W9-CKH-520

SIGNALS, SYSTEMS, AND TRANSFORMS

SIGNALS, SYSTEMS, AND TRANSFORMS

James A. Cadzow
Arizona State University

Hugh F. Van Landingham
Virginia Polytechnic Institute and State University

PRENTICE-HALL, INC., Englewood Cliffs, New Jersey 07632

Library of Congress Cataloging in Publication Data

Cadzow, James A.
Signals, systems, and transforms.

Includes index.
1. Signal theory (Telecommunication) 2. System analysis. 3. Fourier transformations.
I. Van Landingham, Hugh F., 1935– . II. Title.
TK5102.5.C24 1985 003 84-22841
ISBN 0-13-809542-6

Editorial/production supervision: *Raeia Maes*
Cover design: *Debra Watson*
Manufacturing buyer: *Anthony Caruso*

Printed in the United States of America

10 9 8 7 6 5 4 3 2 1

ISBN 0-13-809542-6 01

Prentice-Hall International, Inc., *London*
Prentice-Hall of Australia Pty. Limited, *Sydney*
Editora Prentice-Hall do Brasil, Ltda., *Rio de Janeiro*
Prentice-Hall Canada Inc., *Toronto*
Prentice-Hall Hispanoamericana, S.A., *Mexico*
Prentice-Hall of India Private Limited, *New Delhi*
Prentice-Hall of Japan, Inc., *Tokyo*
Prentice-Hall of Southeast Asia Pte. Ltd., *Singapore*
Whitehall Books Limited, Wellington, *New Zealand*

Contents

PREFACE ix

1 *INTRODUCTION TO SIGNALS AND SYSTEMS 1*

1.1 Introduction to Signals 1
1.2 Continuous to Discrete-Time Signal Conversion 5
1.3 Introduction to Systems 7
1.4 Digital Filtering 9
1.5 Circuit Analysis 9
1.6 Numerical Integration 10
1.7 Numerical Differentiation 14
1.8 Problems 19

2 *DISCRETE-TIME SIGNALS 22*

2.1 Introduction 22
2.2 Changing the Time Variable 25
2.3 Signal Operations 29
2.4 Elementary Operations on Signals 31
2.5 Fundamental Signals 40
2.6 Exponential Signal Generators 48
2.7 Measure of Signal's Size 51

2.8 Modeling by Discrete-Time Operators 52
2.9 Problems 60

3 *CONTINUOUS-TIME SIGNALS 65*
3.1 Introduction 65
3.2 Representation of Continuous-Time Signals 66
3.3 Concepts of Continuity and Differentiability 67
3.4 Changing the Time Variable 72
3.5 Elementary Operations on Signals 77
3.6 Fundamental Signals 82
3.7 Measure of Signal Size 95
3.8 Problems 95

4 *LINEAR OPERATIONS ON SIGNALS 99*
4.1 Introduction 99
4.2 Linear Discrete-Time Signal Operations 100
4.3 Linear Continuous-Time Signal Operations 102
4.4 Linear Operators 103
4.5 Linearity of Discrete- and Continuous-Time Operators 105
4.6 Unit-Impulse Response Determination 106
4.7 Causal Linear Operators 110
4.8 Time Invariance 111
4.9 Operator Stability 112
4.10 Response to Exponential Excitations 114
4.11 Evaluation of the Convolution Operation 116
4.12 Linear Time-Varying Signal Operators 119
4.13 Problems 120

5 *LAPLACE TRANSFORM 125*
5.1 Introduction 125
5.2 Laplace Transform Integral 125
5.3 Region of Absolute Convergence 128
5.4 Laplace Transforms of Basic Signals 129
5.5 Properties of Laplace Transform 134
5.6 Time-Convolution Property 143
5.7 Time-Correlation Property 145
5.8 Poles and Zeros 148

5.9 Inverse Laplace Transform 152
5.10 Rational and Stable Signals 162
5.11 Problems 164
5.12 Appendix 5A: Region of Convergence 168

6 *THE z-TRANSFORM 173*
6.1 Introduction 173
6.2 Basic Properties of the z-Transform 176
6.3 z-Transform Inversion 182
6.4 Sampled Data 195
6.5 Problems 200

7 *TRANSFER FUNCTIONS 202*
7.1 Introduction 202
7.2 The Single-Sided Laplace Transform 203
7.3 Initial-Value Problems 206
7.4 Transfer Functions of Continuous-Time Systems 213
7.5 Discrete-Time Transfer Functions 217
7.6 System Applications 223
7.7 Problems 240

8 *FOURIER SERIES REPRESENTATION 246*
8.1 Introduction 246
8.2 Signal Approximation 256
8.3 Orthogonal Basis Signals 260
8.4 Changing the Time Interval of Approximation 266
8.5 General Fourier Series 267
8.6 Exponential Fourier Series 270
8.7 Periodic Signal Representation 275
8.8 Spectral Content of Periodic Signals 278
8.9 Parseval's Theorem and Signal Power 281
8.10 Fourier Series Representation of Impulse Train 285
8.11 Differentiation of Fourier Series 287
8.12 Response of Linear Systems to Periodic Inputs 289
8.13 Problems 291

9 *THE DISCRETE FOURIER TRANSFORM AND THE FAST FOURIER TRANSFORM ALGORITHM 299*

9.1 Introduction 299
9.2 The Fourier Transform 299
9.3 The Discrete Fourier Transform 303
9.4 Application of the DFT 309
9.5 The Fast Fourier Transform (FFT) Algorithms 318
9.6 DFT Properties and Fast Convolution 327
9.7 Data Windows 333
9.8 Problems 336

INDEX 342

Preface

In very general terms, a *system* is a mechanism that operates upon *signals* (a form of information) to produce other signals. As examples, a stereo system takes a low-level audio signal and produces a high-level sound signal from the system's speakers, a radar tracking system takes radar return signals and produces estimates of where a target will be at the next radar return, and an economical model takes available economic data and predicts future economic behavior. In these diverse situations, the system concept is central and plays an important role in our increasingly quantitative-oriented world. Understanding the *system theory approach* is therefore becoming indispensable in such disciplines as engineering, economics, computer science, modeling, mathematics, and science.

This textbook is concerned with presenting the fundamental aspects of the system theory approach. It is written at a level that is comprehensible to students who have had a course in calculus, have an ability to manipulate complex numbers, and have had some exposure to differential equations. For most electrical engineering students, it should be possible to master the ideas in this text at the junior-year level. Students in other disciplines may be appropriately introduced to the system theory approach upon completion of the aforementioned prerequisites.

Although many of the motivating examples that appear throughout this text are oriented toward electrical engineering, a conscientious effort has been made to incorporate examples from other disciplines as well. The reasoning behind this approach is twofold. First, it is strongly felt that electrical engineering students should be made to appreciate that the tools they use in studying circuits, communication networks, and control systems are directly applicable to a far wider class of interesting applications.

Second, in recent years other quantitative-oriented disciplines are increasingly being exposed to concepts that have been standard to the electrical engineering profession. With this in mind, this textbook seeks to expose the important aspects of system theory from a general viewpoint and to demonstrate their applicability to electrical engineering and other disciplines by means of selected examples and problems.

It is generally possible to classify a given system as being either *discrete-time*, *continuous-time*, or a combination of discrete- and continuous-time. It is widely appreciated that the basic concepts central to discrete-time systems are more easily understood than are their continuous-time counterparts. On the other hand, it can be generally said that to each discrete-time concept there exists an identifiable continuous-time analogy. This being the case, we have here made the pedagogical decision to first develop a discrete-time idea and then immediately follow it with the analogous continuous-time idea.

In the introductory chapter, a philosophical treatment of the notions of signals and systems is undertaken. This includes the essential features distinguishing a discrete- and continuous-time signal and system; the conversion operation of continuous- to discrete-time signals; and motivational applications that include digital filtering, circuit analysis, and numerical integration and differentiation.

The formal development of signal theory is begun in Chapter 2 where discrete-time signals are examined in detail. It is there shown that a discrete-time signal may be viewed as a sequence of numbers. A procedure for making changes in the discrete-time variable is then studied. Furthermore, elementary operations on signals and fundamental signals such as the unit-impulse, unit-step, and sinusoid are explored. The chapter concludes with relevant signal operator applications. A similar treatment of continous-time signals is made in Chapter 3. The analogy between discrete- and continuous-time signals is here emphasized.

Chapter 4 develops the fundamental notion of linear signal operators. Linear signal operators are important due to their widespread usage in various practical applications and to the fact that a rather thorough analysis of such operators is possible. A parallel treatment of linear discrete- and continous-time signal operators is here given in which the similarity between analogous concepts is emphasized. Attention is directed toward the homogeniety and additivity properties of linear operators. In addition, such fundamental notions as operator time-invariance, stability, and transfer function are examined.

In contemporary system theory, the use of signal transformation theory is pervasive. For continous-time signals, the Laplace and Fourier transforms are preeminent. The basic properties and applications of the Laplace transform are studied in Chapter 5. The Laplace transform is shown to be an important tool for the study of signals and the linear operations on signals.

In Chapter 6 the techniques of transformation theory are developed for discrete-time signals. The z-transform is closely related to the Laplace transform; its function is to reduce linear difference equations or equivalently linear discrete-time systems to an algebraic form just as the Laplace transform reduces differential equations to algebraic forms. The applications of both Laplace and z-transform theory is presented in Chapter 7. Here the transfer function concept is used in several analysis contexts: modeling, relating interconnected systems, stability analysis, and others.

Chapter 8 presents the fundamental elements of Fourier series expansions and the corresponding signal approximation techniques leading to the development of Fourier transforms (particularly the computational aspects thereof) in Chapter 9.

The material presented is suitable for junior-level engineering students. A preliminary course in network theory is helpful for understanding the applications in Chapter 7. The book may also be used in a self-study mode for engineers and scientists desiring an introduction to the area of signal analysis.

James A. Cadzow
Hugh F. Van Landingham

SIGNALS, SYSTEMS, AND TRANSFORMS

1

Introduction to Signals and Systems

1.1 INTRODUCTION TO SIGNALS

Contemporary *system theory* has found application in virtually all quantitative disciplines and even in disciplines that were heretofore conceived of as being nonquantitative in nature. Central to the system theory philosophy is the concept of *signal.* In a most fundamental sense, the word *signal* connotes the process of conveying information in some format. This interpretation holds for the most primitive form of information transmittal such as the smoke signal system employed by early-day American Indians to the most sophisticated form of modern-day communication theory.

For our purposes, we use the expression signal to denote a measurement or observation that contains information describing some phenomenon. In order to give our study mathematical structure, we designate signals by means of symbols such as the letters u, x, or y and refer to them as *the signals* u, x, or y. Thus, in a particular situation, the signal x might denote a particular time segment of an audio voltage waveform, the time history of an economic process, the time history of the neurological activity of a muscle system, and so forth.

Information by its very nature implies the notion of being variable or changeable. This is readily demonstrated by the ordinary process of conversation in which information is transferred by auditory signals (words). Thus, the prehistoric cave dweller, who could emit only a series of gruntlike sounds, was able to transmit far less information than is his or her twentieth-century counterpart, who is able to use a complex time sequence of sounds (words and sentences).

A signal in which the information characteristics can change, or fluctuate, is said to be dependent on another variable, which has been classically called time. This independent variable is referred to as time, since, in a large variety of signal theory applications, the underlying independent variable is intrinsically time. For example, the electrocardiogram signal that is displayed on a hospital monitor is seen to have an amplitude (voltage corresponding to the heart's electrical activity) that changes as a function of time. It must be mentioned, however, that there are many situations in which the signal is dependent on a variable other than time (for instance, distance, temperature, or frequency). Thus, when studying the behavior of a vibrating string, the independent variable is a distance measure. The nature of the independent variable is, of course, contingent on the particular measurement or observation being studied. We suffer no real loss in generality, however, by referring to the independent variable as time.

A signal is then very simply an ordinary function of an independent variable. Thus, the value of the signal x at the time instant t is denoted by the symbol $x(t)$. The reader must be careful in distinguishing the difference between the symbol x, which denotes the entire time history of the signal, and the symbol $x(t)$, which specifies the value of the signal at the time instant t. Although this distinction is of importance from a precise mathematical viewpoint, we often use the symbol $x(t)$ to denote a signal whenever there is no danger of misinterpretation. This practice is common in much of signal theory literature.

Continuous- and Discrete-Time Signals

As indicated above, a signal denotes a measurement or observation that contains information relevant to some phenomenon. Generally, the measurement's amplitude changes as time evolves. The manner in which the time variable evolves plays a most profound role in the resultant signal analysis. In many practical situations, the given measurement can change at any instant of time. These signals are called *continuous-time signals*, to reflect the continuous dependence of the signal on time. On the other hand, there exists an important class of processes in which the relevant signals can change value (or are defined) only at specific instants of time. These signals are said to be *discrete-time signals*. These rather abstract concepts are best illustrated by examples.

A sketch of the temperature fluctuation in a room might appear as shown in Fig. 1.1a. Here, signal x specifies the time history of the room temperature with $x(t)$ denoting its value at the specific time instant t. Since the room's temperature is capable of changing at any instant of time, this is clearly a continuous-time signal. In point of fact, many of nature's phenomena are modeled by relationships (differential equations) that are explicitly dependent on continuous-time signals. This is exemplified by Newton's laws of motion, voltage-current relationships in electrical networks, thermodynamic laws, and so forth.

On the other hand, there exists a class of dynamical phenomena that, typically, are man-made in origin and are characterized by discrete-time signals. Examples of this type of signal are abundant in the fields of econometrics, numerical analysis (algorithms), social sciences, operations research, computer sciences, etc. As an illus-

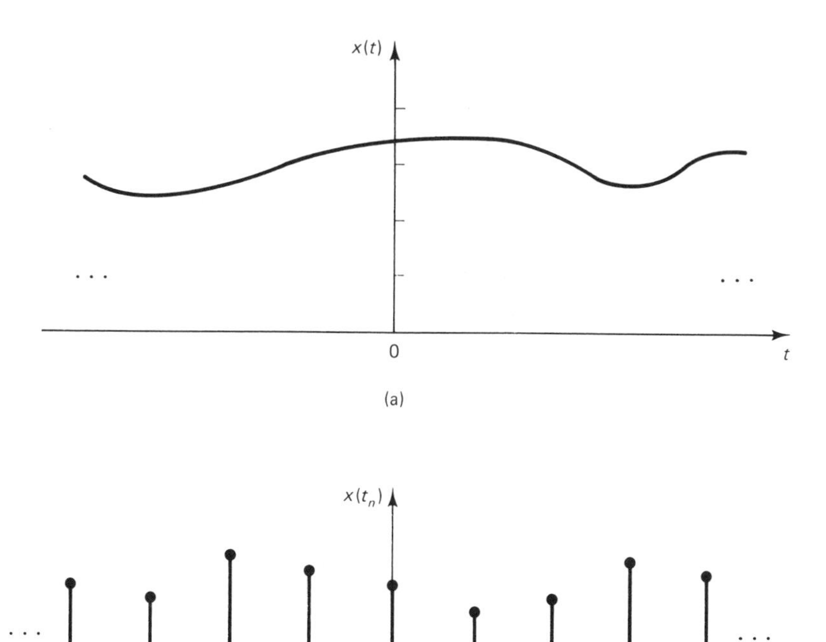

Figure 1.1 Sketch of typical signals that are functions of time: (a) room temperature versus t, and (b) gross national product versus t_n.

tration, consider the determination of our nation's gross national product (GNP). The gross national product is calculated at the end of specific three-month intervals, at which time the various economical components that constitute GNP are determined. A typical plot of GNP might appear as shown in Fig. 1.1b where t_n denotes the end of the given nth three-month period at which time the GNP is to be evaluated. This signal is obviously discrete-time in nature. It is noted that the abscissa axis in this plot is drawn in a continuous manner even though the signal itself has meaning only at the specific time instants t_n. It is for illustrative purposes that we have so displayed the abscissa axis.

Before proceeding further, let us give a more general interpretation to a discrete-time signal. In essence, a discrete-time signal is an ordered set of numbers

$$\ldots, x(t_{-2}), x(t_{-1}), x(t_0), x(t_1), x(t_2), \ldots \tag{1.1a}$$

where the discrete-time variable t_n indicates in which position the number $x(t_n)$ appears in the set of numbers. The three dots to the left of $x(t_{-2})$ and to the right of $x(t_2)$ indicate that the set of numbers continues indefinitely to the left and right, respectively. With this interpretation, it follows that we can think of a discrete-time signal as being a sequence of numbers. For notational convenience, it is desirable to suppress the t in the independent variable t_n and express $x(t_n)$ as $x(n)$ with n being an integer. Therefore, we hereafter interpret a discrete-time signal as being a sequence of numbers in which $x(n)$ denotes the nth member of the sequence. Thus, the discrete-time signal (1.1a) will be hereafter more compactly represented as

$$\ldots, x(-2), x(-1), x(0), x(1), x(2), \ldots \tag{1.1b}$$

In using this shorthand notation, however, it is important to always keep in mind the implicit concept that the integer argument n designates the time instant t_n at which the measure $x(t_n)$ becomes known.

There has recently been a great deal of interest devoted to the study of discrete-time signals. This is obviously a byproduct of the digital computer's development and utilization. The digital computer is a device typically employed to carry out some form of data processing in a rapid manner. Since the computer can essentially only add, subtract, multiply, and divide numbers, the data upon which it operates must be in the format of a sequence of numbers (recorded on magnetic tape, disks, cards, etc.). Therefore, the digital computer is typically used to perform some systematic processing of data which are in the form of a discrete-time signal. Hopefully, this will serve as an adequate motivational stimulus for the further study of discrete-time signals.

The signals displayed in Fig. 1.1 are obviously different in nature. In Fig. 1.1a, the time variable t takes on a continuum of values (that is, values in an interval), and it is for this reason that the corresponding signal is said to be a continuous-time signal. On the other hand, the time variable for the signal displayed in Fig. 1.1b is defined only at discrete-time instants, which results in such signals being referred to as discrete-time signals. Most signals can be classified as being either continuous- or discrete-time in nature as exemplified in Table 1.1.

Since continuous- and discrete-time signals are basically different, it is only natural that different methods have evolved for analyzing their characteristics. Thus we treat these two important classes of signals separately. Wherever possible, however, we point out the many common characteristics shared by each. In the next two chapters, we study some basic properties of discrete-time signals and then extend these concepts to continuous-time signals. This order of presentation reflects the fact that discrete-time signals are inherently easier to characterize and study.

TABLE 1.1 EXAMPLES OF DISCRETE- AND CONTINUOUS-TIME SIGNALS

Signal Description	Signal Type
Monthly new house sales in U.S.A.	Discrete-time
Hourly traffic flow at a highway intersection	Discrete-time
Weekly hotel occupancies	Discrete-time
Daily room temperature at 8:00 A.M.	Continuous-time
Voltage waveform at an amplifier's output terminal	Continuous-time
Speed of a launched rocket	Continuous-time
Electrocardiogram recording	Continuous-time

1.2 CONTINUOUS TO DISCRETE-TIME SIGNAL CONVERSION

In many applications, the underlying descriptive signal(s) being investigated (or used) is inherently continuous-time in nature. If we are to employ the considerable powers of the digital computer for the processing of such signals, however, it is necessary to convert these signals into a format that is compatible with digital computation. Namely, it is necessary to transform the continuous-time signal into a sequence of numbers that may then be manipulated by a digital computer algorithm. This transformation process is commonly referred to as *analog-to-digital (A-to-D) conversion.*

The operation of A-to-D conversion may be conveniently depicted as a switch closing instantaneously at the sample instants t_n. This conceptual model is depicted in Fig. 1.2, where the continuous-time signal $x(t)$ appears at the switch's input terminal and the associated sampled elements $x(t_n)$ appear at the output terminal. A-to-D converters are commonly available hardware items that appear in a variety of computer-based systems as typified by digital controllers and signal processors.

It is possible to provide a rather thorough analysis of the sampling operation. This is particularly true in the case where the sampling instants are equidistant, that is,

$$t_n = nT \qquad \text{for } n = 0, \pm 1, \pm 2, \dots \tag{1.2}$$

in which T is a fixed time interval specifying the *sampling period.* For this uniform sampling scheme, it is readily shown that no information is lost through the sampling process provided that (1) the continuous-time signal is bandlimited and (2) the sampling period T is selected to be smaller than the reciprocal of the highest frequency component of the continuous-time signal. This is a rather startling result since it implies that in such cases, the entire continuous-time signal can be equivalently represented by its sampled values. (See Fig. 1.2.)

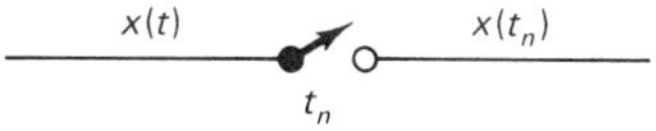

$x(t)$ $x(t_n)$ t_n

Figure 1.2 Continous-time to discrete-time signal conversion.

Example 1.1

Determine the number sequence generated when the continuous-time signal

$$x(t) = \begin{cases} 1 - |t| & \text{for } -1 \le t \le 1 \\ 0 & \text{for all other values of } t \end{cases}$$

is uniformly sampled with sampling period (1) $T = \frac{1}{4}$ s (second), (2) $T = \frac{1}{2}$ s, and (3) $T = 1$ s.

It is beneficial to make a plot of $x(t)$ versus t as shown in Fig. 1.3a in order to visualize the sampling operation. In the following, we shall drop the explicit appearance of sampling period T and write $x(n)$ instead of $x(nT)$ for the sampled signal. Thus, the reader must interpret the sampled signal as a sequence of numbers spaced by T-second intervals, where T is the underlying sampling period.

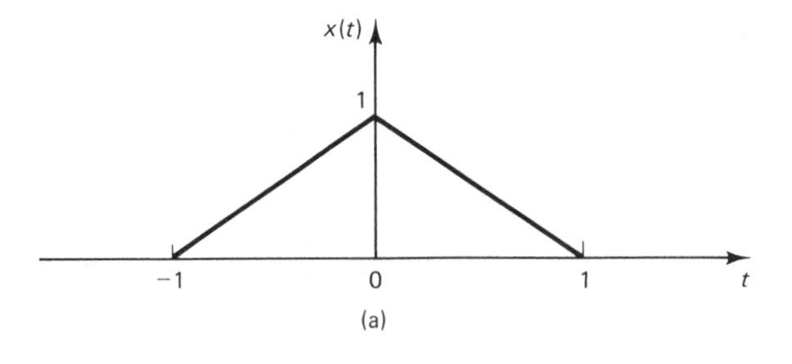

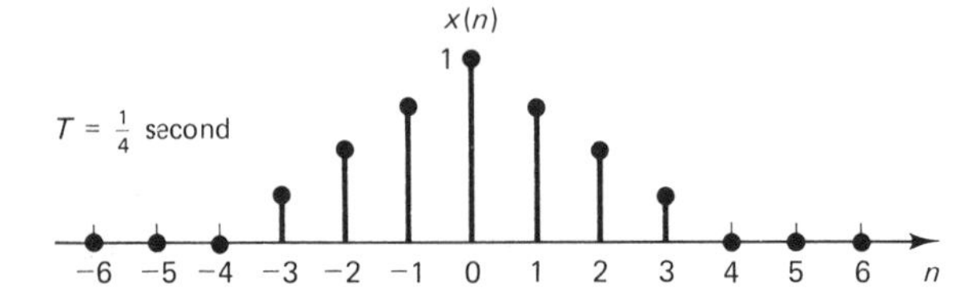

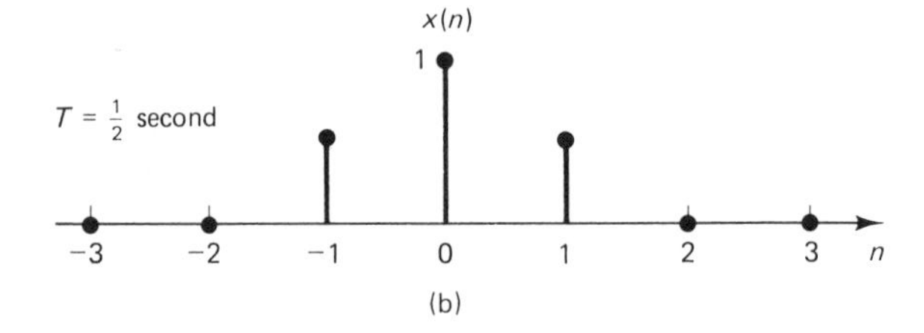

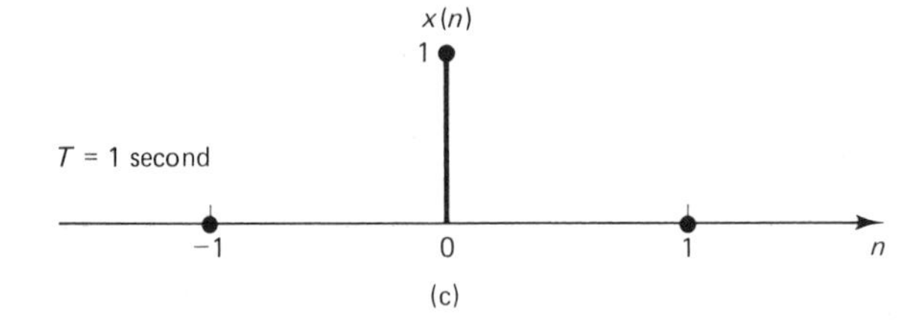

Figure 1.3 Process of uniform sampling with different sampling periods: (a) unsampled waveform (effectively, $T = 0$ s); (b) $T = \frac{1}{4}$ s; (c) $T = \frac{1}{2}$ s; and (d) $T = 1$ s.

Figure 1.3b–d gives a plot of the resultant sampled sequences generated for the three specified sampling periods. Although the same function $x(t)$ is being sampled, it is clear that the sampled sequence obtained depends very critically on the sampling period T. For example, all essential information can be lost by selecting T too large, as is evident in this case for $T = 1$ second.

Example 1.2

Determine the number sequence generated when the continuous-time function

$$x(t) = \begin{cases} 0 & \text{for } t < 0 \\ t + e^t & \text{for } t > 0 \end{cases}$$

is uniformly sampled with sampling period T.

In contrast to the approach taken in Example 1.1, we determine the resultant sampled signal using analytical means. Specifically, the sampled number $x(nT)$ is simply obtained by evaluating the function $x(t)$ at the time instant $t = nT$. For the function above, we then have

$$x(nT) = \begin{cases} 0 & \text{for } n = -1, -2, -3, \ldots \\ nT + e^{nT} & \text{for } n = 0, 1, 2, \ldots \end{cases}$$

and, as is our practice in the remainder of this text, we now drop the explicit appearance of T in the argument of $x(nT)$ to obtain

$$x(n) = \begin{cases} 0 & \text{for } n = -1, -2, -3, \ldots \\ nT + e^{nT} & \text{for } n = 0, 1, 2, \ldots \end{cases}$$

From this expression, it is apparent that the sequence generated depends strongly on the sampling period T.

1.3 INTRODUCTION TO SYSTEMS

Although the study of continuous- and discrete-time signals is important within its own right, we are primarily concerned with investigating procedures whereby a given signal x is changed (transformed) into another signal y in some systematic manner. This transformation procedure is represented by the mathematical notation

$$y = \boldsymbol{T}x \tag{1.3}$$

where $\boldsymbol{T}$ represents some well-defined rule by which the signal x is changed into the signal y. Relationship (1.3) defines a "system" characterization and is depicted as shown in Fig. 1.4. The arrows on the lines leading into and out of the box indicate the direction of signal flow.

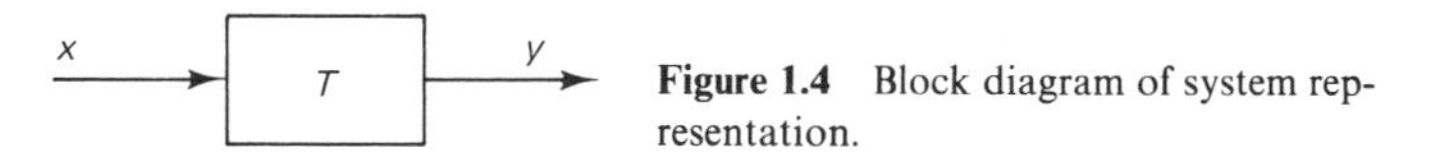

Figure 1.4 Block diagram of system representation.

In this representation, we interpret x as being the system's input signal (or excitation) and y as the system's corresponding output signal (or response). Thus, the excitation signal x is said to generate the response signal y through the characteristic rule $\boldsymbol{T}$. The rule $\boldsymbol{T}$ within the box completely defines the operational characteristic of the system. We are mainly concerned with those situations in which this rule takes the form of a linear differential equation or a linear difference equation. Typical examples of this system's viewpoint now follow.

1. The system is an automobile, the excitation is the accelerator pedal position, and the response is the automobile's velocity.
2. The system is the U.S. economy, the excitation is the prime interest rate, and the response is the inflation rate.
3. The system is an FM stereo receiver, the excitation is an RF signal (to which

the receiver is tuned), and the response is the sound emanating from the receiver speakers.

4. The system is a signal processor (e.g., a filter), the excitation is data which is being operated upon, and the response is the processed data.

It will be subsequently shown that for linear systems, there exists a so-called system *unit-impulse response signal* h, which completely characterizes the system's rule $\boldsymbol{T}$. Thus, for this class of systems, the response signal y may be generated by appropriately convolving the system's input signal x with the system's unit-impulse response signal h. This operational procedure is denoted by

$$y = h * x \tag{1.4}$$

The terms *linear*, *unit-impulse response*, and *convolving* are defined in subsequent chapters. We invoke these concepts at this time to further justify our study of signals, since from relationship (1.4) it is seen that the response signal is completely specified by the convolving of two signals h and x.

The reader may have questioned the relevance of the system theory concept in which there is postulated a pair of signals x and y related to each other by means of a well-defined rule $\boldsymbol{T}$. A little thought will convince the reader, however, that this philosophical notion is used extensively in the modeling of various physical phenomena. For instance, investigators in such diverse fields as engineering, social sciences, econometrics, biology, operational research, and so forth are continually striving to construct *models* that satisfactorily approximate the dynamical behavior of real-world phenomena. Once such models have been generated, the investigator is then able to study the relationships between the model's (system's) input and output signals. Thus, the economist might use a model of the U.S. economy to study the relationships between (1) the variable's prime interest rate, money supply, and so forth, which might serve as the model's input signals and (2) the gross national product, employment, and so forth, which might serve as the corresponding response signals. If the economical model is sufficiently representative, then a rational basis for making satisfactory economical policy is possible (i.e., the correct selection of the input signal in order to achieve a desired response).

It is beyond the scope of this text to develop methodology for generating models of various phenomena. This is a study within itself. In fact, there exists a wealth of excellent texts in such fields as circuit theory, applied mechanics, biology, econometrics, thermodynamics, social sciences, fluid dynamics, and so forth, in which the modeling concept is extensively treated. Invariably, these models are constructed so that the model's input and output signals are related to each other by means of a differential or difference equation.

One of our objectives is, then, to develop procedures for characterizing the dynamical relationships that exist between the input and outputs signals relative to models that are governed by linear differential or difference equations. This development will not be oriented toward any specific discipline since the concepts to be presented are applicable to a variety of disciplines. We shall now provide several examples in which system theory concepts are extensively employed.

1.4 DIGITAL FILTERING

In order to motivate the need for studying discrete-time signals and systems, let us consider a frequently occurring application in which there is given a sequence of N measurements of some phenomenon. It is desired to extract desirable information concerning the phenomenon from this measurement sequence. This objective is often made difficult due to the presence of an additive corruptive *measurement noise.* Namely, the given measurements are assumed to be of the form

$$x(n) = s(n) + w(n), \qquad 1 \leq n \leq N \tag{1.5}$$

in which $s(n)$ represents the useful signal component and $w(n)$ the undesired additive noise component.

The task at hand is to remove the undesired noise component $w(n)$ from the provided measurements without unduly modifying the information-bearing signal. This objective is often effected by using the data measurements as the input to a digital filter. The filter is designed so that it effectively blocks the noise component while allowing the signal component to flow through relatively unimpeded. For many applications of this nature, this signal processing operation may be implemented by means of the linear operator

$$y(n) = h(0)x(n) + h(1)x(n-1) + h(2)x(n-2) + \cdots + h(q)x(n-q) \tag{1.6}$$

in which the constants (coefficients) $h(0), h(1), \ldots, h(q)$ are chosen in order to effect the required filtering. Much of digital filter theory is directed toward the development of systematic procedures for synthesizing (i.e., selecting the $h(k)$ coefficient values) linear filters of this form. As one might suspect, the selection of an appropriate set of filtering coefficients is dependent on the intrinsic time (and frequency) behavior of the signal and additive noise components.

Upon examination of relationship (1.6), it is seen that the filter's output signal $y(n)$ is generated by multiplying the present and past input elements by the filtering coefficients. The required operations of multiplication, addition, and data storage (i.e., retaining the q "old" data values $x(n-1), x(n-2), \ldots, x(n-q)$) to compute $y(n)$ are very effectively carried out on a digital computer. As such, the signal operation as represented by expression (1.6) is very logically called a *digital filter operation.* It constitutes the rule $\boldsymbol{T}$ depicted in Fig. 1.4 that relates to the system's input and output signals. The theory dealing with digital filters is rather extensive and will not be treated separately in this text. The concepts herein developed for discrete-time signals and systems, however, are fundamental to the study of digital filters.

1.5 CIRCUIT ANALYSIS

Most students of electrical engineering have been exposed to the basic ideas of simple circuit analysis by their sophomore year. This theory deals with the characterization of the voltage and current relationships that appear in given circuit

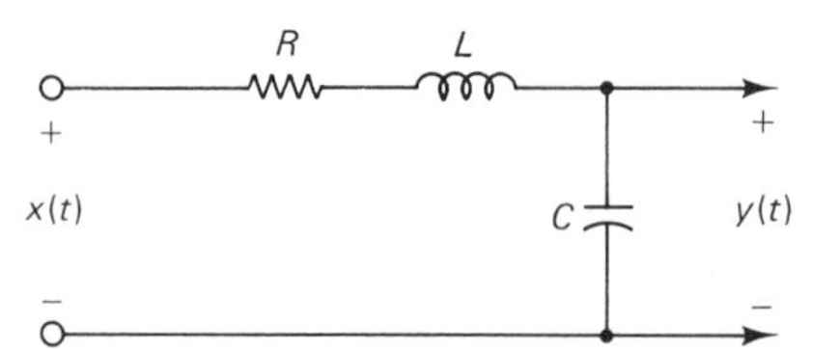

Figure 1.5 A simple circuit.

configurations. A circuit is typically composed of an interconnection of resistors, capacitors, inductors, and active generators. To demonstrate how system theory may be applied to this important discipline, let us consider the circuit shown in Fig. 1.5. Furthermore, let it be desired to evaluate the relationship between the circuit's input voltage $x(t)$ and output voltage $y(t)$. Using standard circuit theory analysis, it is readily shown that this required relationship takes the form

$$\frac{d^2y(t)}{dt^2} + \frac{R}{L}\frac{dy(t)}{dt} + \frac{1}{LC}\,y(t) = \frac{1}{LC}\,x(t) \tag{1.7}$$

Thus, the circuit's input and output voltages are seen to be related by a linear second-order differential equation. In this case, differential equation (1.7) constitutes the well-defined rule $\boldsymbol{T}$ relating the circuit system's input and output signals as operationally depicted in Fig. 1.4. It is to be noted that a similar system interpretation can be given to more complex circuits. Since a circuit's voltages and currents are continuous-time in nature, it is seen that a circuit configuration is an example of a continuous-time system.

1.6 NUMERICAL INTEGRATION*

In many scientific and engineering applications, one is frequently confronted with the task of integrating a function of continuous time. If the function does not have a closed-form integral, it is usually necessary to approximate the integral using numerical techniques. A very popular and convenient technique is now presented.

Suppose we desire to evaluate the integral

$$y(t) = \int_0^t x(\tau)\,d\tau$$

The value of the function $y(t)$ is simply equal to the area under the curve $x(\tau)$ in the time interval $0 \le \tau \le t$. It is this area which we now numerically approximate. Since numerical techniques are to be used, it is expedient to evaluate the integral function $y(t)$ at the time instants

$$t_n = nT \qquad \text{for } n = 0, 1, 2, \ldots$$

*This section taken from James A. Cadzow, *Discrete-Time Systems* (Englewood Cliffs, N.J.: Prentice-Hall, Inc., 1973). Used by permission.

that is,

$$y(nT) = \int_0^{nT} x(\tau)\, d\tau \qquad \text{for } n = 0, 1, 2, \ldots \tag{1.8}$$

How one selects the integration time subinterval T is a most important consideration and is shortly discussed. Let us now break up this time integral as follows:

$$y(nT) = \int_0^{nT-T} x(\tau)\, d\tau + \int_{nT-T}^{nT} x(\tau)\, d\tau \qquad \text{for } n = 1, 2, 3, \ldots$$

The first integral is the area under the curve $x(\tau)$ in the time interval $0 \le \tau \le nT - T$, which by expression (1.8) is denoted by $y(nT - T)$. Therefore,

$$y(nT) = y(nT - T) + \int_{nT-T}^{nT} x(\tau)\, d\tau \qquad \text{for } n = 1, 2, 3, \ldots \tag{1.9}$$

If we knew the value of $y(nT - T)$, then to determine $y(nT)$, we need only evaluate

$$\int_{nT-T}^{nT} x(\tau)\, d\tau$$

which is the area of the curve $x(\tau)$ in the time interval $nT - T < \tau \le nT$. In general, $x(\tau)$ will not have a closed-form integral, which necessitates that we find an approximate value for this area. A possible method is suggested in Fig. 1.6 whereby the area of the shaded rectangle serves as a good approximation to the integral, that is

$$\int_{nT-T}^{nT} x(\tau)\, d\tau \approx \text{area of shaded rectangle } = Tx(nT) \tag{1.10}$$

where $x(nT)$ denotes the value that the function being integrated takes on at time $\tau = nT$. Substituting this approximation into relationship (1.9), we arrive at an iterative process for carrying out numerical integration, namely

$$y_a(nT) = y_a(nT - T) + Tx(nT) \qquad \text{for } n = 1, 2, 3, \ldots$$

where the subscript a in $y_a(nT)$ is used to emphasize the fact that $y_a(nT)$ is an approximation of $y(nT)$ as given by expression (1.8).

At the risk of causing some confusion, we shall now drop the explicit appearance of T in the arguments of $y_a(nT)$, $y_a(nT - T)$, and $x(nT)$ to rewrite the numerical integration algorithm in its more standard iterative form

$$y_a(n) = Tx(n) + y_a(n - 1) \qquad \text{for } n = 1, 2, 3, \ldots \tag{1.11}$$

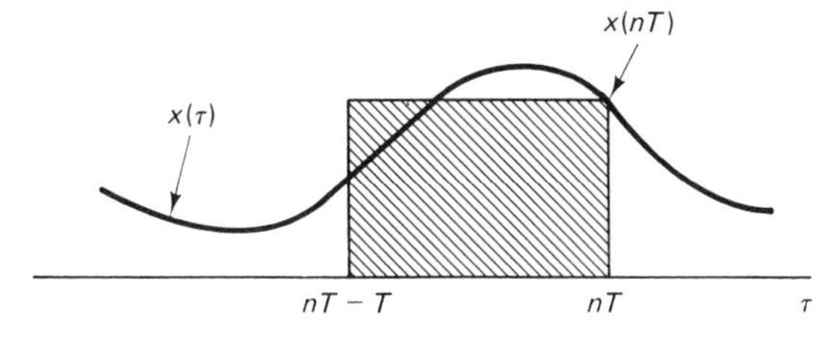

Figure 1.6 Rectangular approximation of an integral.

This expression is recognized as a first-order difference equation with the signal $y_a(n)$ being interpreted as an approximation to integral (1.8). Since the iteration commences at $n = 1$, it follows that $y_a(0)$ plays the role of the initial condition. From expression (1.8), it is apparent that the value of $y_a(0)$ is always zero.[1] Hence, at $n = 1$, we have

$$\begin{aligned} y_a(1) &= Tx(1) + y_a(0) \\ &= Tx(1) \end{aligned}$$

At the second iteration we have

$$y_a(2) = Tx(2) + y_a(1)$$

where the value to be assigned to $y_a(1)$ was computed one iteration previously and then stored in the computer's memory. Continuing in this manner, the numerical integration algorithm (1.11) generates the approximate integral $y_a(n)$ from the sampled values of the function to be integrated, $x(\tau)$. Figure 1.7 depicts the process of integration and its numerical approximation.

Selection of the Integration Time Subinterval T

In selecting the integration time subinterval T, two conflicting factors come into play:

1. Making T small enough so that the area approximation (1.10) is good.
2. Making T large enough so that one has to evaluate expression (1.11) as few times as possible in order to calculate the integral of $x(\tau)$ over a given time interval $0 \leq \tau \leq t_f$. The number of iterations necessary in this case is

$$N = \left[\frac{t_f}{T}\right]$$

where the [ ] symbol means $N = t_f/T$ if t_f/T is an integer; and if not, then N is selected as the next largest integer.

A compromise value of T is normally chosen so that both factors are taken into consideration.

[1]More generally, $y_a(0)$ may be interpreted as the integral of $x(\tau)$ over the interval $-\infty < \tau \leq 0$ when

$$y(t) = \int_{-\infty}^{t} x(\tau)\, d\tau.$$

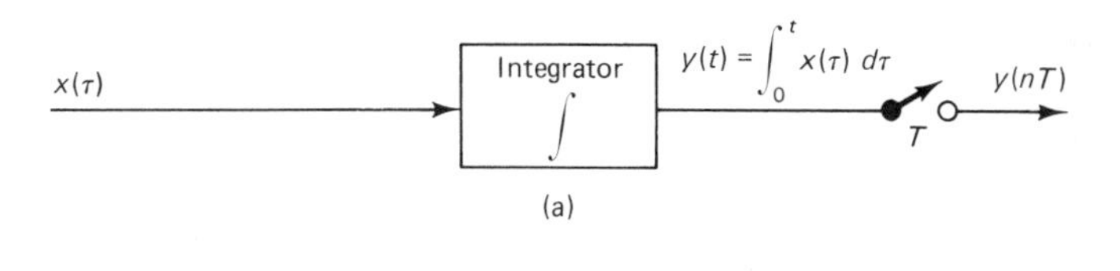

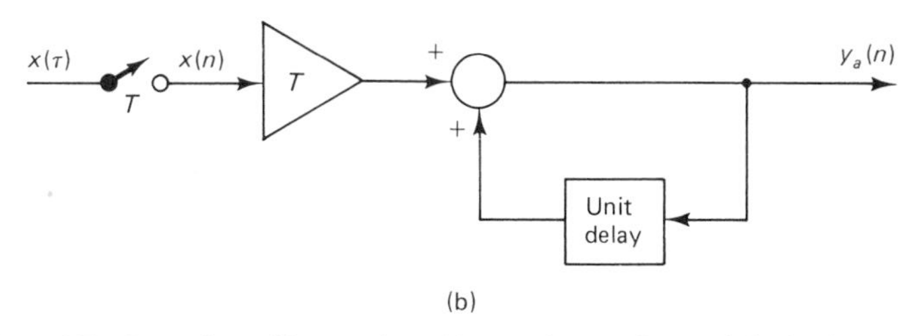

Figure 1.7 Operation of integration: (a) actual operation and (b) implementation of numerical integration algorithm in expression (1.11).

Example 1.3

Approximate the value of the integral

$$\int_0^1 \sqrt{1-\tau^3}\, d\tau$$

using numerical integration. For this example, we are integrating the function $x(\tau) = \sqrt{1-\tau^3}$ with $t_f = 1$ s. The sequence of numbers that results when this function is uniformly sampled is then given by

$$x(n) = \sqrt{1-(nT)^3} \qquad \text{for } n = 1, 2, 3, \ldots$$

A sampling period of $T = 0.1$ s is selected, which yields the sequence shown in the second column of Table 1.2. If this sequence is used as the input signal to the numerical integration algorithm (1.11), there results the response signal $y_a(n)$ given in the third column of Table 1.2. The desired approximate value of the integral is then $y_a(10) = 0.77999$, where discrete time $n = 10$ corresponds to $nT = 1$ s.

TABLE 1.2 NUMERICAL INTEGRATION

n	$x(n)$	$y_a(n)$
1	0.99950	0.09995
2	0.99599	0.19955
3	0.98641	0.29819
4	0.96747	0.39494
5	0.93541	0.48848
6	0.88544	0.57702
7	0.81056	0.65808
8	0.69857	0.72794
9	0.52058	0.77999
10	0.00000	0.77999

1.7 NUMERICAL DIFFERENTIATION*

In many disciplines, the differential equation plays a major role in characterizing the behavior of various phenomena. Numerous examples are to be found in the fields of engineering, economics, business, physiology, physics, and so forth. The study of such phenomena requires that one have the ability to solve differential equations. In most practical situations, however, these solutions are difficult to obtain for a variety of reasons. The investigator is then forced to seek approximate solutions, usually with the aid of an analog or a digital computer.

Obtaining an approximate solution of a differential equation with the use of a digital computer requires that the differential equation be put into a form suitable for digital computation. Specifically, there is a basic incompatibility between a differential equation, which is dependent on continuous-time signals, and digital computation, which involves discrete-time signals. We now present a simple transformation procedure whereby a differential equation is converted into an *equivalent* difference equation.

First-Order Differential Equations

Consider the simple electronic resistor-capacitor network shown in Fig. 1.8. Students of electrical engineering can verify that the relationship between the input voltage, $x(t)$ and the output voltage $y(t)$ is given by

$$\frac{dy(t)}{dt} + \frac{1}{RC}\,y(t) = \frac{1}{RC}\,x(t)$$

where the resistor has a resistance of R ohms and the capacitor a capacitance of C farads. In order to treat a more general class of problems from different disciplines, let us consider a generalization of this relationship as given by

$$\frac{dy(t)}{dt} + \alpha y(t) = \beta x(t) \tag{1.12}$$

where α and β are arbitrary constants (for example, $\alpha = \beta = 1/RC$ for the electronic network of Fig. 1.8). We shall interpret $x(t)$ and $y(t)$ as being the input and output signals, respectively, of the system characterized by this first-order linear differential equation.

This simple differential equation is readily solved using classical methods when the input signal is of a highly specialized form, e.g., $x(t) = 1$ for $t \geq 0$ and zero otherwise. However, in most practical applications, the input signal is usually such as to make the determination of the resultant output signal $y(t)$, at best, extremely difficult to obtain. We are then forced to determine an approximate solution using numerical techniques.

Since numerical techniques are to be used, we normally restrict our attention

*This section taken from James A. Cadzow, *Discrete-Time Systems* (Englewood Cliffs, N.J.: Prentice-Hall, Inc., 1973). Used by permission.

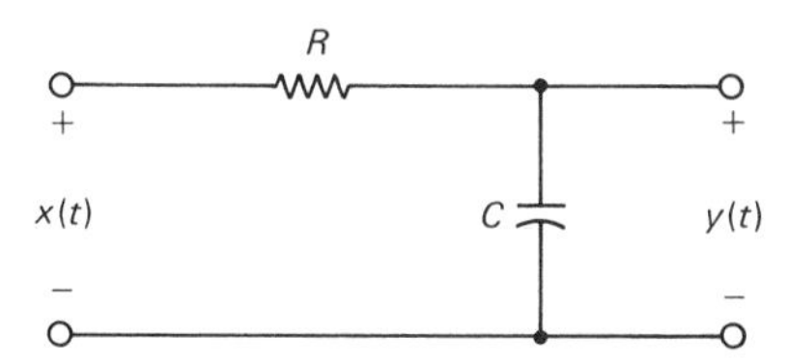

Figure 1.8 Resistor-capacitor network.

to investigating $y(t)$ at the sampling instants nT, where $n = 0, 1, 2, \ldots$, and T is the sampling period. The parameter T must be carefully selected to obtain a good approximate solution efficiently. More is said about how one makes this selection later.

Evaluating relationship (1.12) at $t = nT$, we have

$$\frac{dy(nT)}{dt} + \alpha y(nT) = \beta x(nT) \qquad \text{for } n = 1, 2, 3, \ldots \tag{1.13}$$

From elementary calculus, the derivative of a function $y(t)$ at $t = nT$ is simply the slope of the curve $y(t)$ at $t = nT$. This slope may be approximated by the first difference relationship

$$\frac{dy(nT)}{dt} \approx \frac{y(nT) - y(nT - T)}{T} \tag{1.14}$$

with the approximation being better as the sampling period T is made smaller as is made clear in Fig. 1.9.[2] Inserting this approximation of the derivative into (1.13), we have

$$\frac{y_a(nT) - y_a(nT - T)}{T} + \alpha y_a(nT) = \beta x(nT) \qquad \text{for } n = 1, 2, 3, \ldots$$

where the subscript a on y_a is used to emphasize the fact that $y_a(nT)$ is an approximation of $y(nT)$. Solving this relationship for $y_a(nT)$ gives the desired difference equation relating $y_a(nT)$ with $x(nT)$. That is,

$$y_a(n) = \frac{\beta T}{1 + \alpha T}\, x(n) + \frac{1}{1 + \alpha T}\, y_a(n - 1) \qquad \text{for } n = 1, 2, 3, \ldots \tag{1.15}$$

where we have dropped the explicit appearance of T in the arguments of $y_a(nT)$, $y_a(nT - T)$, and $x(nT)$ for the sake of notational convenience.

This iterative relationship may be utilized in a computer program format to find an approximate solution of the original differential equation (1.12). Typically, when one is required to find the solution of a first-order differential equation of this

[2]Another approximation to the derivative is given by

$$\frac{dy(nT)}{dt} \approx \frac{y(nT + T) - y(nT - T)}{2T}$$

but this is somewhat more complex to implement. Why?

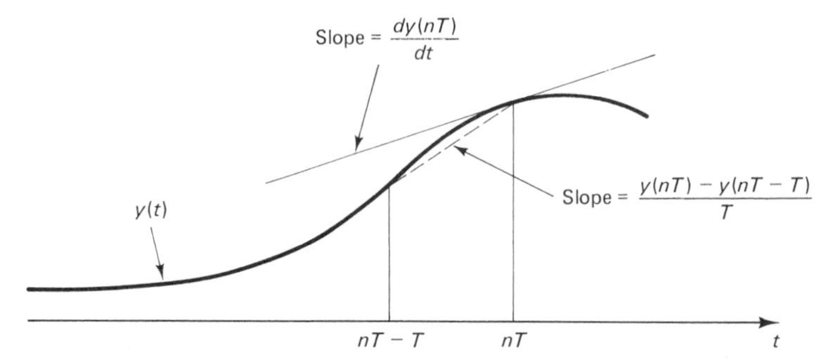

Figure 1.9 Approximation of derivative.

form, a prescribed value for the initial condition $y(0)$ is given. To take this factor into account, one simply begins the iteration of expression (1.15) at $n = 1$, with the initial condition $y_a(0)$ set equal to $y(0)$ (the initial condition associated with the differential equation).

In summary, given the sampled values of the continuous-time signal $x(t)$ and the differential equation's associated initial condition—that is, given $x(n)$ and $y(0)$—expression (1.15) may be used to iteratively determine an approximate solution of the original differential equation (1.12). The resultant response $y_a(k)$, when plotted with its elements spaced T seconds apart, serves as an approximation of the desired solution. Figure 1.10 gives a system interpretation of the prescribed numerical process.

Selection of Sampling Time Interval T

The sampling time parameter T must be selected so that

1. The approximation of the derivative as given by expression (1.14) is good. This, in general, requires that T be made small. How small depends on the time constants of the original differential equation and the frequency content of the input signal (see comments in Section 1.2).
2. The iterative procedure is efficient. This means that we wish to use as few iterations of difference equation (1.15) as possible to obtain a numerical solution of the original differential equation over a given time interval $0 \leq t \leq t_f$.

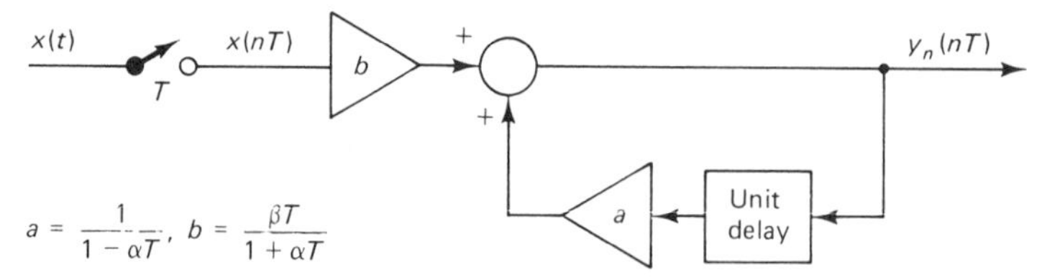

Figure 1.10 Numerical solution of first-order differential equation.

The number of iterations needed for a sampling period of T seconds is

$$N = \left[\frac{t_f}{T}\right]$$

Obviously, for a fixed value of t_f, N is decreased by making T large. A compromise value of T is selected to partially satisfy conditions 1 and 2. How one makes this selection is a rather involved matter and has occupied the interest of numerical analysts for years. If the sampling time T is made too large, the approximation suffers and a bad numerical solution results.

Example 1.4

For the RC network depicted in Fig. 1.8, numerically evaluate this system's unit step response when $R = 10^6$ ohms and $C = 10^{-6}$ farads. The voltage across the capacitor at time $t = 0$ is taken to be zero, that is, $y(0) = 0$.

In this case $\alpha = \beta = 1/RC = 1$; therefore relationship (1.12) becomes

$$\frac{dy(t)}{dt} + y(t) = x(t) \tag{1.16}$$

with initial condition $y(0) = 0$. Since the input signal is specified as being a *continuous-time unit step*, we have

$$x(t) = \begin{cases} 0 & \text{for } t < 0 \\ 1 & \text{for } t \geq 0 \end{cases}$$

This signal is of a simple form, and the solution to the prescribed differential equation may be readily obtained and is given by

$$y(t) = 1 - e^{-t} \qquad \text{for } t \geq 0 \tag{1.17}$$

Readers unfamiliar with the procedures for solving differential equations may verify that this function is the solution by direct substitution. Specifically, we first take the derivative of the postulated solution (1.17), which results in

$$\frac{dy(t)}{dt} = e^{-t} \qquad \text{for } t > 0$$

Substituting this expression and expression (1.17) into the governing differential equation (1.16) yields the identity $1 = 1$ for $t > 0$. Thus the postulated solution satisfies the differential equation and also the initial condition requirement $y(0) = 0$. It is therefore the solution as required.

The main purpose of this example is to obtain an approximate numerical solution of differential equation (1.16) to illustrate the concepts of numerical differentiation just presented. Utilizing relationship (1.15) with $\alpha = \beta = 1$, the equivalent difference equation generated by this differential equation is

$$y_a(n) = \frac{T}{1+T}\,x(n) + \frac{1}{1+T}\,y_a(n-1) \qquad \text{for } n = 1, 2, 3, \ldots \tag{1.18}$$

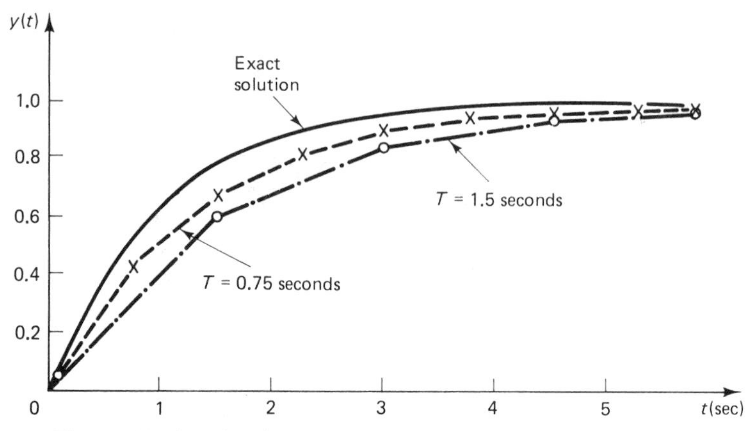

Figure 1.11 Actual and numerical solutions of differential equation (1.16).

The uniform sampled version of the continuous-time step input signal is readily found to be the discrete-time unit step signal:

$$x(n) = \begin{cases} 0 & \text{for } n = -1, -2, -3, \ldots \\ 1 & \text{for } n = 0, 1, 2, \ldots \end{cases}$$

It will be subsequently shown that the solution of difference equation (1.18) to this input is given by

$$y_a(n) = 1 - \left(\frac{1}{1+T}\right)^n \qquad \text{for } n = 0, 1, 2, \ldots \tag{1.19}$$

Again, by substituting this postulated solution into the governing difference equation (1.18), one may verify its correctness.

To illustrate how the selected sampling period T affects the accuracy of the numerical solution, plots of the actual solution and the numerical solutions for $T = 0.75$ s and $T = 1.5$ s are shown in Fig. 1.11. It is apparent that the smaller sampling period yields a more accurate numerical solution as anticipated.

Generalization. It is possible to readily generalize the approach taken above to solve higher-order differential equations. To illustrate this, let it be desired to approximate the second derivative of the signal $\{y(t)\}$ at time $t = nT$. Using the first-difference approximation, we have

$$\frac{d^2y(nT)}{dt^2} \approx \frac{\dfrac{dy(nT)}{dt} - \dfrac{dy(nT-T)}{dt}}{T}$$

Incorporating the previously established approximation (1.14) with appropriate choices of the argument, it follows that

$$\frac{d^2y(nT)}{dt^2} \approx \frac{\dfrac{y(nT) - y(nT - T)}{T} - \dfrac{y(nT - T) - y(nT - 2T)}{T}}{T}$$

$$= \frac{y(nT) - 2y(nT - T) - y(nT - 2T)}{T} \qquad (1.20)$$

This *second-difference* expression may then be substituted for $d^2y(nT)/dt^2$ where required. Clearly, we can continue this process to effect numerical approximations for higher-order derivatives.

1.8 PROBLEMS

1.1. In words, describe how each of the signals shown in Table 1.1 can be considered as a function of either continuous or discrete time.

1.2. Determine whether the following signals are discrete- or continuous-time signals. Give a brief explanation.

(a) Temperature in a room.

(b) Closing price of a stock on the New York Stock Exchange.

(c) Position of the steering wheel of a car in motion.

(d) Weight of an individual taken every day at 5:00 P.M.

1.3. Make a typical plot versus time for each of the signals described in Problem 1.2.

1.4. A vibrating string is fixed between two points so that the ends do not move as shown in Fig. P1.4. When the string is in motion, describe the variables on which the vertical amplitude u depends.

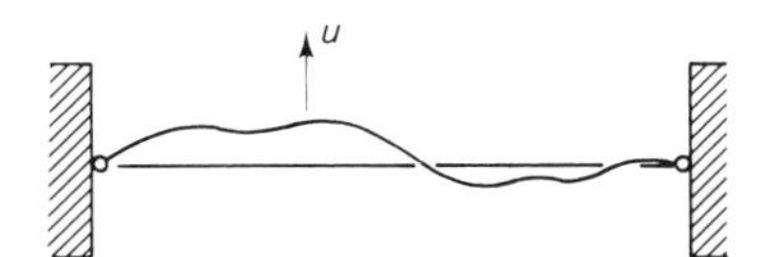

Figure P1.4

1.5. A current signal $i(t)$ amperes is applied to the resistor capacitor combination shown in Fig. P1.5.

(a) Find the operation that relates the current $i(t)$ to the voltage signal $v_c(t)$ across the capacitor.

(b) Calculate $v_c(t)$ for $t \geq 0$ if $i(t) = 1$ ampere (constant) and $v_c(0) = 0$.

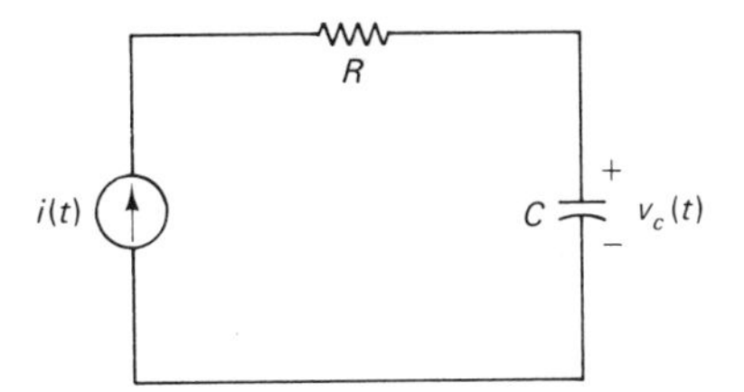

Figure P1.5

1.6. What signals are involved in the body when one is using eye-hand coordination such as picking up an object?

1.7. A mass of m kilograms reacts to a force of F newtons as shown in Fig. P1.7. If x is the distance in meters of the mass from a reference (inertial) frame, state the relation between the signals $F(t)$ and $x(t)$. What force signal corresponds to $x(t) = 10t^2$?

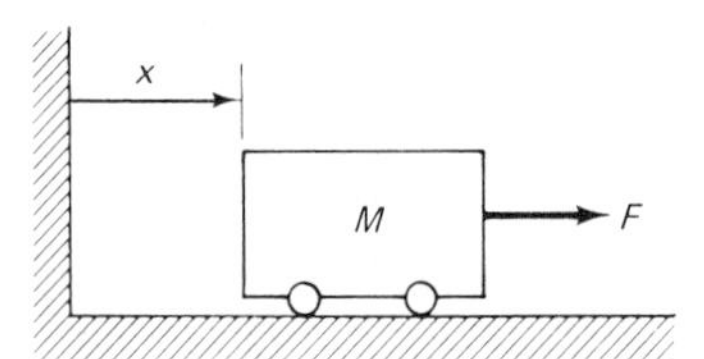

Figure P1.7

1.8. What signals does a driver use to control the operation of a car? State any limits that might be involved with these control signals.

1.9. An inventory system for a large department store uses a computer to calculate which items are to be ordered each week. What signals should enter into this calculation for a particular item?

1.10. For the continuous-time signal

$$x(t) = \begin{cases} 1 & \text{for } -1 \leq t \leq 1 \\ 0 & \text{for } \quad t < -1 \quad \text{and} \quad t > 1 \end{cases}$$

make a plot of the discrete signal that results if this signal is uniformly sampled and the sampling period T is

(a) $T = 0.2$ s

(b) $T = 0.45$ s

In what way and why do the two uniformly sampled versions of the continuous-time signal differ?

1.11. Determine and plot the sequence of numbers generated when the continuous-time signal

$$x(t) = \begin{cases} 1 - t^2 & \text{for } -1 \leq t < 1 \\ 3 - t & \text{for } 1 \leq t \leq 3 \\ 0 & \text{for all other values of } t \end{cases}$$

is uniformly sampled with sampling period T given by

(a) $T = 0.5$ s

(b) $T = 1$ s

(c) $T = 4$ s

Use both the graphical and analytic approaches taken in Examples 1.1 and 1.2. Comment as to why the resultant sequences differ and what might be a proper choice of the sampling period for this continuous-time signal from the selections (a), (b), or (c).

1.12. For the continuous-time signal

$$x(t) = \begin{cases} t \sin 2\pi t & \text{for } -\frac{1}{2} \leq t \leq 2 \\ 0 & \text{for } t < -\frac{1}{2} \quad \text{and} \quad t > 2 \end{cases}$$

make a plot of the discrete signal that results if this continuous-time signal is uniformly sampled with sampling period $T = 0.25$ s.

1.13. Given the continuous-time signal specified by

$$x(t) = \begin{cases} t^3 & \text{for } 0 \le t < 1 \\ \dfrac{3}{2} - \dfrac{t}{2} & \text{for } 1 \le t \le 3 \\ 0 & \text{for all other values of } t \end{cases}$$

determine the resultant sequence of numbers obtained by uniformly sampling $x(t)$ with a sampling period of

(a) $T = \frac{1}{2}$ s

(b) $T = \frac{1}{3}$ s

Use both the graphical and analytic approaches taken in Examples 1.1 and 1.2.

1.14. For the continuous-time signal

$$x(t) = \sin \omega t \qquad \text{for all } t$$

make a plot of the discrete signal that results if this waveform is uniformly sampled with the sampling period

(a) $T = \pi/4\omega$ s

(b) $T = \pi/2\omega$ s

(c) $T = \pi/\omega$ s

From these results, what general comments can be made about the sampling rate needed for sinusoidal signals of radian frequency ω?

1.15. In Fig. P1.15, a continuous-time signal $x(t)$ is shown being sampled ideally so that the exact amplitudes of $x(t)$ are available as a sequence of numbers:

$$x(0),\ x(T),\ x(2T),\ \ldots$$

where T is the sample interval. Discuss how the operation shown approximates an integration on the continuous-time signal $x(t)$, that is, show that

$$y(nT) \approx \int_0^{nT} x(t)\, dt$$

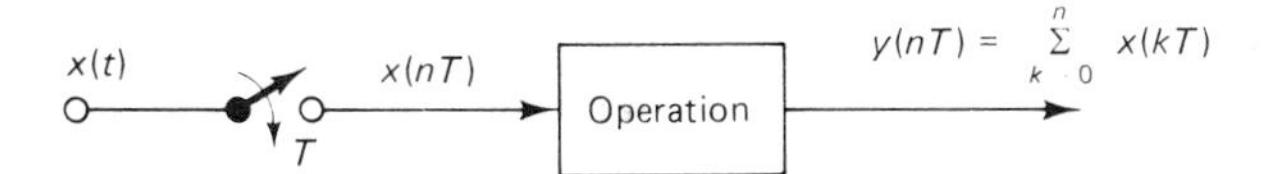

Figure P1.15

2

Discrete-Time Signals

2.1 INTRODUCTION

As indicated in Section 1.1, a discrete-time signal is information that is conveyed, or appears, in the form of a sequence of numbers. In using the term *sequence*, we are in fact stipulating that the numbers that constitute a sequence occur in a well-ordered manner. With this in mind, a sequence can be formally written as

$$\boldsymbol{x} = \{x(n)\} \qquad \text{for all integers } -\infty < n < \infty$$

with $x(n)$ denoting the nth element of the sequence. Thus a sequence is interchangeably denoted by the lower-case boldface italicized symbol $\boldsymbol{x}$ or by the braced symbol $\{x(n)\}$. The integer-order variable n specifies the position in which the particular sequence element $x(n)$ appears. For example, the number $x(4)$ immediately precedes $x(5)$ and immediately follows $x(3)$ in the sequence.

In most practical applications, the values of the individual sequence elements become known on a *sequential time basis*. It is then beneficial to envision a sequence as the result of a process whereby its individual elements are omitted, one at a time, at regularly spaced time instants. Thus, at time t_3 the value $x(3)$ becomes known, then at time t_4 the value $x(4)$ is made known, then $x(5)$ at time t_5, and so forth for all discrete-time instants. In accordance with the convention introduced in Chapter 1, the integer index n is used in place of the time instant t_n at which the element $x(n)$ first becomes known.

From a purely mathematical standpoint, a *sequence* is a function which is defined on the set of integers. Thus, $x(n)$ specifies the value of the function (or sequence) $\boldsymbol{x}$ at the integer argument n. For historical reasons, we will refer to the integer variable n as *discrete-time*. In addition, it is sometimes convenient to refer to the sequence $\boldsymbol{x}$ as *the sequence* $x(n)$, even though, in a strict sense, $x(n)$ is but one element of the sequence.

Methods of Discrete-Time Signal Representation

There are basically three ways in which a sequence can be represented:

1. It may be possible to specify the element values constituting a sequence by means of a convenient mathematical formula (also called a *closed form expression*).
2. We may explicitly list the sequence elements in the tabular format

$$\boldsymbol{x} = \{\ldots, x(-2), x(-1), \underset{\uparrow}{x(0)}, x(1), x(2), \ldots\} \tag{2.1}$$

   where the under arrow is used to locate the element $x(0)$ with the elements $x(1)$, $x(2)$, $x(3)$, etc. occurring immediately to the right and $x(-1)$, $x(-2)$, $x(-3)$, etc. falling immediately to the left of the element $x(0)$.
3. We may graphically display a sequence.

To demonstrate these different representation methods, let us consider the specific sequence whose individual elements are given by the mathematical formula

$$x(n) = \begin{cases} \left(\dfrac{1}{2}\right)^n & \text{for } 0 \le n \le 4 \\ 1 + n(2)^{n+2} & \text{for } -3 \le n \le -1 \\ 0 & \text{otherwise} \end{cases} \tag{2.2}$$

In this expression, it is important to realize that the independent time variable n is *restricted* to be an integer. Thus, the inequality $0 \le n \le 4$ is to be interpreted as n taking on the integer values 0, 1, 2, 3, and 4, exclusively, and not as n taking on all values in this interval (e.g., $n \neq 1/2$). It is also possible to represent this sequence in a tabular array. Specifically, after evaluating expression (2.2) for $-3 \le n \le 4$, the following tabular format is generated

$$\boldsymbol{x} = \left\{\ldots, 0, 0, 0, -\frac{1}{2}, -1, -1, \underset{\uparrow}{1}, \frac{1}{2}, \frac{1}{4}, \frac{1}{8}, \frac{1}{16}, 0, 0, 0, \ldots\right\}$$

The need for an under arrow to locate the $x(0)$ element is made clear in this example. If this point of reference is not specified, the tabular array would be ill-defined, since one would then be unable to ascertain which term corresponded to $n = 0$. Finally, the given sequence may also be depicted graphically as shown in Fig. 2.1. In this graphical representation, we reemphasize again that $x(n)$ is defined only

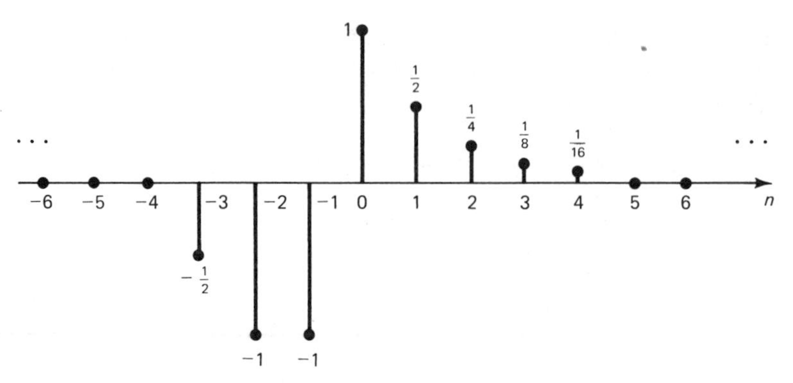

Figure 2.1 Graphical representation of sequence (2.2).

for n as an integer, (for example, $x(1/2)$ is meaningless). It is only for illustrative purposes that the abscissa axis is drawn as a continuous line.

Which of the three sequence representation methods one chooses to use in a given application is very much dependent on the particular sequence being considered. For instance, in some cases it may not be possible to represent a given sequence by a convenient mathematical expression. Of necessity, we must then resort to one of the two remaining representation procedures. Fortunately, in many applications it is usually possible to represent the constituent sequences by means of convenient, simple closed-form expressions.

We now give an interpretation to a sequence expressed in its tabular format

$$\boldsymbol{x} = \{\ldots, x(-2), x(-1), \underset{\uparrow}{x(0)}, x(1), x(2), \ldots\} \tag{2.3}$$

which has important implications relative to our subsequent studies of signals and systems. Namely, we can consider the sequence as specified by representation (2.3) to be a row vector containing an infinite number of components. The reader is undoubtedly familiar with the finite dimensional version of this vector concept (e.g., the three-dimensional vector $\{2, -1, 4\}$). In making this vector interpretation, the following advantages accrue:

1. We may use many of the powerful concepts from linear algebra to further our understanding of signals and systems.
2. The apparently different topics of discrete- and continuous-time signals and systems may be studied from a common setting.
3. We may use the convenient notation of vectors and operators to concisely represent various relationships.

The reader who is unfamiliar with standard vector concepts should not be overly concerned, however, since the fundamental vector space techniques needed are developed herein at a rather relaxed pace. Our plan is that of developing the *classical* and the *vector-space* approaches to the study of signals and systems simultaneously. This is done in appreciation of the fact that the classical approach is usually much easier to initially comprehend. We then build upon this comprehension to aid one's understanding of the more abstract vector-space approach. This is important since the vector-space approach eventually yields a deeper degree of insight and a better understanding of the fundamental concepts of signals and systems.

In terms of notation, we usually denote a sequence by $\{x(n)\}$—or more concisely $x(n)$—when using the classical approach. On the other hand, the boldface italicized symbol $\boldsymbol{x}$ is incorporated when referring to a sequence while taking the vector-space approach. Although the notations $\{x(n)\}$ and $\boldsymbol{x}$ are equivalent representations of a sequence, there is much to be gained in terms of understanding by incorporating these separate notations.

2.2 CHANGING THE TIME VARIABLE

In much of what is to follow regarding our investigation of signals and systems, it is necessary to make a so-called *change in the time variable.* In discrete-time system studies, there are two fundamental changes that can be made: (1) shifting (or translating) the time variable, and (2) transposing the time variable. These two procedures can be used separately or in combination form. The reader is advised to attain a thorough understanding of these procedures since they are frequently used. Many of the more fundamental properties of signals and systems are revealed only after one makes a suitable change in the underlying time variable. Fortunately, there exists a very straightforward method for making such changes that avoids the inherent difficulties that might otherwise arise.

Shifting the Time Variable

Right shifting or delaying the sequence $\{x(n)\}$ by a k-unit increment is accomplished by replacing the time variable n by $n - k$ everywhere it appears in the representation $\{x(n)\}$. Formally, the sequence $\{x(n)\}$ is changed into the sequence $\{x(n-k)\}$ in the following manner:

$$\{x(n-k)\} = \{\ldots, x(-2-k), x(-1-k), \underset{\uparrow}{x(-k)}, x(1-k), x(2-k), \ldots\} \tag{2.4}$$

where k is a fixed integer. A careful interpretation of relationship (2.4) indicates that the sequence $\{x(n-k)\}$ is generated by simply shifting each element of the original sequence $\{x(n)\}$ by k time units to the right. This is made apparent by noting that the order of the elements of the sequence $\{x(n-k)\}$ remains the same as in $\{x(n)\}$, but the $n = 0$ term is now $x(-k)$ instead of $x(0)$.

Let us illustrate this shifting of the time variable for the specific time sequence as characterized by

$$x(n) = \begin{cases} \dfrac{1}{2} & \text{for } -\infty < n \leq -3 \\[2ex] 2 + \dfrac{n}{2} & \text{for } \; -2 \leq n \leq 0 \\[2ex] -\sin \dfrac{\pi n}{2} & \text{for } \quad 1 \leq n \leq 3 \\[2ex] 1 & \text{for } \quad 4 \leq n < \infty \end{cases} \tag{2.5}$$

It is first noted that this sequence is governed by four distinct expressions over the four intervals $-\infty < n \leq -3$, $-2 \leq n \leq 0$, $1 \leq n \leq 3$, and $4 \leq n < \infty$. Thus, the particular formula that is used to determine the element $x(n - k)$ is dependent upon in which interval $n - k$ lies. With this in mind, it follows that

$$x(n-k) = \begin{cases} \dfrac{1}{2} & \text{for } -\infty < n - k \leq -3 \\[2ex] 2 + \dfrac{1}{2}(n-k) & \text{for } \quad 2 \leq n - k \leq 0 \\[2ex] -\sin \dfrac{\pi(n-k)}{4} & \text{for } \quad 1 \leq n - k \leq 3 \\[2ex] 1 & \text{for } \quad 4 \leq n - k < \infty \end{cases}$$

It is next noted that the set of integers n that satisfy the inequality $n_1 \leq n - k \leq n_2$ is identical to the set of integers that satisfy the "equivalent" inequality $n_1 + k \leq n \leq n_2 + k$. Therefore, the desired expression for the shifted sequence is

$$x(n-k) = \begin{cases} \dfrac{1}{2} & \text{for } \quad -\infty < n \leq -3 + k \\[2ex] 2 + \dfrac{1}{2}(n-k) & \text{for } -2 + k \leq n \leq k \\[2ex] -\sin \dfrac{\pi(n-k)}{4} & \text{for } \quad 1 + k \leq n \leq 3 + k \\[2ex] 1 & \text{for } \quad 4 + k \leq n < \infty \end{cases}$$

This sequence is simply a version of the original sequence with a right shift by k time units, as is made apparent in Fig. 2.2, where $\{x(n - k)\}$ is plotted for the cases $k = 0$ (no shift), $k = -1$ (a left shift of one unit), and $k = 2$ (a right shift of two units).

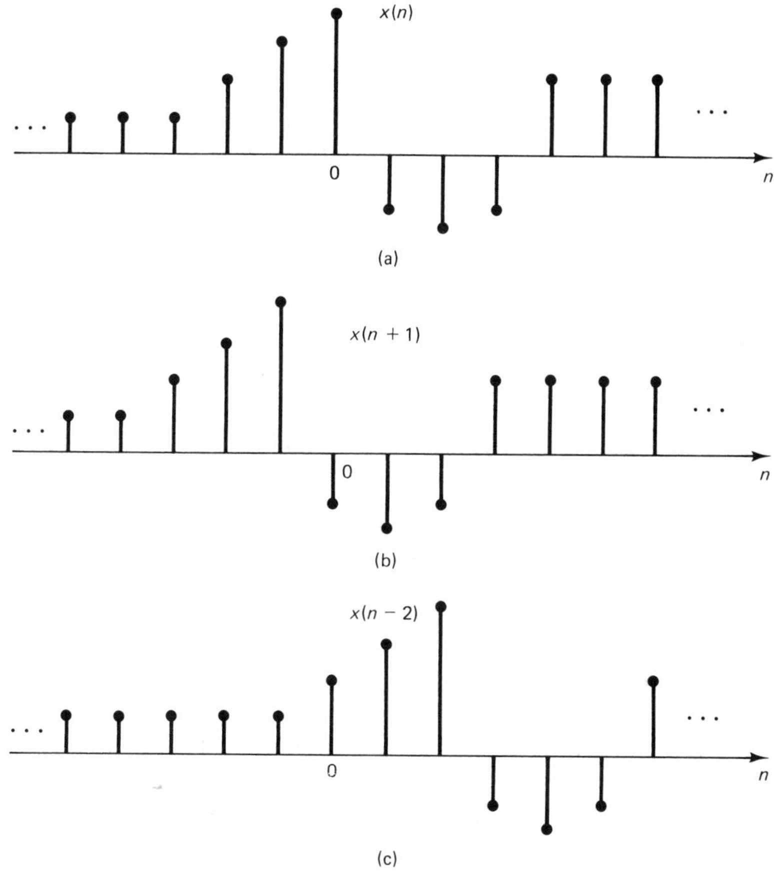

Figure 2.2 Examples of shifting the time variable: (a) $k = 0$, (b) $k = -1$, and (c) $k = 2$.

Transposing the Time Variable

One achieves a transpose of the time variable by replacing the variable n by $-n$ everywhere it appears in the representation of $\{x(n)\}$. Namely, the sequence $\{x(n)\}$ becomes the transposed sequence $\{x(-n)\}$ where

$$\{x(-n)\} = \{\ldots, x(2), x(1), \underset{\uparrow}{x(0)}, x(-1), x(-2), \ldots\} \tag{2.6}$$

If one were to graph the sequences $\{x(n)\}$ and $\{x(-n)\}$ versus n, they would be mirror images of one another with respect to the origin (that is, $n = 0$). This is a direct result of the trivial identity $x(-(-n)) = x(n)$.

To illustrate the transpose of the time-variable operation, let us consider the sequence specified by equation (2.5). As a first step, we replace n by $-n$ everywhere it appears to yield

$$x(-n) = \begin{cases} \dfrac{1}{2} & \text{for } -\infty < -n \leq -3 \\[2ex] 2 - \dfrac{1}{2}n & \text{for } -2 \leq -n \leq 0 \\[2ex] -\sin \dfrac{-\pi n}{4} & \text{for } \quad 1 \leq -n \leq 3 \\[2ex] 1 & \text{for } \quad 4 \leq -n < \infty \end{cases}$$

It is next observed that the set of integers n that satisfy the inequality $n_1 \leq -n \leq n_2$ is identical to $-n_1 \geq n \geq -n_2$. Thus, the desired expression for the transposed sequence is given by

$$x(-n) = \begin{cases} \dfrac{1}{2} & \text{for } \infty > n \geq 3 \\[2ex] 2 - \dfrac{n}{2} & \text{for } \quad 2 \geq n \geq 0 \\[2ex] \sin \dfrac{\pi n}{2} & \text{for } -1 \geq n \geq -3 \\[2ex] 1 & \text{for } -4 \geq n > -\infty \end{cases}$$

The mirror image characterization of the sequences $\{x(n)\}$ and $\{x(-n)\}$ is made apparent in the plots shown in Fig. 2.3.

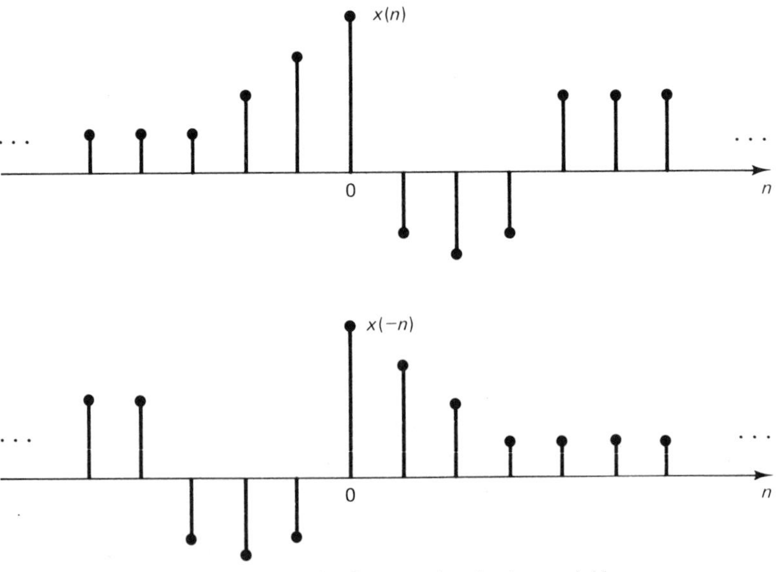

Figure 2.3 Example of transposing the time variable.

Combination of Shifting and Transposing Operations

It is possible to use the approach just taken to make a change of the time variable that combines the shifting and transpose operations. Specifically, the sequence $\{x(-n-k)\}$ is formally given by

$$\{x(-n-k)\} = \{\ldots, x(2-k), x(1-k), \underset{\uparrow}{x(-k)}, x(-1-k), x(-2-k), \ldots\} \tag{2.7}$$

A careful interpretation of the sequence $\{x(-n-k)\}$ indicates that it can be generated by first transposing the sequence $x(n)$ to form $x(-n)$ and then left shifting the elements of $\{x(-n)\}$ each by k time units to generate the desired sequence $\{x(-n-k)\}$.

2.3 SIGNAL OPERATIONS

The digital processing of information that appears in the format of a sequence of numbers entails a systematic procedure whereby the information-bearing sequence $\{x(n)\}$ is transformed into (that is, changed or operated upon) another sequence that is denoted by $\{y(n)\}$. We represent this mathematical transformation of one sequence into another sequence by means of the associational relationship

$$\{x(n)\} \xrightarrow[T]{} \{y(n)\} \tag{2.8a}$$

which is to be read as "sequence $\{x(n)\}$ is transformed into sequence $\{y(n)\}$ by means of a well-defined rule," which is here denoted by the uppercase boldface italicized symbol $\boldsymbol{T}$. In virtually every case of practical interest, the *operator* rule $\boldsymbol{T}$ appears in the form of a mathematical formula relating the elements of sequence $\{y(n)\}$ to those of sequence $\{x(n)$. Thus, given a specific sequence $\{x(n)\}$, one simply applies this characterizing formula to generate the corresponding elements of sequence $\{y(n)\}$.

The concept of transforming one sequence into another by means of a well-defined operator rule appears rather abstract in the general setting as given by expression (2.8a). This representation procedure has been made purposely general in order to include all possible operator rules. In any specific application, however, the meaning is quite clear as the following example demonstrates.

Example 2.1

To illustrate the operator rule concept, let us consider the special case whereby the elements of the sequences $\{y(n)\}$ and $\{x(n)\}$ are related to one another by the so-called *first-difference* operator rule

$$y(n) = x(n) - x(n-1) \qquad \text{for all } n$$

Namely, this operator rule states that the nth element of sequence $\boldsymbol{y}$ is obtained by simply subtracting the $\boldsymbol{x}$ sequence's $n-1$st element from its nth element. Clearly, given any sequence $\boldsymbol{x}$, we can systematically use this rule to generate the corresponding sequence $\boldsymbol{y}$ on an element-by-element basis.

For instance, let us now find the sequence $\boldsymbol{y}$ that results when applying this operator rule to the specific sequence

$$x(n) = \begin{cases} 1 & \text{for } n \geq 0 \\ 0 & \text{for } n < 0 \end{cases}$$

The general nth element $y(n)$ is readily obtained by performing the difference operation $x(n) - x(n-1)$. For the specified sequence $\boldsymbol{x}$, it is seen that the element values $x(n)$ and $x(n-1)$ are identical when n is either positive or negative. This observation then yields

$$\begin{aligned} y(n) &= x(n) - x(n-1) \\ &= 0 - 0 = 0 \qquad \text{for all } n < 0 \end{aligned}$$

and

$$\begin{aligned} y(n) &= x(n) - x(n-1) \\ &= 1 - 1 = 0 \qquad \text{for all } n > 0 \end{aligned}$$

The only other element to be determined is $y(0)$, which is simply given by

$$\begin{aligned} y(0) &= x(0) - x(-1) \\ &= 1 - 0 = 1 \end{aligned}$$

Thus, the sequence $\{y(n)\}$ that results when the given sequence $\{x(n)\}$ is operated upon is found to be

$$y(n) = \begin{cases} 1 & \text{for } n = 0 \\ 0 & \text{for } n \neq 0 \end{cases}$$

or equivalently,

$$\boldsymbol{y} = \{\ldots, 0, 0, 0, \underset{\uparrow}{1}, 0, 0, 0, \ldots\}$$

The rule used in the example above [i.e., $y(n) = x(n) - x(n-1)$], although of an elementary nature, depicts the types of operations that are typically used in discrete-time system studies. It should be apparent that, given any sequence $\{x(n)\}$, it is a conceptually simple matter to generate the corresponding sequence $\{y(n)\}$ by means of the characterizing rule. In general, this is done on an element-by-element basis. Namely, one uses the characterizing rule to generate the individual elements one at a time, that is, $\ldots$, $y(-2)$, then $y(-1)$, then $y(0)$, and so forth, $\ldots$, .

This process of transforming one sequence into another sequence as depicted by the *associational relationship* in expression (2.8a) could also be represented in terms of the equivalent operational equation

$$\{y(n)\} = \boldsymbol{T}\{x(n)\} \quad \text{or} \quad \boldsymbol{y} = \boldsymbol{T}\boldsymbol{x} \tag{2.8b}$$

In this representation, the symbol $\boldsymbol{T}$ is said to be an *operator*, which when applied to the sequence $\boldsymbol{x}$ gives rise to the sequence $\boldsymbol{y}$. Although representations (2.8a) and (2.8b) each depict the same process, we prefer to use the latter for much of what is to follow. This operational equation is shown in Fig. 2.4, where the arrows indicate the

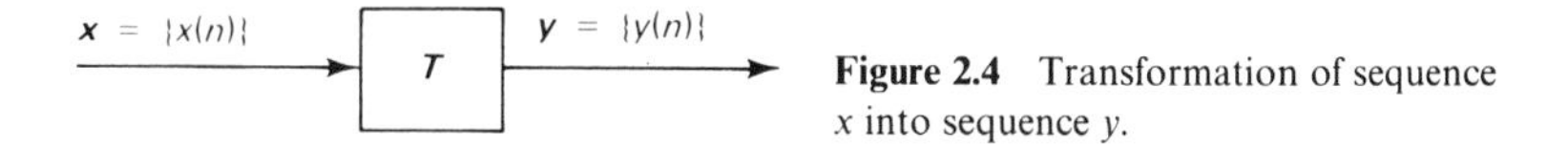

Figure 2.4 Transformation of sequence x into sequence y.

direction of information flow. We refer to $\boldsymbol{x}$ as the *excitation* (or *input*) *sequence* and $\boldsymbol{y}$ as the resultant *response* (or *output*) *sequence*. The input sequence, then, is said to influence the behavior of the response sequence through the operator rule $\boldsymbol{T}$. This cause-and-effect relationship between sequences $\boldsymbol{x}$ and $\boldsymbol{y}$ constitutes the dynamical behavior of the operator rule $\boldsymbol{T}$.

In most practical applications, the values of the input sequence elements are made available on a sequential time basis. Namely, at time $n = 0$ the value of $x(0)$ is specified, then at time $n = 1$ the value of $x(1)$ is specified, and so forth. We can then envision a chain of input sequence elements being supplied in this sequential time basis to the input terminal of operator $\boldsymbol{T}$ as shown in Fig. 2.4. This operator rule is simultaneously applied to the given input sequence to, in turn, generate the response sequence elements in this sequential time manner. Thus, at time $n = 0$ the value of $y(0)$ is generated, at time $(n - 1)$ the value of $y(1)$ is generated, and so forth.

2.4 ELEMENTARY OPERATIONS ON SIGNALS

The transform operation as denoted by expressions (2.8a) and (2.8b) is quite general in that it encompasses all possible procedures for transforming one sequence into another. If there is to be any hope of developing a desirable analytical structure to our study of sequence operators, however, we must of necessity restrict the types of operations (transformations) to be considered. What is so surprising, and at the same time encouraging, is that it is possible to achieve extremely sophisticated forms of discrete-time–signal processing by means of the three elementary sequence operations: (1) shifting, (2) multiplication by a constant, and (3) summation of sequences.

Shift Operation

In the right-shift (or delay) operation, the input sequence $\{x(n)\}$ is operated upon to produce the response sequence $\{y(n)\}$ according to the well-defined rule

$$y(n) = x(n - 1) \tag{2.9}$$

for all integer values of n. This is a particularly simple rule since the value of each element of the response sequence is seen to be dependent on only one element of the input sequence. The right-shift operator is denoted by the suggestive symbol $\boldsymbol{S}$ (that is, $\boldsymbol{S}$ for shift), so that its input and response sequences are operationally related by

$$\left.\begin{aligned} &\boldsymbol{y} = \boldsymbol{S}\boldsymbol{x} \\ \text{where} \qquad & \\ &y(n) = x(n-1) \qquad \text{for all } n \end{aligned}\right\} \tag{2.10a}$$

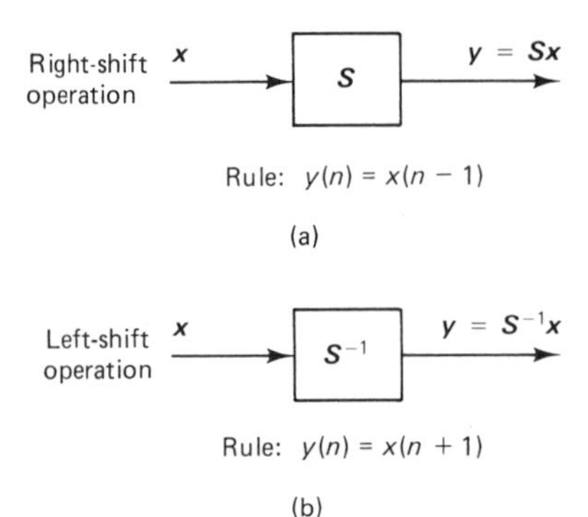

Figure 2.5 Elementary right- and left-shift operators.

The right-shift operation is depicted as shown in Fig. 2.5a. Using relationship (2.10a), it is seen that the sequences $\boldsymbol{x}$ and $\boldsymbol{Sx}$ have the simple tabular array correspondence as given by

$$\left.\begin{aligned} \boldsymbol{x} &= \{\ldots, x(-2), x(-1), \underset{\uparrow}{x(0)}, x(1), x(2), \ldots\} \\ \boldsymbol{Sx} &= \{\ldots, x(-3), x(-2), \underset{\uparrow}{x(-1)}, x(0), x(1), \ldots\} \end{aligned}\right\} \quad (2.10b)$$

Thus, the sequence $\boldsymbol{Sx}$ is seen to be a replication of the sequence $\boldsymbol{x}$ shifted one unit to the right. Alternatively, we could have referred to $\boldsymbol{S}$ as a delay operator, since the response sequence is identical to the input sequence except delayed by a one time unit.

Example 2.2

Determine the sequence that results when the sequence expressed in the tabular array

$$\boldsymbol{x} = \{\ldots, 0, 0, -2, \underset{\uparrow}{4}, 1, 1, 3, 0, 0, 0, \ldots\}$$

is operated upon by the right-shift operator S. Applying relationship (2.10a), it follows that

$$\boldsymbol{y} = \boldsymbol{Sx} = \{\ldots, 0, 0, 0, -2, \underset{\uparrow}{4}, 1, 1, 3, 0, 0, \ldots\}$$

These two sequences are shown in Fig. 2.6, where the right-shift characteristic is again in evidence.

It is possible to generalize the right-shift operator concept in an obvious manner. This is readily demonstrated by considering the task of generating a sequence $\boldsymbol{w}$ that is identical to a sequence $\boldsymbol{x}$ except shifted to the right by two time units, that is, $w(n) = x(n-2)$. One may achieve this operation by simply right shifting the sequence $\boldsymbol{Sx}$, that is,

$$\boldsymbol{w} = \boldsymbol{S}[\boldsymbol{Sx}]$$

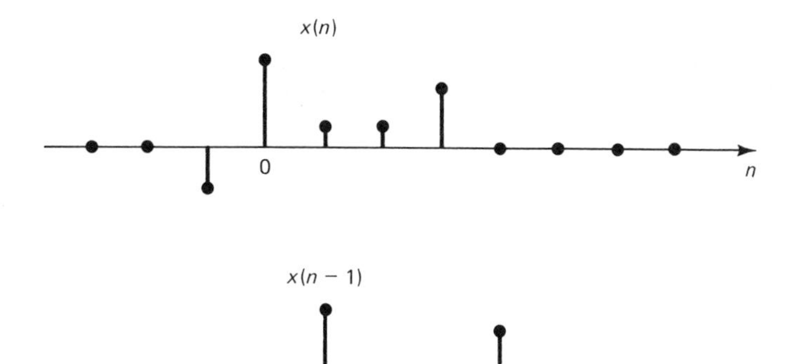

Figure 2.6 Sequence and its right-shifted version.

To demonstrate the validity of this assertion, we will use the tabular array representation of the sequence $\boldsymbol{Sx}$

$$\boldsymbol{Sx} = \{\ldots, x(-3), x(-2), x(-1), \underset{\uparrow}{x(0)}, x(1), x(2), \ldots\}$$

so that

$$\boldsymbol{S}[\boldsymbol{Sx}] = \{\ldots, x(-4), x(-3), x(-2), \underset{\uparrow}{x(-1)}, x(0), x(1), \ldots\}$$

Clearly, the sequence $\boldsymbol{S}[\boldsymbol{Sx}]$ possesses the desired property for a right shift by two time units in which $w(n) = x(n-2)$.

In this procedure, it was necessary to apply the right-shifting operator $\boldsymbol{S}$ first to the sequence $\boldsymbol{x}$ and then to the sequence $\boldsymbol{Sx}$. It is notationally convenient to represent this compound operation as

$$\boldsymbol{w} = \boldsymbol{S}[\boldsymbol{Sx}] = \boldsymbol{S}^2\boldsymbol{x} \tag{2.11}$$

where the exponent 2 over the right-shift operator $\boldsymbol{S}$ implies a twofold application of operator $\boldsymbol{S}$. This double-shift operation may be depicted as shown in Fig. 2.7, where the sequence element values that appear at time n are shown below the input and output terminals of each of the shift operators.

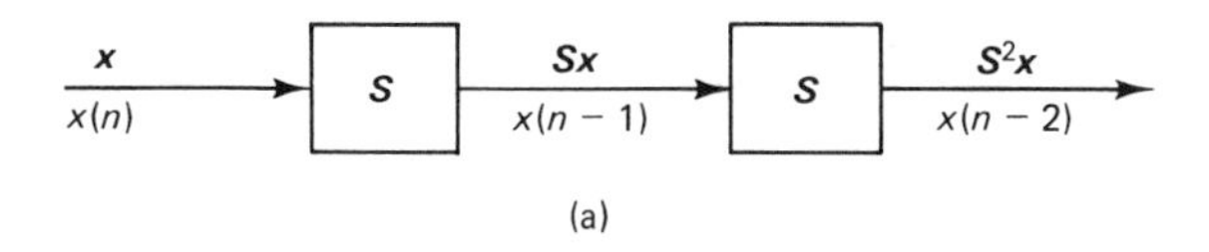

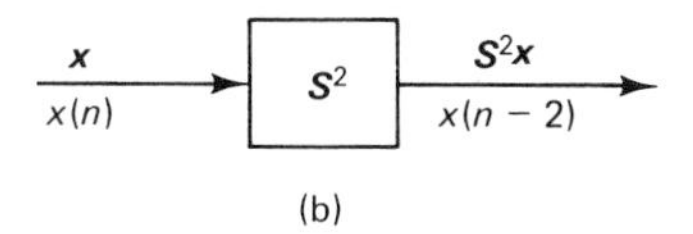

Figure 2.7 Equivalent representations of a double right-shift operation: (a) actual configuration of two right-shift operators and (b) mathematical equivalent.

Continuing on in an obvious manner, it is apparent that a k-fold application of the right-shift operator $\boldsymbol{S}$ yields a right shift of k time units. Therefore, if the sequences $\boldsymbol{x}$ and $\boldsymbol{y}$ are operationally related to one another by

$$\left.\begin{aligned} &\boldsymbol{y} = \boldsymbol{S}^k\boldsymbol{x} \\ \text{Then} \\ &y(n) = x(n-k) \qquad \text{for all } n \end{aligned}\right\} \tag{2.12}$$

where k is a fixed integer. In this representation, the operator $\boldsymbol{S}^k$ denotes the compound operation

$$\boldsymbol{S}^k = \underbrace{\boldsymbol{S}\ \ \boldsymbol{S}\ \ \boldsymbol{S}\ \ \cdots\ \ \boldsymbol{S}}_{k \text{ times}}$$

The integer k in relationship (2.12) has been conceived of as being strictly positive, so that the operator $\boldsymbol{S}^k$ is said to impose a right shift (or delay) of k time units to the sequence upon which it operates. There is nothing, however, that restricts us from considering k to be a negative integer. In fact, a literal interpretation of expression (2.12) for k negative would imply that the sequence $\boldsymbol{S}^k\boldsymbol{x}$ is a replication of sequence $\boldsymbol{x}$ shifted (or advanced) to the left by $-k$ time units. With this in mind, the left-shift operator $\boldsymbol{S}^{-1}$ is introduced where

$$\left.\begin{aligned} &\boldsymbol{y} = \boldsymbol{S}^{-1}\boldsymbol{x} \\ \text{implies} \\ &y(n) = x(n+1) \qquad \text{for all } n \end{aligned}\right\} \tag{2.13a}$$

The sequences $\boldsymbol{x}$ and $\boldsymbol{S}^{-1}\boldsymbol{x}$ are then seen to be related in the left-shifted fashion as given by

$$\left.\begin{aligned} \boldsymbol{x} &= \{\ldots, x(-1), \underset{\uparrow}{x(0)}, x(1), x(2), \ldots\} \\ & \qquad\quad \swarrow \quad \swarrow \quad \swarrow \\ \boldsymbol{S}^{-1}\boldsymbol{x} &= \{\ldots, x(0), \underset{\uparrow}{x(1)}, x(2), x(3), \ldots\} \end{aligned}\right\} \tag{2.13b}$$

The left-shift operator is depicted as shown in Fig. 2.5b. It is also apparent from this development that the compound operator $\boldsymbol{S}^{-k}$ (that is, the operator $\boldsymbol{S}^{-1}$ applied k times) imparts a left shift of k time units to the sequence upon which it operates.

There is an obviously close relationship existent between the right-shift $\boldsymbol{S}$ and left-shift $\boldsymbol{S}^{-1}$ operators. This is made apparent by noting that if the operator $\boldsymbol{S}^{-1}$ is applied to the sequence $\boldsymbol{Sx}$, then the original sequence $\boldsymbol{x}$ is obtained, that is,

$$\boldsymbol{x} = \boldsymbol{S}^{-1}[\boldsymbol{Sx}]$$

Namely, the right shift imposed upon $\boldsymbol{x}$ by $\boldsymbol{S}$ is effectively removed by the left-shift operator $\boldsymbol{S}^{-1}$ to generate the original sequence $\boldsymbol{x}$. This validity of this property is readily established by the following two-step operation:

$$\boldsymbol{x} = \{\ldots, x(-2), x(-1), \underset{\uparrow}{x(0)}, x(1), x(2), \ldots\}$$

$$\boldsymbol{Sx} = \{\ldots, x(-3), x(-2), \underset{\uparrow}{x(-1)}, x(0), x(1), \ldots\}$$

$$\boldsymbol{S}^{-1}[\boldsymbol{Sx}] = \{\ldots, x(-2), x(-1), \underset{\uparrow}{x(0)}, x(1), x(2), \ldots\} = x$$

In a similar fashion, one can establish the relationship

$$\boldsymbol{S}\,\boldsymbol{S}^{-1}\boldsymbol{x} = \boldsymbol{x}$$

which implies that the left shift imposed by operator $\boldsymbol{S}^{-1}$ upon $\boldsymbol{x}$ is effectively removed by a right-shift operation. The right- and left-shift operators are then said to be the "inverses" of each other in the sense that one restores the changes created by the other (i.e., $\boldsymbol{S}\,\boldsymbol{S}^{-1}\boldsymbol{x} = \boldsymbol{S}^{-1}\,\boldsymbol{Sx} = \boldsymbol{x}$). It then follows that when either of the two compound operators $\boldsymbol{S}\,\boldsymbol{S}^{-1}$ and $\boldsymbol{S}^{-1}\,\boldsymbol{S}$ is applied to any sequence $\boldsymbol{x}$, the resultant response sequence is identical to $\boldsymbol{x}$. This observation is operationally denoted by

$$\boldsymbol{S}\,\boldsymbol{S}^{-1} = \boldsymbol{S}^{-1}\,\boldsymbol{S} = \boldsymbol{I} \tag{2.14}$$

where $\boldsymbol{I}$ is the *identity operator* that leaves any sequence unchanged, that is,

$$\boldsymbol{Ix} = \boldsymbol{x} \qquad \text{for all sequences of } \boldsymbol{x} \tag{2.15}$$

An operator algebra can be now generated strictly based on the right- and left-shift operators. For instance, if the sequences $\boldsymbol{x}$ and $\boldsymbol{y}$ are related by

$$\left.\begin{aligned} \boldsymbol{y} &= \boldsymbol{Sx} \\ \text{then} \qquad & \\ \boldsymbol{x} &= \boldsymbol{S}^{-1}\boldsymbol{y} \end{aligned}\right\} \tag{2.16}$$

and vice versa. This relationship is readily proven by first operating on each side of the sequence identity $\boldsymbol{y} = \boldsymbol{Sx}$ by the operator $\boldsymbol{S}^{-1}$ to yield $\boldsymbol{S}^{-1}\boldsymbol{y} = \boldsymbol{S}^{-1}\,\boldsymbol{Sx}$. It is next noted that $\boldsymbol{S}^{-1}\,\boldsymbol{Sx} = \boldsymbol{x}$, thereby giving the desired relationship $\boldsymbol{S}^{-1}\boldsymbol{y} = \boldsymbol{x}$. This result can also be obtained by noting that the first expression in equation (2.16) implies $y(n) = x(n-1)$, whereas the second requires $x(n) = y(n+1)$ or, equivalently, $x(n-1) = y(n)$. These two relationships are then seen to be identical so that one implies the other.

It can also be shown that the shift operators $\boldsymbol{S}$ and $\boldsymbol{S}^{-1}$ satisfy normal polynomial relationships. Specifically, using the readily proven identity $\boldsymbol{S}^{p-q}\boldsymbol{x} = \boldsymbol{S}^{p}\,\boldsymbol{S}^{-q}\boldsymbol{x}$, it follows that

$$\boldsymbol{S}^{p}\,\boldsymbol{S}^{-q} = \boldsymbol{S}^{p-q} \tag{2.17}$$

for all integer values of p and q. This relationship will be of fundamental importance in our subsequent studies of linear discrete-time systems.

In summary, the compound operator $\boldsymbol{S}^k$ is a shift operator that imparts a right shift of k time units upon the sequence to which it is applied. If the integer k is positive (negative), then a right (left) shift of k (or $-k$) time units is imparted. On the other hand, when k is set to zero, the response sequence $\boldsymbol{S}^0\boldsymbol{x}$ is seen to be simply $\boldsymbol{x}$ as indicated by expression (2.12). Thus, the operator $\boldsymbol{S}^0$ is, in fact, equal to the identity operator $\boldsymbol{I}$.

Multiplication Operation by a Constant

The multiplication operation by a constant has a particularly simple meaning. Namely, the sequence $\boldsymbol{y}$ is said to be equal to the sequence $\boldsymbol{x}$ multiplied by the scalar α as denoted by

$$\left.\begin{aligned} \boldsymbol{y} &= \alpha\boldsymbol{x} \\ \text{if} \qquad & \\ y(n) &= \alpha x(n) \qquad \text{for all } n \end{aligned}\right\} \tag{2.18a}$$

Thus, the sequence $\alpha\boldsymbol{x}$ is seen to have elements that are simply scalar multiples of the corresponding elements of sequence $\boldsymbol{x}$ as given by

$$\alpha\boldsymbol{x} = \{\ldots, \alpha x(-2), \alpha x(-1), \underset{\uparrow}{\alpha x(0)}, \alpha x(1), \alpha x(2), \ldots\} \tag{2.18b}$$

This operation is depicted in Fig. 2.8a.

Example 2.3

Determine the sequence $\boldsymbol{y} = 0.5\boldsymbol{x}$ when the sequence $\boldsymbol{x}$ is given by

$$\boldsymbol{x} = \{\ldots, 3, 4, \underset{\uparrow}{1}, 7, 6, \ldots\}$$

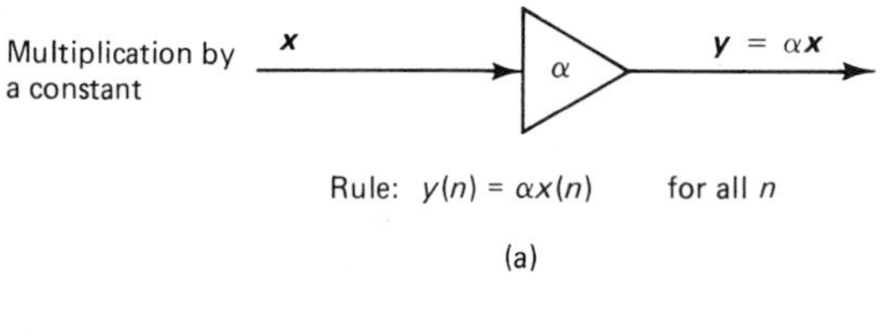

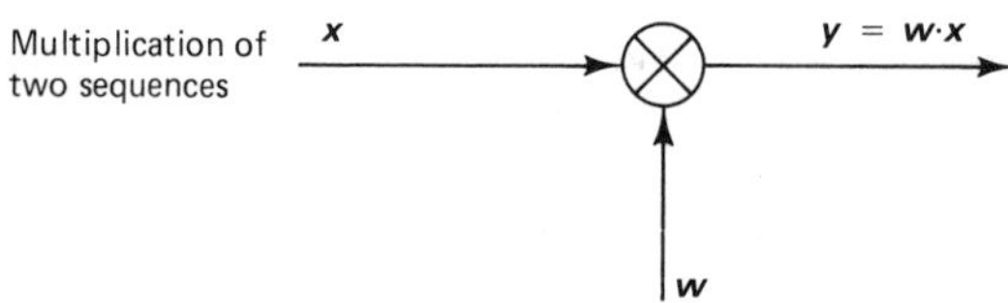

Figure 2.8 Elementary operations of (a) multiplying by a constant and (b) multiplication of two sequences.

Using the defining relationships (2.18a) we have

$$y = \frac{1}{2}x = \{\ldots, 1.5, -2, \underset{\uparrow}{0.5}, 3.5, 3, \ldots\}$$

The multiplication operation has been carried out on the specific elements $x(-2)$, $x(-1)$, $x(0)$, $x(1)$, and $x(2)$ with the clear implication that the remaining elements are similarly multiplied by $\frac{1}{2}$.

Example 2.4

Determine the sequence $\boldsymbol{y} = 0.5\boldsymbol{x}$ for the sequence $\boldsymbol{x}$ given by

$$x(n) = \begin{cases} (0.25)^n & \text{for } n \geq 0 \\ 0 & \text{for } n < 0 \end{cases}$$

In this case, the sequence $\boldsymbol{x}$ being operated upon is characterized by a convenient formula. The sequence $\boldsymbol{y} = \frac{1}{2}\boldsymbol{x}$ is therefore also similarly characterized and is given by (after using definition (2.18a))

$$y(n) = (0.5)\ x(n) = \begin{cases} (0.25)^{n+2} & \text{for } n \geq 0 \\ 0 & \text{for } n < 0 \end{cases}$$

In the operational expression $\boldsymbol{y} = \alpha\boldsymbol{x}$, it is important to realize that $\boldsymbol{x}$ and $\boldsymbol{y}$ are sequences and that α is a constant (not a sequence). Whenever this operation is symbolically used, the scalar multiplier appears either as a Greek letter (e.g., $\alpha\boldsymbol{x}$) or as a number (e.g., $2\boldsymbol{x}$). There exists another related operation which is used occasionally in our studies, namely, the product of the two sequences $\boldsymbol{w}$ and $\boldsymbol{x}$, which is defined as another sequence denoted by

$$\left.\begin{aligned} &\boldsymbol{y} = \boldsymbol{w} \cdot \boldsymbol{x} \\ &\text{where} \\ &y(n) = w(n) \cdot x(n) \qquad \text{for all } n \end{aligned}\right\} \quad (2.19)$$

The multiplication of two sequence operations is illustrated as shown in Fig. 2.8b. It is further noted that if the sequence $\boldsymbol{w}$ has all its elements equal to a constant α, then the operations (2.18) and (2.19) are identical.

It is possible to combine the two elementary operations—the shift and multiplication by a constant—to achieve related sequence operations. Let us illustrate this by considering the following sequence operation:

$$\left.\begin{aligned} &\boldsymbol{y} = \alpha\boldsymbol{S}\boldsymbol{x} \\ &\text{which has the time domain interpretation} \\ &y(n) = \alpha x(n-1) \qquad \text{for all } n \end{aligned}\right\} \quad (2.20)$$

Namely, the sequence operation $\alpha\boldsymbol{S}$ is seen to impart a right shift as well as scalar multiplication to the sequence upon which it operates. We can implement this

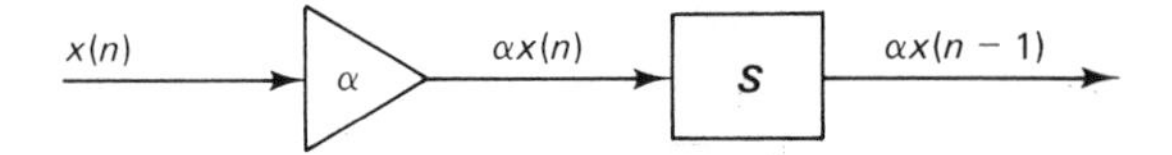

Figure 2.9 Compound operation αS.

compound operation using elementary operations as indicated in Fig. 2.9, where the sequence element values that appear at time n are shown. In a straightforward extension, the more general compound operator $\alpha \boldsymbol{S}^k$ is defined by

$$\left.\begin{aligned} &\boldsymbol{y} = \alpha \boldsymbol{S}^k \boldsymbol{x} \\ \text{where} \quad & \\ &y(n) = \alpha x(n-k) \qquad \text{for all } n \end{aligned}\right\} \tag{2.21}$$

Summation Operation

The last elementary sequence operation to be considered is that in which two or more sequences are additively combined to form the so-called sum sequence. For example, the two sequences $\boldsymbol{w}$ and $\boldsymbol{x}$ are summed to form the sum sequence $\boldsymbol{y}$ as denoted by

$$\left.\begin{aligned} &\boldsymbol{y} = \boldsymbol{w} + \boldsymbol{x} \\ \text{where} \quad & \\ &y(n) = w(n) + x(n) \qquad \text{for all } n \end{aligned}\right\} \tag{2.22}$$

The elements of the sum sequence are then given in the row vector format:

$$\boldsymbol{y} = \{\ldots, w(-1) + x(-1), \underset{\uparrow}{w(0) + x(0)}, w(1) + x(1), \ldots\} \tag{2.23}$$

where $w(n)$ and $x(n)$ denote the nth elements of sequences $\boldsymbol{w}$ and $\boldsymbol{x}$, respectively. The summation operation is depicted in Fig. 2.10a.

It is possible to generalize the concept of sequence summation to include those cases where more than two sequences are summed. Namely, the sum of the m sequences $\boldsymbol{x}_1, \boldsymbol{x}_2, \ldots, \boldsymbol{x}_m$ is denoted by

$$\left.\begin{aligned} &\boldsymbol{x} = \boldsymbol{x}_1 + \boldsymbol{x}_2 + \cdots + \boldsymbol{x}_m \\ \text{where} \quad & \\ &x(n) = x_1(n) + x_2(n) + \cdots + x_m(n) \qquad \text{for all } n \end{aligned}\right\} \tag{2.24}$$

Thus, the elements of the sum sequence x are simply obtained by summing the corresponding elements of the individual sequences forming the sum. This operation is as depicted in Fig. 2.10a except with m arrows entering the sum node.

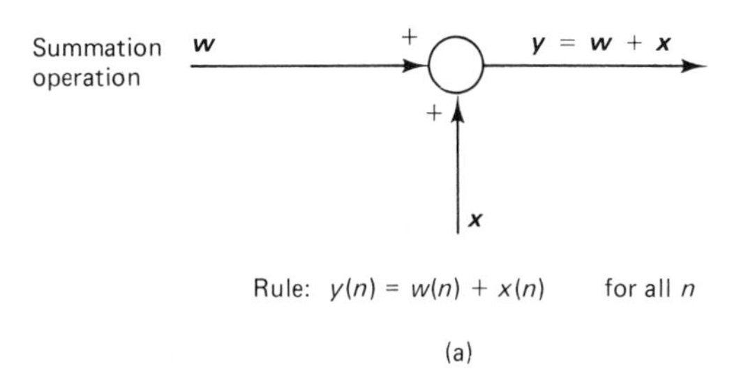

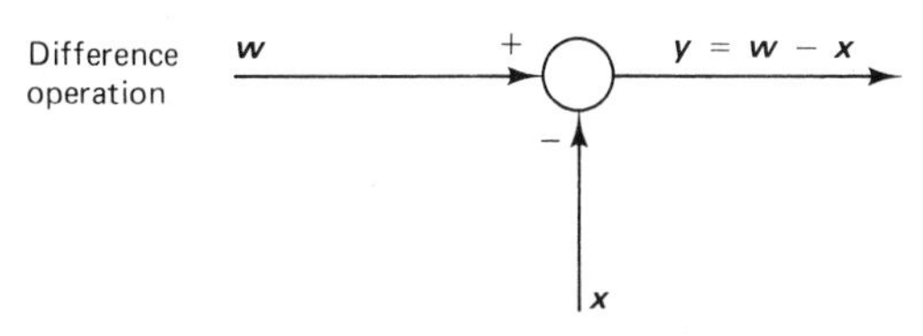

Figure 2.10 Elementary operations of sequence summation and difference.

Example 2.5

Determine the sum of the three sequences

$$x_1 = \{\ldots, -1, 2, \underset{\uparrow}{7}, 3, 1, \ldots\}$$

$$x_2 = \{\ldots, 4, -5, \underset{\uparrow}{1}, 2, 2, \ldots\}$$

$$x_3 = \{\ldots, 2, 1, \underset{\uparrow}{3}, 0, 4, \ldots\}$$

Applying relationships (2.24), it is found that the sum sequence $x = x_1 + x_2 + x_3$ is given by

$$x = \{\ldots, 5, -2, \underset{\uparrow}{11}, 5, 7, \ldots\}$$

It is frequently necessary to generate a sequence that is the so-called difference of two given sequences. The difference sequence $y = w - x$ is defined as being

$$\left.\begin{aligned} y &= w - x \\ \text{where} \quad & \\ y(n) &= w(n) - x(n) \qquad \text{for all } n \end{aligned}\right\} \tag{2.25}$$

This operation is illustrated in Fig. 2.10b, where it is observed that a negative sign appears at the x terminal.

In summary, we have considered the three elementary operations—shift, multiplication by a constant, and summation of sequences—as well as some variants of these operations. It should be apparent that each of these three elementary sequence operations is readily implemented on a digital computer. For instance, the right-shift operator is in effect a memory operation whereby a number $x(n-1)$ is stored at time $n-1$ and recalled at time n to form $y(n) = x(n-1)$. Similarly, multiplication by a constant is simply the operation of multiplying each element of a sequence, applied sequentially, by a constant stored in the computer's memory. Finally, the summation operation corresponds to the adding of the component values of two or more sequences whose values are presented to the computer in a sequential manner.

It is noted that the operations of right shift and multiplication by a constant each require the use of one memory unit in the digital computer. This is important since it will subsequently be shown that many forms of contemporary digital signal processing can be implemented by an appropriate combination of the three elementary operations herein described. The cost of this computer implementation in terms of hardware and computation time is dependent on the required computer memory capability. Thus, in synthesizing a specific digital signal processing operation, it is best to use as few right shift and multiplication by a constant operations as possible to achieve the required processing objective.

2.5 FUNDAMENTAL SIGNALS

There are three basic signals, each governed by a simple mathematical formula, that play dominant roles in our studies of discrete-time signals and systems. They are the unit-impulse (Kronecker delta), the unit-step, and the complex-exponential sequences.

Unit-Impulse Sequence

The simplest and most fundamental sequence used in the theory of linear discrete-time systems is the unit-impulse (or Kronecker delta) sequence as defined by

$$\delta(n) = \begin{cases} 1 & \text{for } n = 0 \\ 0 & \text{for } n \neq 0 \end{cases} \tag{2.26a}$$

or in its equivalent tabular (vector) array

$$\boldsymbol{\delta} = \{\ldots, 0, 0, 0, \underset{\uparrow}{1}, 0, 0, 0, \ldots\} \tag{2.26b}$$

The unit-impulse sequence is seen to be a sequence that is identically zero for all nonzero values of its argument and is one when its argument is zero. This sequence plays such a vital role in our studies that we have reserved the special notation $\boldsymbol{\delta}$ for it.

With the defining rule above for the unit-impulse sequence, it then follows that the related sequence $\{\delta(n-k)\}$ must be characterized by

$$\delta(n-k) = \begin{cases} 1 & \text{for } n = k \\ 0 & \text{for } n \neq k \end{cases} \tag{2.27a}$$

where k is a fixed integer. This sequence is seen to be identical to the unit-impulse sequence except that its *point of application* is at time $n = k$ rather than zero. The sequences $\{\delta(n)\}$ and $\{\delta(n-k)\}$ are seen to be operationally related by

$$\{\delta(n-k)\} = \boldsymbol{S}^k\boldsymbol{\delta} \tag{2.27b}$$

where it is recalled that $\boldsymbol{S}^k$ is the compound k right-shift operator. The unit-impulse sequence $\boldsymbol{\delta}$ and its shifted versions $\boldsymbol{S\delta}$ and $\boldsymbol{S}^{-2}\boldsymbol{\delta}$ are displayed in Fig. 2.11.

One of the most important applications of the unit-impulse sequence arises from the fact that *any* sequence can be expressed as a summation of appropriately weighted and shifted unit-impulse sequences. The necessary steps needed to accomplish this objective are best demonstrated by first considering the specific sequence as given in the tabular (vector) array

$$\boldsymbol{x} = \{\ldots, 0, 0, 0, 5, 0, \underset{\uparrow}{2}, -3, 0, 0, 0, \ldots\}$$

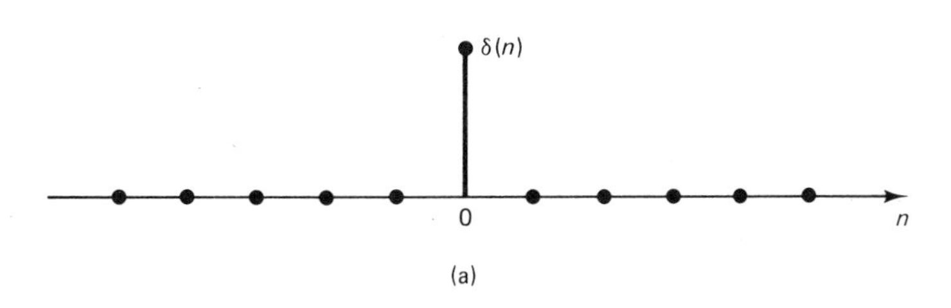

(a)

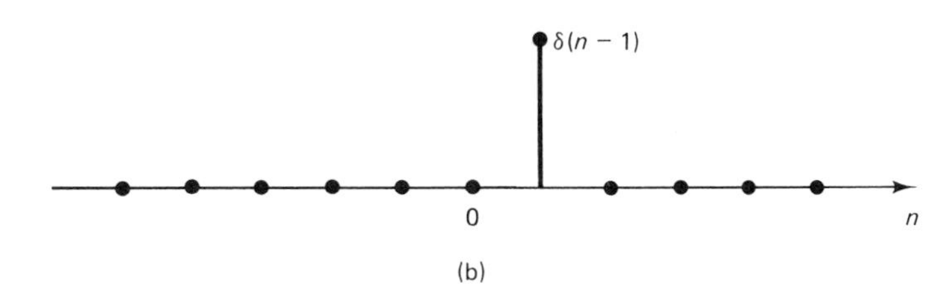

(b)

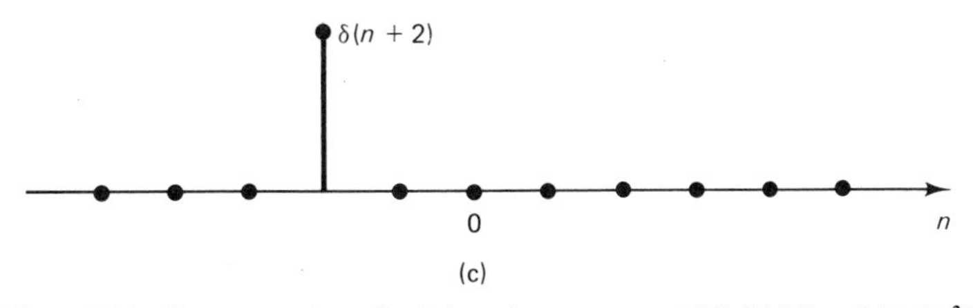

(c)

Figure 2.11 Representation of unit-impulse sequences (a) $\boldsymbol{\delta}$, (b) $\mathrm{S}\boldsymbol{\delta}$, and (c) $\mathrm{S}^{-2}\boldsymbol{\delta}$.

Using the elementary operations of multiplication by a constant and summation, this sequence can be decomposed into the weighted sum of the unit-impulse sequences $\boldsymbol{S}^{-2}\boldsymbol{\delta}$, $\boldsymbol{\delta}$, and $\boldsymbol{S\delta}$, that is,

$$\begin{aligned}\boldsymbol{x} &= 5\{\ldots, 0, 0, 1, 0, \underset{\uparrow}{0}, 0, 0, 0, 0, \ldots\} \\ &\quad + 2\{\ldots, 0, 0, 0, 0, \underset{\uparrow}{1}, 0, 0, 0, 0, \ldots\} \\ &\quad - 3\{\ldots, 0, 0, 0, 0, \underset{\uparrow}{0}, 1, 0, 0, 0, \ldots\} \\ &= 5\boldsymbol{S}^{-2}\boldsymbol{\delta} + 2\boldsymbol{\delta} - 3\boldsymbol{S\delta}\end{aligned}$$

which accomplishes the desired representation. It is sometimes convenient to express this sequence relationship in the time domain by equating the general nth element of each side to yield

$$x(n) = 5\delta(n+2) + 2\delta(n) - 3\delta(n-1)$$

Let us now generalize this approach in order to represent the arbitrary sequence

$$\boldsymbol{x} = \{\ldots, x(-2), x(-1), x(0), x(1), x(2), \ldots\}$$

by an appropriately weighted sum of shifted unit-impulse sequences. This is readily accomplished in an analogous manner and results in[1]

$$\begin{aligned}\boldsymbol{x} &= \cdots + x(-2)\boldsymbol{S}^{-2}\boldsymbol{\delta} + x(-1)\boldsymbol{S}^{-1}\boldsymbol{\delta} + x(0)\boldsymbol{\delta} + x(1)\boldsymbol{S\delta} + x(2)\boldsymbol{S}^{2}\boldsymbol{\delta} + \cdots \\ &= \sum_{k=-\infty}^{\infty} x(k)\boldsymbol{S}^{k}\boldsymbol{\delta} \qquad (2.28a)\end{aligned}$$

The corresponding time-domain characterization is obtained by equating the nth elements of each side of expression (2.28a) to yield

$$x(n) = \sum_{k=-\infty}^{\infty} x(k)\delta(n-k) \qquad (2.28b)$$

It is noted that each summand term, $x(k)\delta(n-k)$, in the right side summation is identically zero for all k except at $k = n$ where it equals $x(n)\delta(0)$. This then yields the desired identity $x(n) = x(n)$.

Although relationships (2.28a) and (2.28b) each yield a procedure for representing a general sequence $\boldsymbol{x} = \{x(n)\}$ in terms of weighted, shifted unit-impulse sequences, they are conceptually very different. Namely, representation (2.28b) is a formula that specifies the general nth element of sequence $\boldsymbol{x}$, whereas representation (2.28a) expresses the sequence $\boldsymbol{x}$ as a weighted [by $x(k)$] sum of shifted, unit-impulse

[1]This result follows from an extended version of the elementary summation operator rule (2.24) in which an infinite number of sequences are summed.

sequences $S^k\delta$. The ability to represent any sequence in the format (2.28a) has important implications in our study of linear discrete-time systems.

Example 2.6

To further demonstrate the unit-impulse method of sequence representation, one may readily verify that the sequence shown in Fig. 2.1 may be concisely expressed as

$$x(n) = -\frac{1}{2}\,\delta(n+3) - \delta(n+2) - \delta(n+1) + \delta(n) + \frac{1}{2}\,\delta(n-1) + \frac{1}{4}\,\delta(n-2) + \frac{1}{8}\,\delta(n-3) + \frac{1}{16}\,\delta(n-4) \qquad \text{for all } n$$

or equivalently

$$\boldsymbol{x} = -\frac{1}{2}\boldsymbol{S}^{-3}\boldsymbol{\delta} - \boldsymbol{S}^{-2}\boldsymbol{\delta} - \boldsymbol{S}^{-1}\boldsymbol{\delta} + \boldsymbol{\delta} + \frac{1}{2}\boldsymbol{S}\boldsymbol{\delta} + \frac{1}{4}\boldsymbol{S}^{2}\boldsymbol{\delta} + \frac{1}{8}\boldsymbol{S}^{3}\boldsymbol{\delta} + \frac{1}{16}\boldsymbol{S}^{4}\boldsymbol{\delta}$$

Unit-Step Sequence

In the study of discrete-time systems, it frequently happens that a given system is suddenly subjected to an input sequence that abruptly changes from a zero level to some other level of constant amplitude. One example of this is a savings account system in which one begins a savings plan and thereafter makes a sequence of constant deposits on a monthly basis. In order to represent the characteristics of this sequence type, we now introduce the so-called unit-step sequence as defined by

$$u(n) = \begin{cases} 1 & \text{for } n \geq 0 \\ 0 & \text{for } n < 0 \end{cases} \tag{2.29a}$$

or in its equivalent tabular (vector) array format

$$\boldsymbol{u} = \{\ldots, 0, 0, 0, \underset{\uparrow}{1}, 1, 1, 1, \ldots\} \tag{2.29b}$$

As in the case of the unit-impulse sequence, we have elected to use a special notation $\boldsymbol{u}$ for the unit-step sequence in order to reflect its importance and frequent use.

It is noted that the unit-step element $u(n)$ is identically equal to one for nonnegative n and zero otherwise. With this in mind, it then follows that the related shifted sequence $\{u(n-k)\}$ has element values given by

$$u(n-k) = \begin{cases} 1 & \text{for } n \geq k \\ 0 & \text{for } n < k \end{cases} \tag{2.30a}$$

where k is a fixed integer. We may equivalently represent this shifted unit-step sequence as

$$\{u(n-k)\} = \boldsymbol{S}^k\boldsymbol{u} \tag{2.30b}$$

Sketches of the sequences $\boldsymbol{u}$, $\boldsymbol{Su}$, and $\boldsymbol{S}^{-2}\boldsymbol{u}$ are shown in Fig. 2.12.

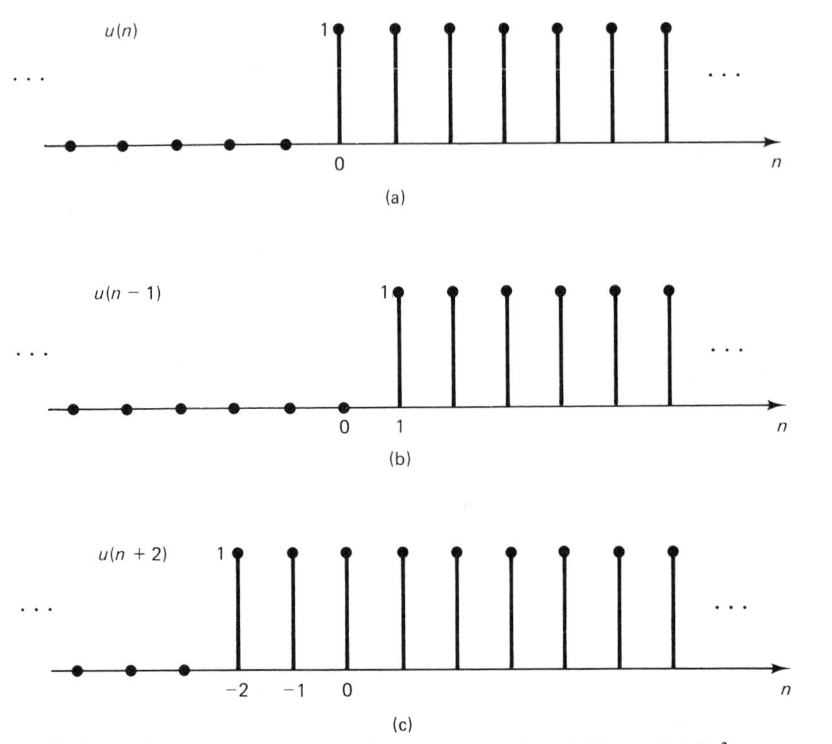

Figure 2.12 Representation of unit-step sequences (a) **u**, (b) **Su**, and (c) $S^{-2}\mathbf{u}$.

The unit-step sequence may also be expressed as a sum of shifted unit-impulse sequences as suggested by representation (2.28a), that is,

$$\boldsymbol{u} = \sum_{k=0}^{\infty} \boldsymbol{S}^k \boldsymbol{\delta} \tag{2.31}$$

by noting that the kth element of sequence $\boldsymbol{S}^k\boldsymbol{\delta} = 1$, whereas all its other elements are zero. If we then sum these shifted unit-impulse sequences as k goes fròm zero to infinity, the result is a sequence whose nonnegative elements are all equal to one and whose negative elements are zero (that is, the unit-step sequence). Conversely, the unit-impulse sequence may be expressed as

$$\begin{aligned} \boldsymbol{\delta} &= \boldsymbol{u} - \boldsymbol{S}\boldsymbol{u} \\ &= (\boldsymbol{I} - \boldsymbol{S})\boldsymbol{u} \end{aligned} \tag{2.32}$$

by observing that the nth element of the sequence $\boldsymbol{u} - \boldsymbol{S}\boldsymbol{u}$ is $u(n) - u(n-1)$. One may readily show that the term $u(n) - u(n-1)$ is identically equal to zero for all n except $n = 0$, 'where it equals one (see Example 2.1). This result can also be obtained by noting that the unit-impulse sequence results when the sequence $\boldsymbol{S}\boldsymbol{u}$ is subtracted from sequence $\boldsymbol{u}$ where these sequences are shown in Fig. 2.12.

Complex Exponential Sequences

In studies related to linear discrete-time systems, it is necessary to characterize those sequences $\boldsymbol{x}$ that satisfy the operational relationship

$$z\boldsymbol{S}\boldsymbol{x} = \boldsymbol{x}$$

or, equivalently,

$$\boldsymbol{S}\boldsymbol{x} = z^{-1}\boldsymbol{x} \tag{2.33}$$

where z is a fixed complex number and $\boldsymbol{S}$ is the right-shift operator considered in the last section. Note that we are here looking for a sequence $\boldsymbol{x}$, which when operated upon by $\boldsymbol{S}$ will give rise to a scalar multiplied version of itself, that is, $z^{-1}\boldsymbol{x}$. Using standard vector-space terminology, it is said that any sequence $\boldsymbol{x}$ that satisfies relationship (2.33) is an *eigensequence* of operator $\boldsymbol{S}$ that has the corresponding *eigenvalue* z^{-1}.

Operational relationship (2.33) may be equivalently expressed in the time domain by equating the general nth components of each side to yield

$$zx(n-1) = x(n) \qquad \text{for all } n \tag{2.34}$$

Thus, the ratio of the eigensequence's nth element to its $n-1$st element—that is, $x(n)/x(n-1)$—is equal to the constant z. It is therefore apparent that *any* sequence that satisfies this time-domain relationship must be a scalar multiplied version of the sequence whose elements are given by[2]

$$x(n) = z^n \qquad \text{for all } n \tag{2.35a}$$

or in its vector format

$$\boldsymbol{x} = \{\ldots, z^{-2}, z^{-1}, \underset{\uparrow}{z^0}, z^1, z^2, z^3, \ldots\} \tag{2.35b}$$

To demonstrate the validity of this eigensequence solution, it is seen that the sequence $\boldsymbol{S}\boldsymbol{x}$ is given by

$$\boldsymbol{S}\boldsymbol{x} = \{\ldots, z^{-3}, z^{-2}, \underset{\uparrow}{z^{-1}}, z^0, z^1, z^2, \ldots\}$$
$$= z^{-1}\{\ldots, z^{-2}, z^{-1}, \underset{\uparrow}{z^0}, z^1, z^2, z^3, \ldots\} = z^{-1}\boldsymbol{x}$$

and operational relationship (2.33) is therefore satisfied.

The sequence as characterized by expression (2.35a) forms the important class of *complex exponential sequences*. It will be subsequently shown that the primary reason for the complex exponential sequence's importance is that it is the *only* sequence which when operated upon by the right-shift operator $\boldsymbol{S}$ will give rise to a scalar multiplied version of itself. This observation in conjunction with the fact that

[2] Any solution to relationship (2.34) is of the form αz^n, where α is an arbitrary constant.

the operator $\boldsymbol{S}$ plays such a fundamental role in linear discrete-time systems should serve as convincing initial evidence of the exponential sequence's significance.

The complex exponential sequence as represented by equation (2.35) is seen to have a time-domain behavior that is completely characterized by the complex number z. This characterization is best revealed by expressing z in its polar representation

$$z = re^{j\omega} \tag{2.36}$$

where r denotes the magnitude and ω the angle of the complex number z. It is possible to represent a large class of sequences by an appropriate linear combination of complex exponential sequences (i.e., a Fourier representation). To gain an insight as to why this is so, let us now examine various possibilities for selecting the parameters r and ω which characterize the complex exponential sequence rewritten as

$$x(n) = r^n e^{j\omega n} \qquad \text{for all } n \tag{2.37}$$

Let us first consider the case where z is restricted to be real, which requires that ω equals zero or π. This results in the class of *real exponential sequences* as given by

$$x(n) = r^n \qquad \text{for all } n \tag{2.38}$$

when ω is taken to be zero, or $x(n) = (-r)^n$ when $\omega = \pi$. A plot of two possible real exponential sequences is shown in Fig. 2.13. It is observed that if the positive parameter r is greater (less) than one, the sequence monotonically increases (decreases) as time increases, whereas the sequence is identically equal to one when $r = 1$. Moreover, the rate at which a real exponential sequence increases (or de-

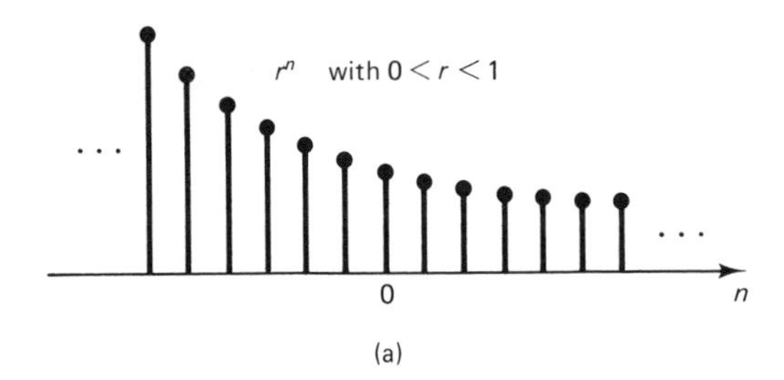

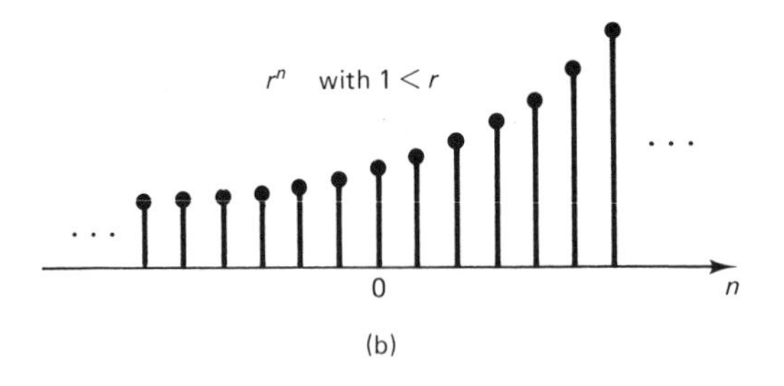

Figure 2.13 Examples of (a) a decaying and (b) a growing real exponential sequence.

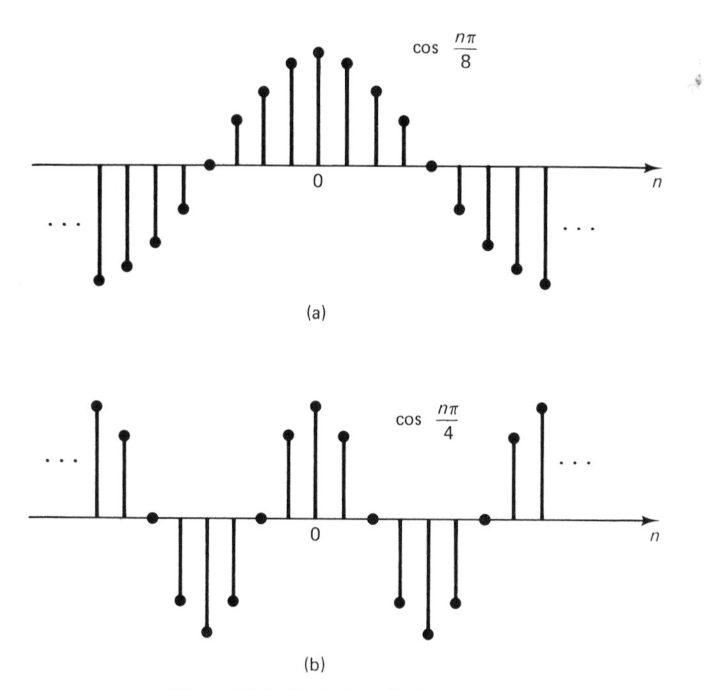

Figure 2.14 Real sinusoidal sequences.

creases) for increasing time is seen to be dependent on how close r is to one. The further r is from one, the larger this rate of change, that is, $x(n)/x(n-1) = r$.

If the magnitude parameter r is fixed at one and ω is other than zero or π, the class of complex sinusoidal sequences is thereby generated, that is,[3]

$$x(n) = e^{jn\omega}$$
$$= \cos n\omega + j \sin n\omega \qquad \text{for all } n \tag{2.39}$$

This sequence is seen to be composed of complex numbers and therefore does not play a direct role in real-world applications, which invariably involve sequences consisting of real numbers. It is possible, however, to suitably combine complex sinusoidal sequences in order to generate real sinusoidal-type sequences. This is easily demonstrated by incorporating the Euler identity to obtain the *real sinusoidal* sequence

$$\cos(n\omega + \theta) = \frac{e^{j(n\omega+\theta)} + e^{-j(n\omega+\theta)}}{2} \tag{2.40}$$

where ω and θ are the sinusoidal sequence's associated *frequency* and *phase* parameters, respectively. A plot of a typical real sinusoidal sequence is shown in Fig. 2.14 with $\omega = \pi/8$ and $\pi/4$, and $\theta = 0$. The frequency parameter ω is seen to control the

[3]The relationship $e^{j\theta} = \cos\theta + j\sin\theta$ is known as the *Euler identity*. It will play a key role in much of our studies and should be committed to memory.

rate at which the oscillations take place. One must be careful in interpreting this observation, however, since the two sinusoidal sequences

$$\{\cos(n\omega + \theta)\} \quad \text{and} \quad \{\cos[n(\omega + 2\pi k) + \theta]\}$$

are found to be identical when k is an integer. This assertion is easily proven using elementary trigonometric identities. Thus, one cannot distinguish the difference between two sinusoidal sequences whose frequency parameters differ by an integer multiple of 2π. With this in mind, we can effectively confine our study to those real sinusoidal sequences whose frequencies lie in the range $0 \leq \omega \leq 2\pi$.

As a further generalization of real exponential and real sinusoidal sequences, we next introduce the *expontenial sinusoidal* sequence as defined by

$$x(n) = r^n \cos(n\omega + \theta) \tag{2.41}$$

This sequence is seen to have the combined characteristics of the real exponential and real sinusoidal sequences.

In any practical application involving a sequence, the nonzero elements of the sequence commence at some finite value of n (i.e., not at $n = -\infty$). This is in recognition of the obvious fact that any dynamical process that generates sequences has not been in operation for the infinite past. With this in mind, let us now define the class of one-sided complex exponential sequences as given by

$$x(n) = \begin{cases} 0 & \text{for } n < 0 \\ z^n & \text{for } n \geq 0 \end{cases} \tag{2.42a}$$

or

$$\boldsymbol{x} = \{\ldots, 0, 0, 0, \underset{\uparrow}{z^0}, z^1, z^2, z^3, \ldots\} \tag{2.42b}$$

where z is a fixed, complex number. It should be apparent that all of the characterizations we have made for the complex exponential sequences (2.35) have analogous interpretations for their one-sided version.

2.6 EXPONENTIAL SIGNAL GENERATORS

In this section, an efficient algorithmic procedure for generating the one-sided exponential sequence

$$y(n) = z^n u(n) \tag{2.43}$$

is given. It is desirable to accomplish this since one-sided exponential sequences play such a prominent role in the characterization of linear discrete-time systems and since they also appear frequently in computer simulations of various dynamical processes. For any positive value of n, the above element $y(n)$ may be directly generated by multiplying z by itself $n - 1$ times to obtain z^n. Unfortunately, for large values of n, the computer time necessary to carry out these $n - 1$ multiplications becomes prohibitive, thereby rendering this direct approach impractical. A significant savings in computation time can be achieved, however, by noting that the elements of this one-sided exponential sequence are related by

$$y(n) = z \cdot y(n-1) \qquad \text{for } n \geq 1 \tag{2.44}$$

with $y(0) = 1$ and $y(n) \equiv 0$ for n negative. Using this simple *algorithm*, one may generate the element $y(n)$ by simply multiplying the previous element value $y(n-1)$ by the constant z. This requires only one multiplication in comparison to $n-1$ multiplications needed in the direct approach. This algorithmic procedure then yields an extremely efficient procedure for generating the desired one-sided exponential.

Let us now put this exponential sequence algorithm into an operator rule format. Specifically, it will be now shown that the response of the operator

$$y(n) = x(n) + z \cdot y(n-1) \tag{2.45}$$

to the unit-impulse sequence (i.e., $x(n) = \delta(n)$) is the desired one-sided exponential sequence (2.43). In using this rule, we are of course restricting $y(n)$ to be identically zero for negative n as is required for one-sided sequences. Substituting $\delta(n)$ for $x(n)$ in relationship (2.45) yields

$$y(n) = \delta(n) + z \cdot y(n-1) \tag{2.46}$$

The only value of n for which $\delta(n)$ is other than zero occurs at $n = 0$ where the response element is

$$\begin{aligned} y(0) &= \delta(0) + z \cdot y(-1) \\ &= 1 + z \cdot 0 = 1 \end{aligned}$$

We have used the fact that $y(-1)$ is zero in arriving at this result. For values of n greater than zero, the term $\delta(n)$ is identically zero, so that relationship (2.46) simplifies to

$$y(n) = z \cdot y(n-1) \qquad \text{for } n \geq 1$$

with $y(0) = 1$. This iterative procedure is then seen to yield the same expression as the one-sided exponential sequence algorithm (2.44). Thus, the unit-impulse response of operator rule (2.46) is seen to yield a highly efficient exponential sequence generator. A block diagram representation for this rule and its elementary operator implementation are shown in Fig. 2.15.

It is beneficial to express the one-sided exponential sequence generator rule (2.46) in the following equivalent sequence relationship

$$\boldsymbol{y} - z\boldsymbol{S}\boldsymbol{y} = \boldsymbol{\delta} \tag{2.47}$$

where

$$\boldsymbol{y} = \{\ldots, 0, 0, 0, \underset{\uparrow}{1}, z, z^2, z^3, \ldots\}$$

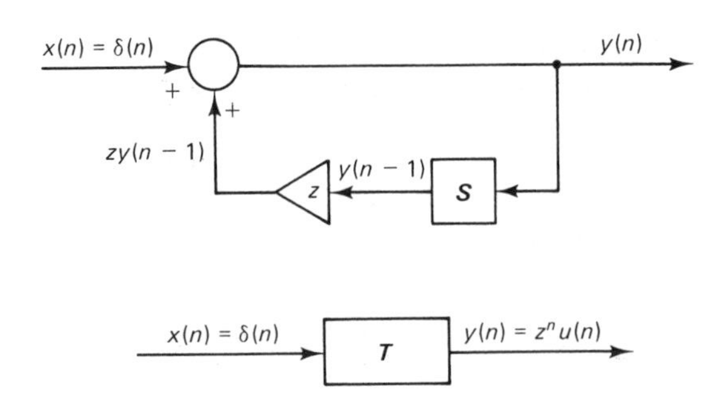

Operator rule: $y(n) = x(n) + z \cdot y(n-1)$ with $x(n) = \delta(n)$ and $y(-1) = 0$

Figure 2.15 One-sided exponential sequence generator.

This equivalence is readily verified by equating the nth components of the right-hand (i.e., $y(n) - zy(n-1)$) and left-hand (i.e., $\delta(n)$) sides of expression (2.47) to arrive at relationship (2.46). This operator relationship indicates that when the compound operator $\boldsymbol{I} - z\boldsymbol{S}$ is applied to the one-sided exponential sequence (2.43), the response sequence is the unit-impulse sequence. This result is of fundamental importance in future considerations.

Example 2.7

Determine the operator rule that generates the specific one-sided exponential sequence

$$y = \left\{\ldots, 0, 0, 0, \underset{\uparrow}{1}, \frac{1}{2}, \frac{1}{4}, \frac{1}{8}, \ldots\right\}$$

In this case, the exponential sequence's characterizing constant z equals $\frac{1}{2}$, so that the required operator rule is

$$y(n) = x(n) + \frac{1}{2}\, y(n-1) \qquad \text{with} \quad x(n) = \delta(n) \quad \text{and} \quad y(-1) = 0$$

Example 2.8

Determine the operator rule that generates the one-sided cosine sequence

$$y(n) = [\cos \omega n]u(n)$$

where ω is a given constant. This cosine sequence may be expressed as a sum of exponential sequences, that is,

$$y(n) = \frac{1}{2}\, e^{j\omega n}u(n) + \frac{1}{2}\, e^{-j\omega n}u(n)$$

We can now generate the desired cosine sequence by first generating the individual exponential sequences $\{x_1(n)\} = \{e^{j\omega n}u(n)\}$ and $\{x_2(n)\} = \{e^{-j\omega n}u(n)\}$ and then performing the operation

$$y(n) = \frac{1}{2}\, x_1(n) + \frac{1}{2}\, x_2(n)$$

The required exponential sequences are generated by using the following operator rules:

$$x_1(n) = \delta(n) + e^{j\omega}x_1(n-1) \qquad \text{with } x_1(-1) = 0$$

$$x_2(n) = \delta(n) + e^{-j\omega}x_2(n-1) \qquad \text{with } x_2(-1) = 0$$

Since each of these operator rules involves multiplication of complex numbers, it is necessary to incorporate complex-number arithmetic in any digital computer program used to generate these complex exponential sequences. It is shown in a later chapter, however, that the desired cosine sequence may be generated by the following equivalent algorithm

$$y(n) = x(n) - [\cos \omega]x(n-1) + [2 \cos \omega]y(n-1) - y(n-2)$$

with $x(n) = \delta(n)$ and $y(-1) = y(-2) = 0$. This algorithm is seen to involve only real-number arithmetic.

2.7 MEASURE OF SIGNAL'S SIZE

If one is to make a meaningful analysis of sequences, it quickly becomes apparent that the notion of *sequence size* is needed. Very simply, a sequence size measure is a rule (called a *norm*) that assigns a nonnegative number to a sequence that in some way reflects the size of that sequence. This rule must satisfy certain attributes usually associated with the concept of size. There are many different size measures which one can assign to a sequence, but the three presented here are the most fundamental.

ℓ_1 Norm

The first measure of sequence size is referred to as the *ℓ_1 norm* of sequence $\boldsymbol{x}$ and will be denoted by the symbol $\|\boldsymbol{x}\|_1$ where

$$\|\boldsymbol{x}\|_1 = \sum_{n=-\infty}^{\infty} |x(n)| \tag{2.48}$$

This norm is seen to be a rule that associates a nonnegative number $\|\boldsymbol{x}\|_1$ equal to the sum of the magnitudes of the elements of sequence $\boldsymbol{x}$. There are various applications in which it is desirable to evaluate $\|\boldsymbol{x}\|_1$. For instance, it is shown in Chapter 4 that the stability characteristics of a linear system are completely determined by the ℓ_1 norm of the system's unit-impulse response. Even more fundamental, it is shown in Chapter 9 that a sequence has a well-defined Fourier transform if it has a finite ℓ_1 norm. Finally, in digital control theory, the ℓ_1 norm gives a measure of the fuel content of a sequence.

ℓ_2 Norm

Another classical measure of sequence size is given by the ℓ_2 norm, which is denoted by $\|\boldsymbol{x}\|_2$, where

$$\|\boldsymbol{x}\|_2 = \left[\sum_{n=-\infty}^{\infty} |x(n)|^2\right]^{1/2} \tag{2.49}$$

Thus, the ℓ_2 norm of sequence $\boldsymbol{x}$ is seen to be equal to the square root of the sum of its elements' magnitudes squared. In regard to physical interpretations, we often refer to the square of $\|\boldsymbol{x}\|_2$ as the *energy content* of sequence $\boldsymbol{x}$, that is,

$$\mathcal{E}(\boldsymbol{x}) \triangleq \sum_{n=-\infty}^{\infty} |x(n)|^2 \tag{2.50}$$

Therefore, if two sequences $\boldsymbol{x}_1$ and $\boldsymbol{x}_2$ are such as to have $\mathcal{E}(\boldsymbol{x}_1) < \mathcal{E}(\boldsymbol{x}_2)$, we say that sequence $\boldsymbol{x}_2$ has more energy than $\boldsymbol{x}_1$.

ℓ_∞ Norm

Finally, in introducing the concept of system stability, it is necessary to introduce the notion of the boundedness of a sequence. In simple terms, a sequence is said to be bounded if its constituent members, that is, the $x(n)$ all have magnitude less than some arbitrary, but fixed, nonnegative number. This concept may be formulated by first introducing the ℓ_∞ norm of sequence $\boldsymbol{x}$ as denoted by $\|\boldsymbol{x}\|_\infty$, where

$$\|\boldsymbol{x}\|_\infty = \sup_{-\infty < n < \infty} |x(n)| \tag{2.51}$$

For most sequences, we can replace "sup" (*supremum*) by *maximum.* We may then interpret $\|\boldsymbol{x}\|_\infty$ as being equal to the largest member of the sequence $\boldsymbol{x}$ in the magnitude sense. It should be noted that there exist many sequences for which $\|\boldsymbol{x}\|_\infty$ is infinity (e.g., $x(n) = n$). With this in mind, a sequence is said to be *bounded* if $\|\boldsymbol{x}\|_\infty$ is finite, as this measure serves as a bound on the elements $x(n)$. That is, if

$$|x(n)| \le \|\boldsymbol{x}\|_\infty \qquad \text{for all } n \tag{2.52}$$

Example 2.9

Let us evaluate the norms $\|\boldsymbol{x}\|_1$, $\|\boldsymbol{x}\|_2$ and $\|\boldsymbol{x}\|_\infty$ for the sequence

$$\boldsymbol{x} = \{\ldots, 0, 0, 0, 1, -1, \underset{\uparrow}{2}, -3, 1, 0, 0, 0, \ldots\}$$

Applying definitions (2.48), (2.49), and (2.52), one readily determines that

$$\|\boldsymbol{x}\|_1 = 8, \quad \|\boldsymbol{x}\|_2 = 4, \quad \text{and} \quad \|\boldsymbol{x}\|_\infty = 3$$

2.8 MODELING BY DISCRETE-TIME OPERATORS*

In forming a model for a given phenomenon, basic assumptions concerning the phenomenon's behavior are first made. These assumptions are then used to generate mathematical expressions that relate signals that characterize the phenomenon being analyzed. This important process is commonly referred to as *system modeling.*

*This section taken from James A. Cadzow, *Discrete-Time Systems* (Englewood Cliffs, N.J.: Prentice-Hall, Inc., 1973). Used by permission.

To illustrate the steps typically used in implementing a mathematical model, we shall now consider three examples.

Amortization

When the purchase of a car, house, or similar item is made, the assumed debt is normally paid for by means of a process known as *amortization*. Under this plan, a debt is repaid by a sequence of periodic payments. A portion of each payment reduces the outstanding principal while the remaining portion is for interest on the loan. Suppose that the original debt to be paid is d dollars and that interest charges are compounded at the rate of r (or $+100r\%$) per payment period. To formulate the amortization process, let $y(n)$ = the outstanding principal after the nth payment, and $x(n)$ = the amount of the nth payment. During the nth payment period, the outstanding debt increases by the interest due on the previous principal $y(n-1)$. After the nth payment, the outstanding principal is then given by

$$y(n) = y(n-1) + ry(n-1) - x(n)$$
$$= (1 + r)y(n-1) - x(n) \tag{2.53}$$

which is a linear first-order difference equation. A configuration consisting of basic units that implements the process of amortization is shown in Fig. 2.16.

In a typical situation, the periodic payments are made in equal amounts:

$$x(n) = p \qquad \text{for } n = 1, 2, \ldots, N$$

where N is the number of periods required to repay the debt. The loaner now wishes to determine the value of p so that the loan is repaid in N periods at the compound rate of r. This is accomplished by finding the response of system (2.53) to the input as specified above. Commencing the iteration at $n = 1$, we have

$$y(1) = (1 + r)y(0) - x(1)$$

and since the initial debt is d, and $x(1) = p$, we have

$$y(1) = (1 + r)d - p$$

At $n = 2$, expression (2.53) gives

$$y(2) = (1 + r)y(1) - p$$

which, after substituting the value of $y(1)$ determined at iteration one, becomes

$$y(2) = (1 + r)[(1 + r)d - p] - p$$
$$= (1 + r)^2 d - p(1 + r) - p$$

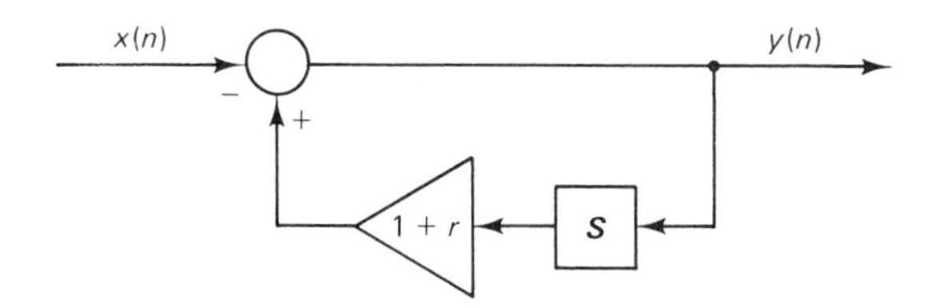

Figure 2.16 Model of amortization.

Following a similar procedure for $n = 3$, we obtain

$$y(3) = (1 + r)^3 d - p(1 + r)^2 - p(1 + r) - p$$

A pattern is seen to evolve, namely, the outstanding principal after the nth payment is given by

$$\begin{aligned} y(n) &= (1 + r)^n d - p(1 + r)^{n-1} - p(1 + r)^{n-2} - \cdots - p(1 + r) - p \\ &= (1 + r)^n d - p \sum_{i=0}^{n-1} (1 + r)^i \qquad \text{for } n = 1, 2, \ldots, N \end{aligned}$$

The summation in this representation may be put into a closed form by using the geometric summation[4] with $\alpha = 1 + r$. This yields

$$y(n) = (1 + r)^n d - p \frac{(1 + r)^n - 1}{r} \qquad \text{for } n = 1, 2, \ldots, N \tag{2.54}$$

To determine the amount of the required payment p so that the debt is reduced to zero after N payments, we let $n = N$ in expression (2.54) and set $y(N) = 0$.

$$0 = y(N) = (1 + r)^N d - p \frac{(1 + r)^N - 1}{r}$$

Solving for p gives

$$p = \frac{r(1 + r)^N}{(1 + r)^N - 1} d$$

which is the required periodic payment needed to reduce the outstanding debt to zero in N periods.

Example 2.10

To demonstrate that expression (2.54) is a solution to difference equation (2.53) for the given input, we must show that the proposed solution satisfies the given difference equation. If it does not, the hypothesized solution is incorrect. Carrying out the prescribed substitution of expression (2.54) into difference equation (2.53) results in

$$\begin{aligned} &(1 + r)^n d - p \frac{(1 + r)^n - 1}{r} \\ &\qquad = (1 + r)\left[(1 + r)^{n-1} d - p \frac{(1 + r)^{n-1} - 1}{r}\right] - p \qquad \text{for } n = 1, 2, 3, \ldots \end{aligned}$$

[4]The geometric summation is specified by

$$\sum_{k=0}^{n} \alpha^k = \frac{1 - \alpha^{n+1}}{1 - \alpha}$$

and holds for all nonnegative integers n and scalars α. Moreover for $|\alpha| < 1$, we have

$$\sum_{k=0}^{+\infty} \alpha^k = \frac{1}{1 - \alpha}$$

with the right side simplifying to

$$= (1 + r)^n d - p \frac{(1 + r)^n - 1 - r}{r} - p$$

$$= (1 + r)^n d - p \frac{(1 + r)^n - 1}{r}$$

and the identity of right and left sides is seen to follow. This demonstrates that the postulated solution (2.54) does satisfy difference equation (2.53).

Model of Rabbit Population

As indicated in Chapter 1, a phenomenon may be properly analyzed only after a representative model of the phenomenon is made. An investigator frequently postulates a number of properties he or she believes the phenomenon to possess and next generates a model that satisfies these properties. The goodness of the model is then judged by how well the model matches the phenomenon's observed behavior.

To illustrate this notion, we shall now generate a model that represents the phenomenon of rabbit population. The idealized properties that rabbit population is hypothesized to have are

1. A pair (one male and one female) of rabbits is born to each pair of adult rabbits at the end of every month.
2. A newborn pair produces their first offspring at two months of age.
3. Once paired, a pair of rabbits remain true to each other and indefinitely produce rabbits according to properties 1 and 2.

Based on these three properties, it is possible to determine a mathematical model that describes rabbit population.

To generate a model, it is first assumed that a pair of newborn rabbits are placed on an unpopulated (with rabbits) island. Next, we define $y(n)$ as the number of pairs of rabbits present at the end of the nth month after placement of the initial pair. This totality of pairs of rabbits, $y(n)$, is composed of either newborn rabbits or those that are one month or older. From property 2 above, it follows that the number of newborn pairs at the end of month n must be equal to $y(n - 2)$, and the number of rabbits that are one or more months old at the end of month n is obviously $y(n - 1)$. Therefore, we must have

$$y(n) = \underbrace{y(n - 1)}_{\substack{\text{One or more}\\ \text{months old at}\\ \text{conclusion of}\\ \text{month } n}} + \underbrace{y(n - 2)}_{\substack{\text{Newborn at}\\ \text{end of } n\text{th}\\ \text{month}}} \tag{2.55}$$

This is a linear second-order difference equation that is a function only of the output signal $y(n)$. To determine the history of rabbit population under the hypothesized model, we take the two initial conditions to be

$$y(0) = 1, \quad y(-1) = 0$$

which reflect the fact that a newborn pair of rabbits was placed on the island initially. Now, letting $n = 1$ in relationship (2.55), we have

$$y(1) = y(0) + y(-1) = 1$$

After the second month ($n = 2$), the number of rabbit pairs is

$$y(2) = y(1) + y(0) = 2$$

Similarly, we can readily determine that

$$y(3) = 3, \quad y(4) = 5, \quad y(5) = 8, \quad y(6) = 13$$

and so forth.

In Chapter 8, methods are developed that enable us to determine that the general solution of difference equation (2.55) with $y(0) = 1$ and $y(-1) = 0$ is given by

$$y(n) = \frac{1}{\sqrt{5}}\left[\left(\frac{1+\sqrt{5}}{2}\right)^{n+1} - \left(\frac{1-\sqrt{5}}{2}\right)^{n+1}\right] \quad \text{for } n = -1, 0, 1, 2, \ldots \tag{2.56}$$

To demonstrate that this expression is indeed the solution, we must show that it satisfies relationship (2.55) and, in addition, that the initial conditions $y(0) = 1$ and $y(-1) = 0$ are satisfied. This is readily shown (see Problem 2.29). The sequence of numbers generated by expression (2.56) is the well-known *Fibonacci sequence.*

We may further generalize the given model. Suppose that, at the end of each month, we place pairs of newborn rabbits on the island in addition to the pairs already present. If $x(n)$ = the number of pairs of newborn rabbits placed on the island at the end of the nth month, we may argue as before that

$$y(n) = \underbrace{x(n)}_{\substack{\text{Newborn at end}\\ \text{of } n\text{th month}}} + \underbrace{y(n-2) + y(n-1)}_{\substack{\text{One or more}\\ \text{months old at}\\ \text{conclusion of}\\ \text{month } n.}} \tag{2.57}$$

It should be apparent that this more general model contains the model in expression (2.55) as a special case. Specifically, if in this model we let the input signal be the unit-impulse signal and the initial conditions are taken to be zero [i.e., $y(-1) = y(-2) = 0$], then the response signal of system (2.57) is given by expression (2.56) for $n = 0, 1, 2, \ldots$. A possible implementation of this model is shown in Fig. 2.17. The model of rabbit population given above is based on three idealized properties that do not reflect the known behavior of this phenomenon. It will therefore typically yield a poor representation and is of general academic interest only.

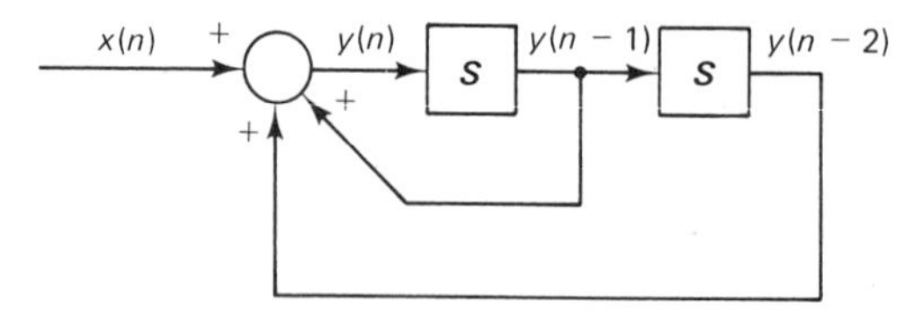

Figure 2.17 Model of rabbit population.

Radar Tracking

Radar tracking offers an excellent example of the application of signal processing techniques. We now give a very simplistic description of how a range radar system operates. The radar system transmits, via its antenna, a pulsed sinusoid of very large amplitude and a very narrow width as shown in Fig. 2.18a. This pulse propagates through space at the speed of light, denoted by c (nominally $c = 3 \times 10^8$ meters/s), until it hits the object being tracked. Some of the pulse's energy is absorbed by the object, and the remainder is reflected in various directions. A portion is reflected back to the radar antenna where it is received and detected. In the ideal situation, the received signal is of the same waveform as the transmitted signal as shown in Fig. 2.18b.

Comparing leading edges of the transmitted and received pulses, we note that a total of Δt seconds is needed for the transmitted pulse to travel to the object being tracked, be reflected, and return to the antenna. Since the pulse propagates at a speed of c meters/s in space, it follows that the object being tracked is located a distance of

$$x = c\,\frac{\Delta t}{2} \quad \text{meters} \tag{2.58}$$

from the radar antenna.

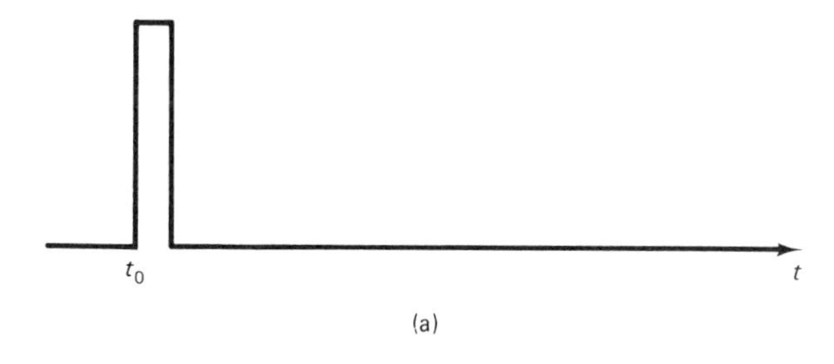

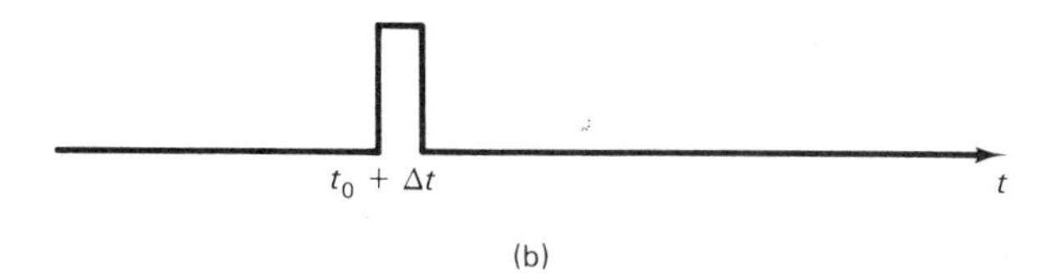

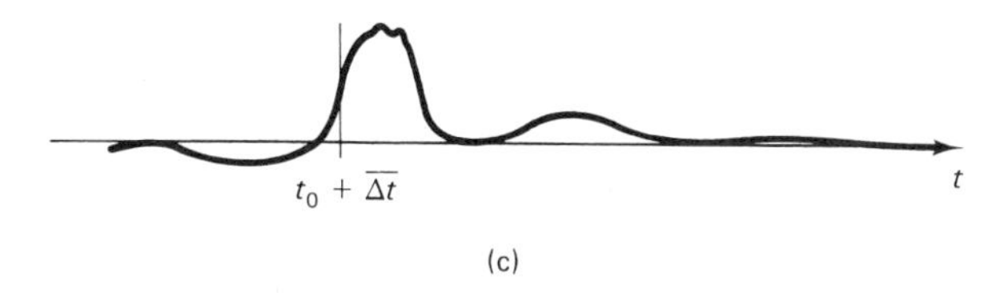

Figure 2.18 (a) Ideal transmitted signal, (b) ideal received signal, and (c) typical received signal.

Therefore, if the value of Δt is determined by means of signal processing in the radar system, expression (2.58) provides a convenient method for determining the distance an object is located from the radar system. Unfortunately, the reflected signal received by the antenna will not be of the ideal shape due to factors such as target size and shape, noise, and multiple objects being tracked. A typical received signal might appear as in Fig. 2.18c. In this realistic situation, the determination of the pulse travel time, Δt, is not directly evident. It is then necessary to build a detector subsystem that will yield an estimate of Δt, denoted by $\overline{\Delta t}$. In general, $\overline{\Delta t} \neq \Delta t$, so that an error in travel time results. Thus, instead of determining the object's true range as given by expression (2.58), we instead obtain

$$\bar{x} = c\,\frac{\overline{\Delta t}}{2}$$

so that we have an error in range given by

$$\Delta x = c\,\frac{\overline{\Delta t} - \Delta t}{2} \tag{2.59}$$

For very accurate range determination, this error in the single-pulse range is normally too large. To obtain greater accuracy, one usually makes periodic measurements of the travel time. This amounts to periodically transmitting pulses every T seconds and obtaining successive measurements of range. Specifically, we acquire a sequence of error-contaminated range measurements that are denoted by

$$x(0),\ x(1),\ x(2),\ \ldots$$

with measurement $x(n)$ being obtained from the nth radar pulse return. A specific procedure for obtaining accurate range determination from these measurements is now outlined. In range data processing, two conflicting requirements are usually made:

1. Accurate range estimation from noise-contaminated measurements of travel time.
2. Fast maneuver-following capability.

Requirement 1 is postulated because of the inherent inaccuracies present in the unprocessed estimates of range. On the other hand, in tracking airborne objects, the dynamics of the signal processor must be capable of tracking an object that is not stationary in space (that is, the object being tracked is moving), and so requirement 2 is necessary. Unfortunately, as one makes the signal processing more noise insensitive, the maneuver-following capabilities suffer, and vice versa. A tradeoff between requirements 1 and 2 must then be made.

In many radar-tracking applications, the signal processor seeks to generate

1. A good estimate of the object's present range.
2. A good estimate of the object's present range velocity (rate of change of range).

3. A good estimate of the object's range one radar pulse time in the future.

These estimates are made by processing the raw data, which are the noise-contaminated measurements on the object's range obtained from the reflected radar pulses. To illustrate how this information processing might proceed, let

$x(n)$ = noise-contaminated measurement of the object's range obtained from the nth radar pulse return

$y(n)$ = estimate of the object's range at the nth radar pulse after signal processing

$\dot{y}(n)$ = estimate of the object's range velocity at the nth radar pulse after signal processing

$y_p(n)$ = prediction of the object's range at nth radar pulse obtained at $(n-1)$th radar pulse after signal processing

If the estimated values of range and range velocity at the $(n-1)$th radar pulse time have been calculated, then a plausible prediction of the range at the next radar pulse time would be

$$y_p(n) = y(n-1) + T\dot{y}(n-1) \tag{2.60}$$

This is the range that we anticipate the object being tracked will be at the nth radar pulse. At the nth measurement of range, we then calculate the difference $x(n) - y_p(n)$. A positive difference implies that the predicted range $y_p(n)$ was underestimated. On the other hand, if the difference is negative, an apparent overestimate of $y_p(n)$ was made. In either case, an update on the estimated range is called for in order that this difference be reduced to zero. The update

$$y(n) = y_p(n) + \alpha[x(n) - y_p(n)]$$

tends to correct either of the observed errors in the difference $x(n) - y_p(n)$, where α is selected to be positive. In a similar fashion, the velocity estimate is updated by the relationship

$$\dot{y}(n) = \dot{y}(n-1) + \frac{\beta}{T}[x(n) - y_p(n)]$$

where T is the period at which radar pulses are transmitted and $\beta > 0$. This set of relationships

$$\left.\begin{aligned} y_p(n) &= y(n-1) + T\dot{y}(n-1) \\ y(n) &= y_p(n) + \alpha[x(n) - y_p(n)] \\ \dot{y}(n) &= \dot{y}(n-1) + \frac{\beta}{T}[x(n) - y_p(n)] \end{aligned}\right\} \tag{2.61}$$

is known as the *alpha-beta* (α-β) *tracking equations* and forms a very simple but highly effective form of range signal processing. Procedures for determining appropriate values for the α and β parameters have been extensively studied.

2.9 PROBLEMS

2.1. Write the sequence of numbers obtained at the output of an ideal sampler with input $x(t) = 10 \sin 2\pi t$. Assume that the first sample is at $t = 0$ and that the sample rate is

(a) 5 Hz

(b) 2 Hz

(c) π Hz

2.2. Given the time sequence $x(n)$ defined by

$$x(n) = \begin{cases} 0 & \text{for } -\infty < n \leq -2 \\ n & \text{for } -1 \leq n \leq 5 \\ 0 & \text{for } 6 \leq n \end{cases}$$

(a) Graph the sequence from $n = -10$ to $n = 10$.

(b) On a separate graph show $x(-n)$.

(c) On a separate graph show $x(3 - n)$.

(d) On a separate graph show $x(n - 3)$.

2.3. A discrete-time signal is given by

$$s(n) = \begin{cases} 1 - e^{-n} & \text{for } 0 \leq n \leq 10 \\ 0 & \text{for other } n \end{cases}$$

(a) Graph $s(n)$ versus n.

(b) Graph the amplified signal $2s(n)$.

(c) Graph the right-shifted signal $s(n-3)$.

(d) Graph the transposed signal $s(-n)$.

(e) Graph the result of all three operations, that is, first amplify s by 2, right shift by 3, and then transpose. Is the result the same in any order?

2.4. In numerical analysis various approximations for the derivative of a function are made. A commonly used one is given by

$$\frac{dx(t)}{dt} \cong \frac{x(t + T) - x(t)}{T}$$

For the signal

$$x(t) = \begin{cases} t & \text{for } t \geq 0 \\ 0 & \text{for other } t \end{cases}$$

(a) Construct a sequence of sampled values, $x(nT)$, $n = 0, 1, 2, \ldots$.

(b) Apply the approximate derivative operation given above to the sequence obtained in part (a). Does the approximation result in a sampled signal that corresponds to dx/dt? Are there any discrepancies?

2.5. Write the approximate derivative operation of Problem 2.4 in terms of the shift operator $\boldsymbol{S}$.

2.6. In many digital communication systems, time synchronization is achieved by preceding a message with a code word that the receiver looks for as an indication that a message is beginning.

(a) If the code word is given by the sequence $\{1, 1, 1, -1, 1\}$, how many shifts (from the arrow) before the five previous (binary) digits correspond to the code word in the following received data string?

$$\begin{array}{c} 1\ \ -1\ \ 1\ \ 1\ \ -1\ \ -1\ \ 1\ \ -1\ \ 1\ \ 1\ \ 1\ \ -1\ \ 1\ \ -1\ \ 1\ \ -1\ \ -1\ \ 1\ \ 1\ \ -1\ \ 1\ \ 1 \end{array}$$

↑ (under the fifth number)

(b) Starting with the first five received numbers, multiply the code word place by place and add the results; then shift right one place and repeat the procedure until the hidden code word is passed. For example, the result of the first calculation is -1, and the result after one shift is $+1$. Plot your results and note what happened when the code word is *found*.

2.7. Using the shift-operator notation $\boldsymbol{S}$, prove

(a) $\boldsymbol{w} = \boldsymbol{S}^n[\boldsymbol{S}^m\boldsymbol{x}] = \boldsymbol{S}^{m+n}\boldsymbol{x}$

(b) $\boldsymbol{x} = \boldsymbol{S}^{-n}[\boldsymbol{S}^n\boldsymbol{x}]$

where $\boldsymbol{S}^n$ is an n-shift-right operation and $\boldsymbol{S}^{-n}$ is an n-shift-left operation.

2.8. If $\boldsymbol{x}$ and $\boldsymbol{y}$ represent two sequences, and

$$x(n) = \begin{cases} (0.5)^n & \text{for } n \geq 0 \\ 0 & \text{for } n < 0 \end{cases}$$

determine $\boldsymbol{y}$ such that $y(n) = 4x(n-1)$. Write out the first few terms of each sequence beginning with $n = 0$.

2.9. Determine the output sequence $\boldsymbol{y}$ in Fig. P2.9 if $\alpha = 1$ and

$$\boldsymbol{x} = \{\ldots 0, 0, \underset{\uparrow}{1}, 2, 3, 2, 1, 0, 0, \ldots\}$$

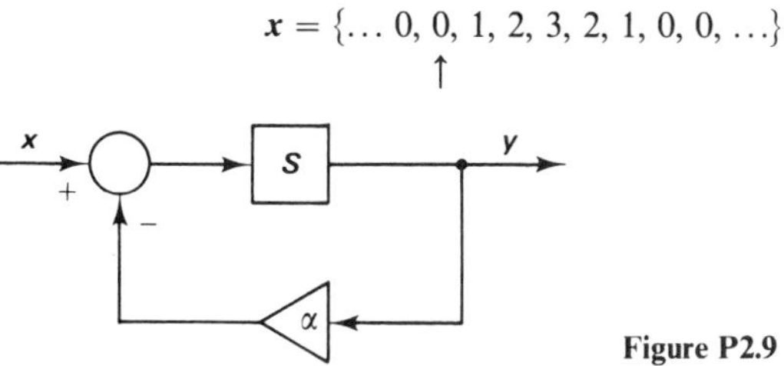

Figure P2.9

2.10. Write the time-domain expression relating the sequences $\{x(n)\}$ and $\{y(n)\}$ for the operation shown in Fig. P2.10.

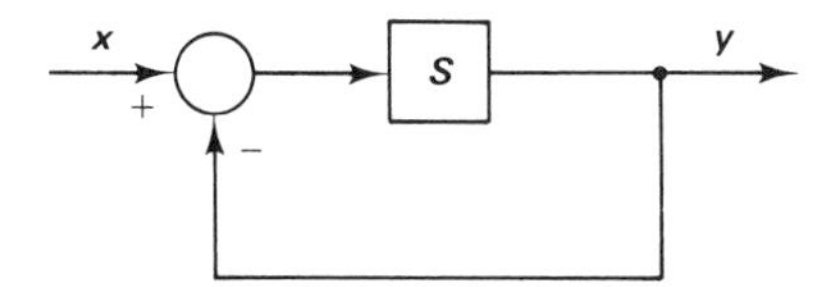

Figure P2.10

2.11. Perform the indicated operations. Sketch the results.

(a) $y(n) = 5x(3-n)$

(b) $y(n) = x(n) \cdot w(n)$

(c) $y(n) = x(n) + w(n)$

where

$$\boldsymbol{x} = \{\ldots, 0, 1, \underset{\uparrow}{2}, 3, 4, 0, 0, \ldots\}$$

and

$$\boldsymbol{w} = \{\ldots, -1, 1, -1, \underset{\uparrow}{1}, -1, 1, -1, 1, \ldots\}$$

2.12. Express the following sequence as a sum of Kronecker delta signals

$$x(n) = \begin{cases} 2 & \text{for } n = -2 \\ -1 & \text{for } n = 1 \\ 5 & \text{for } n = 2 \\ 7 & \text{for } n = 5 \\ 0 & \text{for otherwise} \end{cases}$$

2.13. Show that

$$\sum_{k=-\infty}^{n-1} u(k) = nu(n)$$

2.14. Evaluate and plot the discrete-time signal $x(n) = x_1(n) + x_2(n)$ where $x_1(n) = (-1)^{n^2-1}u(n)$ and $x_2(n) = \frac{1}{2}[1 + (-1)^n]u(n)$.

2.15. Determine the signal $\{x_2(n)\}$ which when added to the signal

$$x_1(n) = \begin{cases} 0 & \text{for } n < 0 \\ 2 & \text{for } n = 0, 2, 4, 6, \ldots \\ -3.5 & \text{for } n = 1, 3, 5, 7, \ldots \end{cases}$$

is such that the sum signal is a step signal of amplitude 2, that is, $x_1(n) + x_2(n) = 2u(n)$.

2.16. Using the unit-impulse $\boldsymbol{\delta}$, show that an arbitrary sequence $\boldsymbol{x}$ can be represented as the sum

$$\boldsymbol{x} = \sum_{n=-\infty}^{\infty} x(n)\boldsymbol{S}^n\boldsymbol{\delta}$$

2.17. Sketch the following sequences:

(a) $x_1(n) = 3\delta(n + 2) + 2\delta(n + 1) + \delta(n) + 2\delta(n - 3)$

(b) $\boldsymbol{x}_2 = \sum_{n=0}^{5} n\boldsymbol{S}^n\boldsymbol{\delta}$

(c) $\boldsymbol{x}_3 = \sum_{n=-3}^{3} n^2\boldsymbol{S}^{-n}\boldsymbol{\delta}$

2.18. Find the sequence $\boldsymbol{y}$ in Fig. P2.10 if $\boldsymbol{x} = \boldsymbol{\delta}$.

2.19. A digital control program implements the following equation:

$$c(k) = e(k) + c(k - 1) \qquad \text{for } k = 1, 2, 3, \ldots$$

where $\boldsymbol{c}$ is the control signal and $\boldsymbol{e}$ is the feedback error signal.

(a) Show that Fig. P2.10 is a diagram of the control program where $\boldsymbol{x} = \boldsymbol{e}$ and $\boldsymbol{y} = \boldsymbol{c}$.

(b) For a unit-step error signal $\boldsymbol{e} = \boldsymbol{u}$, find the control signal $\boldsymbol{c}$. Assume that $c(0) = 0$.

2.20. One way of analyzing a discrete-time signal is to take successive differences. This method generates an array of values, and with luck a simple signal can be recognized at some stage of the process. To form the first difference of the sequence $\boldsymbol{x}$ (call this new sequence $\Delta\boldsymbol{x}$), we define

$$\Delta\boldsymbol{x} = (\boldsymbol{I} - \boldsymbol{S})\boldsymbol{x}$$

Thus, the second-difference sequence is

$$\Delta^2 x = (I - S)\,\Delta x = (I - S)^2 x$$

and so on for higher differences.

(a) Generate an array showing rows x, Δx, $\Delta^2 x, \ldots,$ for the sequence

$$x = \{\ldots\ 0, 12, 13, 20, 39, 76, 137, \ldots\}$$

(b) What is the next entry of the sequence x following 137?

2.21. Find an "exponential" sequence x that satisfies

$$Sx = \lambda x$$

and has the value $x(0) = 1$ if

(a) $\lambda = 2$

(b) $\lambda = \frac{1}{2}$

(c) $\lambda = e^{j\pi/2}$

(d) $\lambda = 2e^{j\pi/4}$

2.22. Determine the parameter z in expression (2.46) so that with $y(-1) = 0$, y will be the one-sided exponential sequence

$$y = \{\ldots, 0, 0, 0, \underset{\uparrow}{1}, 0.1, 0.01, 0.001, \ldots\}$$

2.23. Show that samples of the signal $\sin 2\pi t$ can be generated every 0.1 second using the sine generator given below with $\omega = 2\pi$ and $T = 0.1$, by calculating the first 10 values of the sequence y if $y(-1) = y(-2) = 0$. Plot your results.

$$y(k) = (\sin \omega T)\delta(k-1) + (2\cos \omega T)y(k-1) - y(k-2) \qquad \text{for } k = 0, 1, 2, \ldots$$

2.24. Calculate the l_1, l_2, and l_∞ norms of each of the following sequences:

(a) $x_a = \{\ldots, 0, 0, 0, \underset{\uparrow}{-1}, 1, 2, 5, -3, -1, 0, 0, \ldots\}$

(b) $x_b(n) = \begin{cases} 0 & \text{for } n < 0 \\ \dfrac{1}{2} & \text{for } n \geq 0 \end{cases}$

(c) $x_c = \{\ldots, 0, 0, 0, \underset{\uparrow}{1}, 2, 3, 4, 5, \ldots, 10, 0, 0, \ldots\}$

(d) $x_d = 20\delta + 30S\delta + 10S^{-1}\delta$

2.25. A debt of \$240 is to be amortized by equal payments of \$80 at the end of each month, plus a final partial payment one month after the last \$80 payment. If interest is imposed at the annual rate of 12% compounded monthly, construct an amortization schedule to show the required payments.

2.26. Suppose John wishes to double his money in six years by loaning it out at a rate of r, compounded yearly. The borrower is to pay the loan back in one lump payment at the completion of six years. Find the necessary value of r.

2.27. Give a closed-form relationship for the response of the system as governed by expression (2.53) to the unit Kronecker delta input. The initial condition $y(-1)$ is taken to be zero.

2.28. In the model of rabbit reproduction, suppose that properties 1, 2, and 3 hold as specified in Section 2.8 and that three newborn pairs of rabbits are deposited on an isolated island. Determine the resultant rabbit population.

2.29. Verify that expression (2.56) satisfies difference equation (2.55).

2.30. In a model of rabbit reproduction, suppose that at the end of the nth month a total of $x(n)$ one-month-old rabbit pairs are deposited on an island with $n = 0, 1, 2, \ldots$. If properties 1, 2, and 3 as specified in Section 2.8 are assumed to hold, determine the difference equation that governs the resultant rabbit population.

2.31. Elaborate on reasons why the model of rabbit population as governed by relationship (2.55) does not conform to the real world. For example, what happens to $y(n)$ as specified by equation (2.56) as n gets large?

2.32. Determine the first four iterations of the α-β tracking system as specified by relationship (2.61) when $\alpha = 0.4$, $\beta = 0.1$, and $T = 1$ s, while

$$x(n) = 2000 + 1000n \qquad \text{for } n \geq 0$$

The initial conditions are $y(-1) = \dot{y}(-1) = 0$.

2.33. Suppose it is desired to determine the range that a stationary target in space (e.g., synchronized satellite) is located from an α-β tracking system. If r denotes this fixed range, then the input signal to the tracker is

$$x(n) = \begin{cases} 0 & \text{for } n = 1, 2, 3, \ldots \\ r & \text{for } n = 0, 1, 2, \ldots \end{cases}$$

Verify that the α-β tracker yields desirable results in the steady state (i.e., large values of n) for this situation. (Hint: Let $y(n) \approx y_p(n) \approx r$ and $\dot{y}(n) \approx 0$ for large n.)

3

Continuous-Time Signals

3.1 INTRODUCTION

As indicated in Chapter 1, a *signal* is an ordinary function of an independent variable t. This independent variable may take on all values in the function's *domain of definition*. For continuous-time signals, this domain of definition is of an interval form that, unless otherwise noted, is taken to be the entire real axis $-\infty < t < \infty$.[1] We have agreed to call the independent variable *time* in order to indicate that many physical signals naturally evolve as a function of time.

A continuous-time signal is generally denoted by a boldface italicized letter symbol such as $\boldsymbol{x}$ whereas its value at the specific time instant t is signified by $x(t)$. We take some mathematical liberties, however, and sometimes refer to a signal as $x(t)$ instead of $\boldsymbol{x}$. This is clearly a mathematically improper notation since $x(t)$ is, in fact, but one value of the signal $\boldsymbol{x}$ at the fixed, but arbitrary, time t. We have elected to use this imprecise notation since it often clarifies matters and it also conforms with the notation incorporated in most signal theory textbooks.

It will be soon made apparent that many of the concepts used for discrete-time signals have corresponding analogies in continuous-time signal theory. Thus, we will introduce such continuous-time operations as "multiplying a signal by a constant" and "adding two or more signals." The continuous-time operation of differentiation, however, does not have an exact analogy and raises many subtle and basic issues that must be satisfactorily resolved if a true understanding of continuous-time signal theory is to result. A major portion of this chapter is thus devoted to a study of the differentiation operator and its effect on continuous-time signals.

[1]This is in contrast to discrete-time signals for which the domain of definition is the set of integers.

3.2 REPRESENTATION OF CONTINUOUS-TIME SIGNALS

One of the primary considerations in our study of signal theory is that of appropriately characterizing signals. If a signal's principal attributes can be extracted from its basic time behavior description, then much has been gained in terms of an understanding of its intrinsic structure. One goal of this text is to effectively characterize a signal's structure. We begin this by first describing the signal itself in its natural time-domain setting. There are basically two methods by which a continuous-time signal may be described in the time domain:

1. It may be possible to define a continuous-time signal by means of a mathematical formula, that is, a closed-form expression.
2. It is always conceptually possible to graphically display the time behavior in the form of a plot of $x(t)$ versus t.

Let us illustrate these two representation procedures by considering a specific signal that has the following closed form representation:

$$x(t) = \begin{cases} 2 - \dfrac{t}{2} & \text{for } 0 \leq t \leq 1 \\ 2 & \text{for } -3 \leq t < 0 \\ \dfrac{1}{2} & \text{otherwise} \end{cases}$$

This continuous-time signal can also be represented by its graphical equivalence as shown in Fig. 3.1.

Which of the two representation procedures one uses is, of course, dependent on the specific signal being considered. In many cases, there may not exist a convenient formula by which a given signal can be described. One is then forced to use a graphical display in such situations (e.g., the recording of an electrocardiogram

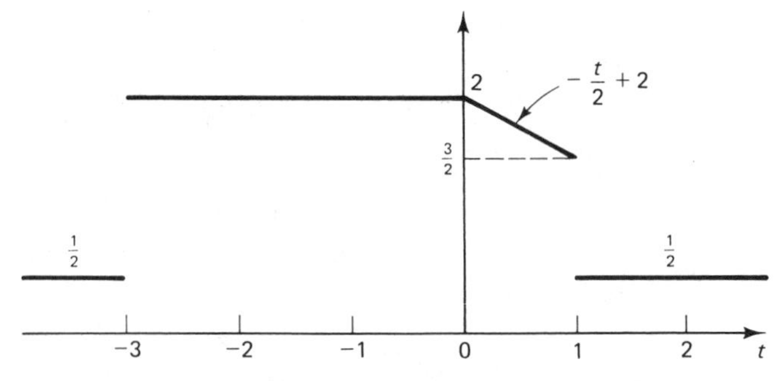

Figure 3.1 Graphical display of a continous-time signal.

signal, the voltage-time waveform of a radar signal, and so forth) or to represent the signal implicitly as the solution or output of some relation such as a differential equation. It must be mentioned, however, that important classes of continuous-time signals do have convenient mathematical formuli. In fact, the elementary study of continuous-time signals and linear systems is primarily concerned with such signals. As a final observation, it is noted that unlike discrete-time signal representations, we are unable to generate a table in which the values of the signal $x(t)$ are explicitly listed as a function of t over the interval of definition. This follows since it can be shown that in any interval, no matter how small, there are an uncountable, infinite number of values of the independent variable t.

3.3 CONCEPTS OF CONTINUITY AND DIFFERENTIABILITY

In studying continuous-time signals, the concepts of signal continuity and differentiability play fundamental roles. Since these ideas are well dealt with in college calculus courses, we will only briefly summarize their salient features. This is effectively begun by defining the quantities

$$x(t_0^+) = \lim_{\varepsilon \to 0^+} x(t_0 + \varepsilon) \tag{3.1}$$

$$x(t_0^-) = \lim_{\varepsilon \to 0^-} x(t_0 + \varepsilon) \tag{3.2}$$

where $\varepsilon \to 0^+$ (or 0^-) denotes the limiting process of ε approaching zero from the right (or left) but never actually equaling zero. In essence, $x(t_0^+)$ then designates the value of the signal $x(t)$ as t approaches t_0 from the right, while $x(t_0^-)$ gives the signal's value as t approaches t_0 from the left. With this in mind, the quantities $x(t_0^+)$ and $x(t_0^-)$ are referred to as the *right* and *left limits*, respectively, of the signal $x(t)$ at time $t = t_0$.

Example 3.1

The reader may readily verify that for the signal shown in Fig. 3.2, we have

$$x\left(\frac{1}{4}\right)^+ = x\left(\frac{1}{4}\right)^- = \frac{7}{4}$$

and

$$x(1^+) = \frac{1}{2} \quad \text{but} \quad x(1^-) = \frac{3}{2}$$

Thus, when t_0 is set equal to $\frac{1}{4}$, the right and left limits are equal. However, this is not the case when t_0 is set equal to 1. This distinctly different behavior of the signal at times $t = \frac{1}{4}$ and 1 is of such importance that mathematicians have introduced the concept of continuity to describe it.

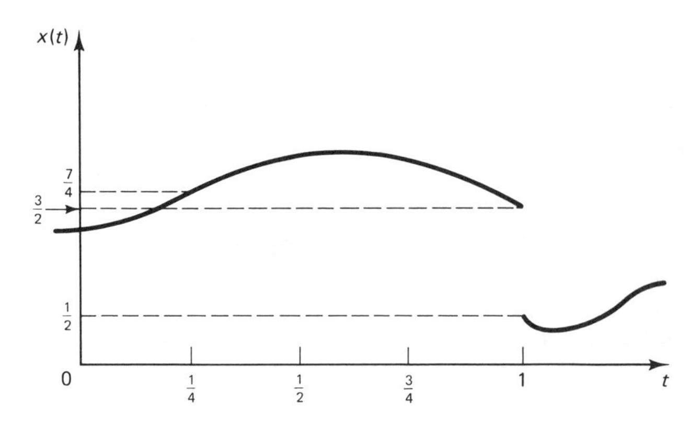

Figure 3.2 Continuous-time signal with a discontinuity of time $t = 1$.

Continuity

A signal x is said to be continuous at the time instant t_0 if the following limiting operation holds:

$$\lim_{\varepsilon \to 0} x(t_0 + \varepsilon) = x(t_0) \tag{3.3}$$

independent of the manner in which ε approaches zero (that is, from the right or left). In other words, continuity at t_0 implies that the signal $x(t)$ does not undergo abrupt changes in value (that is, has jumps) when t is in an *arbitrarily close proximity* of t_0. For example, the signal depicted in Fig. 3.2 is found to be continuous at $t_0 = \frac{1}{4}$ since $x(t)$ is well-behaved at that point of time. On the other hand, if the limiting operation as defined by expression (3.3) does not hold at t_0, then the signal is said to lack continuity or to be *discontinuous* at t_0. Thus, the signal shown in Fig. 3.2 is said to be discontinuous at $t_0 = 1$ since it undergoes an abrupt change in value from $\frac{3}{2}$ to $\frac{1}{2}$ as time goes from 1^- to 1^+.

In virtually all studies related to signals characterizing physical phenomena, the concept of a discontinuous signal is without meaning. As an example, let us consider the situation whereby a d.c. voltage of amplitude E volts is suddenly applied (switched on) to an electronic system at the time instant t_0. We usually approximate this action by the idealized waveform shown in Fig. 3.3a where the signal $\boldsymbol{x}$ jumps instantaneously from zero to E at time instant t_0 (i.e., $\boldsymbol{x}$ is discontinuous at t_0). When one applies the physical laws that govern this switching process, however, it follows that this instantaneous voltage-level change could never be achieved in practice due to inductance in the switch circuit and other reasons. As a matter of fact, if one examined the actual signal waveform on an expanded time scale by taking a microscopic view, the applied voltage might appear as shown in Fig. 3.3b. The actual signal is seen to rise from zero to E in a continuous manner, and it undergoes this transition in a nonzero Δt-second interval. If this transition time interval Δt is sufficiently small relative to the time constants characterizing the system to which $\boldsymbol{x}$ is applied, one can show that the system's response to this actual input signal is virtually identical to the system's response to the approximating ideal signal shown in Fig. 3.3a. It should be readily apparent, however, that the ideal

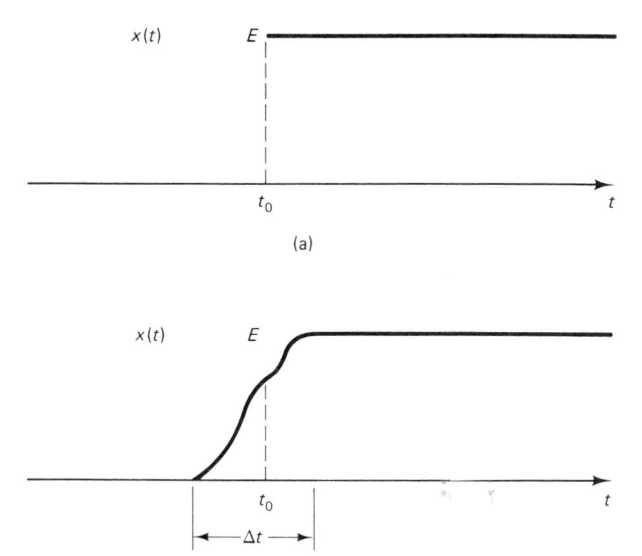

Figure 3.3 (a) Idealized signal, and (b) actual signal of a much expanded time axis.

signal is much simpler to describe (i.e., by the two parameters E and t_0). Moreover, it will be shortly shown that there exists a straightforward procedure for determining a system's response to this ideal input signal (to be developed in subsequent chapters). This being the case, we will often approximate signals that undergo sudden changes in values by signals that have discontinuities. Generally, this is done in order to make the resultant analysis more tractable with the realization that the approximation leads to an acceptably accurate result.

Derivatives

One of the tasks of signal theory is that of analyzing the characteristic features of signals. At a particular time instant t_0, the most important informational quantity that describes the signal is its value as denoted by $x(t_0)$. There exist other measures, however, that yield additional information relative to the intrinsic nature of the signal at t_0. For example, the trend behavior as measured by the manner in which the signal is changing at the given time instant yields a most valuable addition to its characterization. This trend behavior is generally estimated by evaluating the ratio

$$\frac{x(t_0 + \Delta) - x(t_0)}{\Delta}$$

where Δ is a small perturbation. The terms constituting this ratio are shown in Fig. 3.4, for Δ positive where it is seen that $x(t_0 + \Delta) - x(t_0)$ is the amount by which the signal changes in the Δ-second interval. A knowledge of this ratio's value together with $x(t_0)$ enables us to roughly estimate the behavior of the signal $\boldsymbol{x}$ in a neighborhood of the time instant t_0.

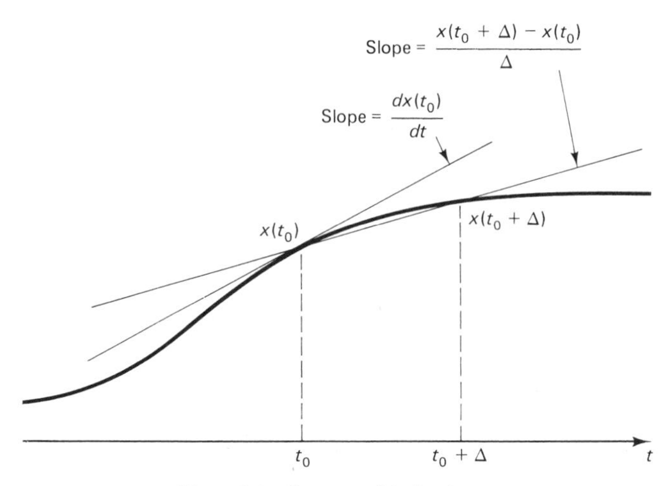

Figure 3.4 Concept of derivative.

We are particularly interested in the nature of this ratio when the width parameter Δ is made arbitrarily close to zero. When this ratio tends to a constant value as Δ is made small in magnitude, the signal is said to have a *derivative* at the time instant as formally defined by the limiting process

$$\frac{dx(t_0)}{dt} = \lim_{\Delta \to 0} \frac{x(t_0 + \Delta) - x(t_0)}{\Delta} \tag{3.4}$$

It is important to realize that the ratio must approach a constant, denoted by $dx(t_0)/dt$, independent of the manner in which Δ approaches zero, (for example, from the left or right) in order for the derivative to exist. If this should not be the case, then the signal is said not to possess a derivative at time t_0.

A careful interpretation of the derivative definition (3.4) and its graphical depiction as given in Fig. 3.4 indicates that $dx(t_0)/dt$ (when it exists) is equal to the slope of the graph of $x(t)$ at t_0. Thus, the derivative yields the desired information relative to the trend behavior of a signal at any given time instant. It must be cautioned, however, that a signal may fail to have a derivative at the time instant t_0. This generally arises because either (1) the signal $x(t)$ is discontinuous at t_0, thereby making the limit operation (3.4) unbounded (i.e., plus or minus infinity), or (2) the ratio equals different constants depending on how Δ approaches zero. Examples of these two possibilities are shown in Fig. 3.5, where each signal fails to possess a derivative at the time instant t_0. The following example will illustrate case (2).

Example 3.2

Determine whether the signal as given by

$$x(t) = \begin{cases} -t + 1 & \text{for } t \leq 0 \\ \cos t & \text{for } t > 0 \end{cases}$$

has a derivative at $t_0 = 0$. To demonstrate that it does not, let us evaluate the so-called

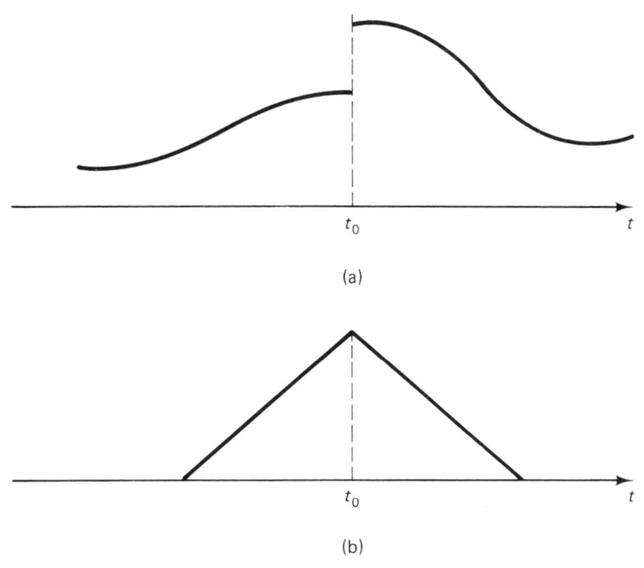

Figure 3.5 Examples of signals that do not have a derivative at t_0.

left derivative as defined by

$$\frac{dx(0^-)}{dt} = \lim_{\Delta \to 0^-} \frac{x(0 + \Delta) - x(0)}{\Delta}$$

$$= \lim_{\Delta \to 0^-} \frac{(-\Delta + 1) - 1}{\Delta}$$

$$= -1$$

and the right derivative

$$\frac{dx(0^+)}{dt} = \lim_{\Delta \to 0^+} \frac{x(0 + \Delta) - x(0)}{\Delta}$$

$$= \lim_{\Delta \to 0^+} \frac{\cos \Delta - 1}{\Delta}$$

$$= 0$$

where use of l'Hospital's rule has been made in determining $dx(0^+)/dt$.[2] Since the ratio

[2]L'Hospital's rule: Let $f(t)$ and $g(t)$ be real and differentiable functions for $a < t < b$ with $dg(t)/dt \neq 0$ for all t in $a < t < b$ where $-\infty \leq a < b \leq +\infty$. Let the limit

$$\lim_{t \to a} \frac{df(t)/dt}{dg(t)/dt} = \rho$$

exist. If $f(t)$ and $g(t)$ each approach zero as t approaches a, or if $g(t)$ approaches $+\infty$ as t approaches a,

$$\lim_{t \to a} \frac{f(t)}{g(t)} = \rho$$

An analogous statement holds for t approaching b or with $g(t)$ approaching $-\infty$ as t approaches b.

that defines the derivative approaches a different value depending on whether Δ approaches zero from positive or negative values, it follows that the given signal does not have a derivative at $t_0 = 0$.

As we have just seen, a signal fails to possess a derivative at time t_0 if the signal is itself discontinuous at t_0. Unfortunately, signals that are modeled with discontinuities tend to arise frequently in practical applications. Even more disquieting, it is necessary to take the derivatives of such signals. How do we then take the derivatives of such signals at their points of discontinuities? From a classical mathematical viewpoint, the derivatives fail to exist at these points, and a dilemma arises. This difficulty is circumvented, however, by the introduction of the so-called unit-impulse signal in Section 3.6.

3.4 CHANGING THE TIME VARIABLE

In many practical situations involving our study of continuous-time signals and systems, it is necessary to make a change in the independent time variable. There are basically three procedures to change the continuous-time variable that we will use over and over again: (1) scaling the time variable, (2) shifting (or translating) the time variable, and (3) transposing the time variable. These three operations can be used individually or in a combination form. For example, it is subsequently shown that the important operation of convolution involves both the shifting and transposing of time-variable operations. The importance of making a change in the time variable cannot be stressed enough. Without a thorough understanding of the essential aspects of this procedure, one cannot hope to fully understand the many subtle signal and system theoretical developments that employ a change of time-variable operation. Fortunately, there exists a relatively simple and well-established procedure that may always be followed when making a change in the time variable. By doing this, many of the pitfalls that might otherwise arise are avoided.

Scaling the Time Variable

A scaling of the independent time variable is achieved by replacing t by αt everywhere it appears in the defining equation for $x(t)$, that is,

$$x(t) \quad \text{becomes} \quad x(\alpha t) \tag{3.5}$$

where α is restricted to be a real and positive constant. We now show that if the signals $x(t)$ and $x(\alpha t)$ are plotted versus t, they will appear identical except for a contraction (i.e., when $\alpha > 1$) or an expansion (i.e., when $\alpha < 1$) of the time axis. As an example, let us consider the specific time signal

$$x(t) = \begin{cases} -1 & \text{for } -\infty < t < -1 \\ 1 + 2t & \text{for } -1 \le t \le 0 \\ \cos \pi t & \text{for } 0 < t \le 2 \\ 1 & \text{for } 2 < t < \infty \end{cases} \tag{3.6}$$

as depicted in Fig. 3.6a. To determine an expression for $x(\alpha t)$, it is first noted that $x(t)$ is governed by a different formula in each of the four time intervals $-\infty < t < -1$, $-1 \leq t \leq 0$, $0 < t \leq 2$, $2 < t < \infty$. Thus, to ascertain the value of $x(\alpha t)$ for a specific value of αt, it is necessary to determine in which interval αt lies

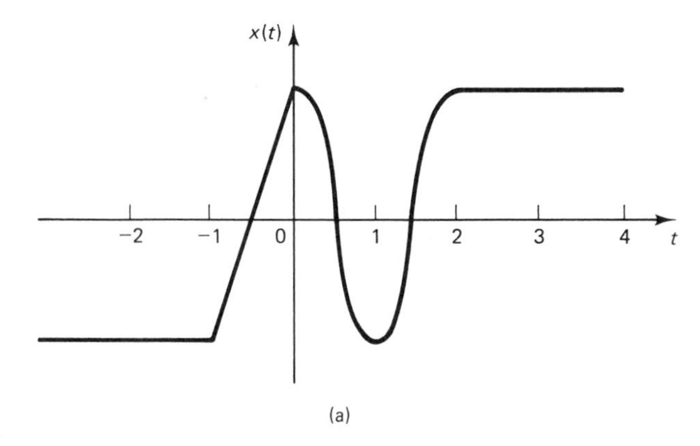

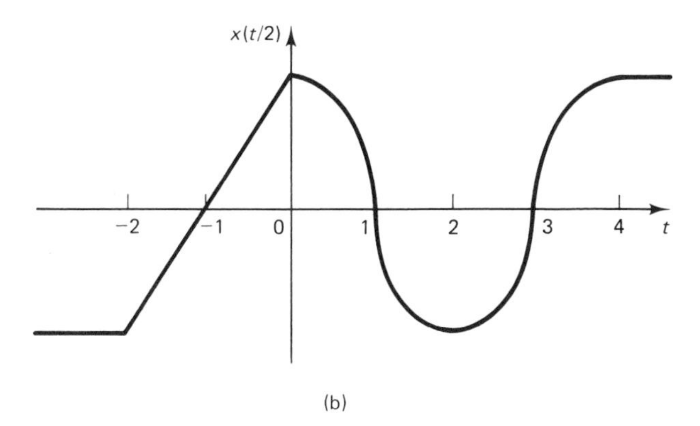

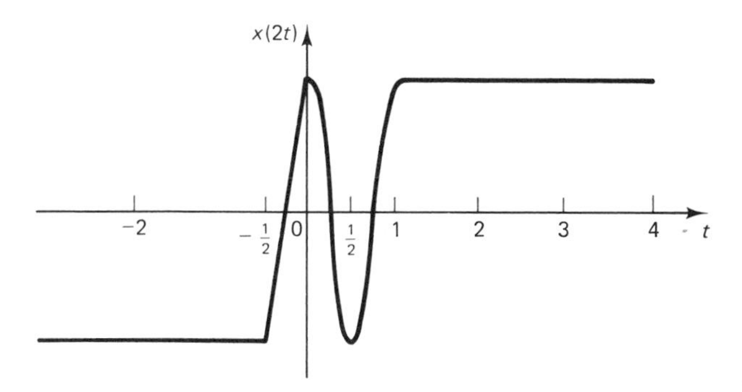

Figure 3.6 Example of scaling the time variable.

and then use the appropriate formula governing that interval. With this in mind, it must follow that

$$x(\alpha t) = \begin{cases} -1 & \text{for } -\infty < \alpha t < -1 \\ 1 + 2\alpha t & \text{for } -1 \leq \alpha t \leq 0 \\ \cos \pi\alpha t & \text{for } 0 < \alpha t \leq 2 \\ 1 & \text{for } 2 < \alpha t < \infty \end{cases}$$

where t has been replaced by αt *everywhere* it appears in representation (3.6). This expression for $x(\alpha t)$ may be put into a more standard form by noting that the interval inequalities $t_1 \leq \alpha t \leq t_2$ and $t_1/\alpha \leq t \leq t_2/\alpha$ are identical when α is a positive real number. Thus, the desired characterization of the signal $x(\alpha t)$ is given by

$$x(\alpha t) = \begin{cases} -1 & \text{for } -\infty < t < -1/\alpha \\ 1 + 2\alpha t & \text{for } -1/\alpha \leq t \leq 0 \\ \cos \alpha\pi t & \text{for } 0 < t \leq 2/\alpha \\ 1 & \text{for } 2/\alpha < t < \infty \end{cases} \tag{3.7}$$

Plots of the signals $x(t/2)$ and $x(2t)$ are shown in Fig. 3.6. It is apparent that $\alpha = \frac{1}{2}$ (or 2) has the effect of expanding (or contracting) the time axis. This is a direct consequence of the fact that any given interval $t_1 \leq t \leq t_2$ in the original graph of $x(t)$ versus t becomes $t_1/\alpha \leq t \leq t_2/\alpha$ in the scale-changed graph of $x(\alpha t)$ versus t. Thus, a choice of $\alpha < 1$ (or > 1) has the effect of expanding (or contracting) the time axis in a plot of $x(\alpha t)$ versus t.

Shifting the Time Variable

A shifting of the time variable by an amount t_0 is accomplished by replacing t by $t - t_0$ everywhere it appears in the representation for $x(t)$, that is,

$$x(t) \quad \text{becomes} \quad x(t - t_0) \tag{3.8}$$

where t_0 is a fixed real number. If we plot the signals $x(t)$ and $x(t - t_0)$ versus t, they would be identical in all aspects except that $x(t - t_0)$ would be a version of $x(t)$ that is shifted to the right by t_0 seconds.

To illustrate this time-shifting operation, let us generate $x(t - t_0)$ for the signal given by expression (3.6). The signal $x(t - t_0)$ is obtained by straightforwardly substituting $t - t_0$ for t *everywhere* it appears in this governing expression, to obtain

$$x(t - t_0) = \begin{cases} -1 & \text{for } -\infty < t - t_0 < -1 \\ 1 + 2(t - t_0) & \text{for } -1 \leq t - t_0 \leq 0 \\ \cos \pi(t - t_0) & \text{for } 0 < t - t_0 \leq 2 \\ 1 & \text{for } 2 < t - t_0 < \infty \end{cases}$$

In order to depict this signal in the standard format, note that the inequality $t_1 \leq t - t_0 \leq t_2$ may be equivalently expressed as $t_1 + t_0 \leq t \leq t_2 + t_0$. With this in mind, the signal $x(t - t_0)$ may be written in the desired format

$$x(t - t_0) = \begin{cases} -1 & \text{for } -\infty < t < -1 + t_0 \\ 1 + 2(t - t_0) & \text{for } -1 + t_0 \leq t \leq t_0 \\ \cos \pi(t - t_0) & \text{for } t_0 < t \leq 2 + t_0 \\ 1 & \text{for } 2 + t_0 < t < \infty \end{cases} \tag{3.9}$$

Clearly, the intervals defining the signal $x(t - t_0)$ are each shifted to the right by t_0 seconds relative to the intervals defining $x(t)$. Since the same formula pertains to each of these corresponding intervals, it then must follow that the graph of $x(t - t_0)$ is obtained by simply shifting the graph of $x(t)$ by t_0 seconds to the right. This shifting process is illustrated in Fig. 3.7, where the graphs of $x(t)$, $x(t + \frac{1}{2})$, and $x(t - 1)$ are sketched.

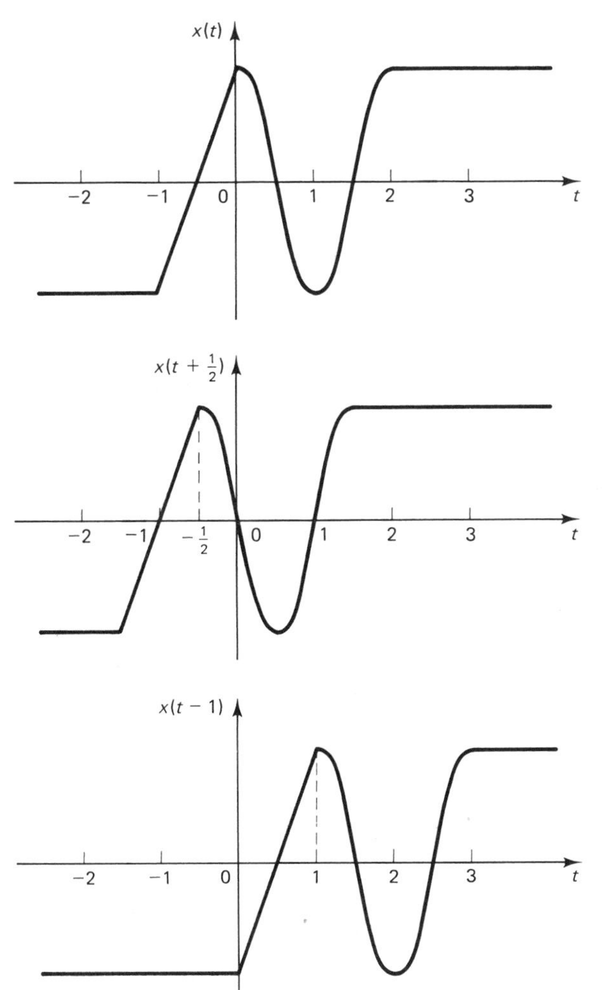

Figure 3.7 Example of shifting the time variable.

Transposing the Time Variable

In transposing the time variable, one simply replaces t by $-t$ everywhere it appears in the representation of $x(t)$, that is,

$$x(t) \quad \text{becomes} \quad x(-t) \tag{3.10}$$

If we graphed $x(t)$ and $x(-t)$, they would be mirror images of each other about the point $t = 0$. Let us illustrate the transpose of the time-variable operation for the signal given by relationship (3.6). This is begun by replacing t by $-t$ *everywhere* it appears in the characterizing expression for $x(t)$. This results in

$$x(-t) = \begin{cases} -1 & \text{for } -\infty < -t < -1 \\ 1 - 2t & \text{for } -1 \le -t \le 0 \\ \cos(-\pi t) & \text{for } 0 < -t \le 2 \\ 1 & \text{for } 2 < -t < \infty \end{cases}$$

In order that the time variable t appears in the desired positive form in the inequalities defining the four intervals over which $x(-t)$ is defined, we use the fact that the interval $t_1 \le -t \le t_2$ is equivalent to $-t_1 \ge t \ge -t_2$. Thus, the desired form for the transposed signal is given by

$$x(-t) = \begin{cases} -1 & \text{for } \infty > t > 1 \\ 1 - 2t & \text{for } 1 \ge t \ge 0 \\ \cos \pi t & \text{for } 0 > t \ge -2 \\ 1 & \text{for } -2 > t > -\infty \end{cases} \tag{3.11}$$

A graph of $x(-t)$ is sketched in Fig. 3.8, where the image relationship between $x(t)$ and $x(-t)$ is apparent.

Let us conclude this section by making a change of time variable that incorporates each of the three basic procedures just described. For the signal $x(t)$ given by representation (3.6), let us properly characterize the signal $x(-3t + 2)$. This first requires the replacement of t by $-3t + 2$ everywhere it appears in relationship (3.6), that is,

$$x(-3t + 2) = \begin{cases} -1 & \text{for } -\infty < -3t + 2 < -1 \\ 1 + 2(-3t + 2) & \text{for } -1 \le -3t + 2 \le 0 \\ \cos \pi(-3t + 2) & \text{for } 0 < -3t + 2 \le 2 \\ 1 & \text{for } 2 < -3t + 2 < \infty \end{cases}$$

To simplify this expression, we use the easily derived fact that the set of t values that satisfy the two inequalities

$$t_1 \le -3t + 2 \le t_2 \quad \text{and} \quad \frac{2 - t_1}{3} \ge t \ge \frac{2 - t_2}{3} \tag{3.12}$$

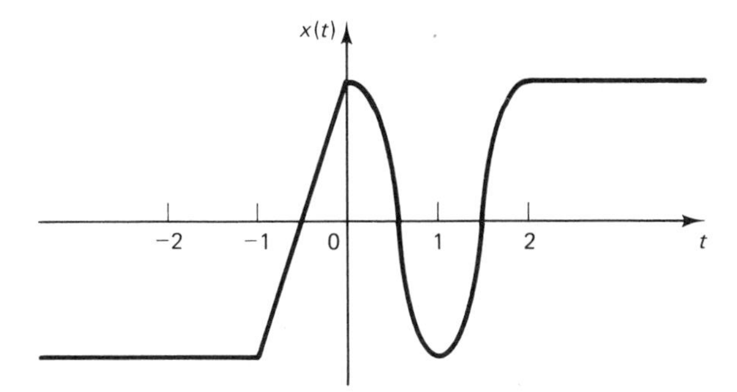

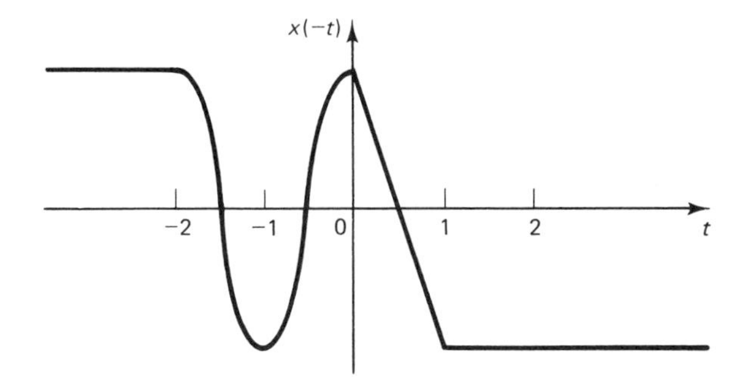

Figure 3.8 Operation of time transposition.

are identical. Thus, the desired form for the signal $x(-3t + 2)$ is then

$$x(-3t+2) = \begin{cases} -1 & \text{for } \infty > t > 1 \\ 5 - 6t & \text{for } 1 \geq t \geq \dfrac{2}{3} \\ \cos 3\pi t & \text{for } \dfrac{2}{3} > t \geq 0 \\ 1 & \text{for } 0 > t > -\infty \end{cases} \tag{3.13}$$

A plot of $x(-3t + 2)$ versus t would result in a version of the original signal $x(t)$ that is contracted by a factor of 3, transposed, and shifted to the left by $\frac{2}{3}$ seconds.

3.5 ELEMENTARY OPERATIONS ON SIGNALS

The processing of data (information) that appear in the form of a continuous-time signal entails a systematic procedure whereby the given signal is transformed (or changed) into another signal by means of a well-defined rule. This transformation procedure is symbolically expressed as

$$y = \boldsymbol{T}x \tag{3.14}$$

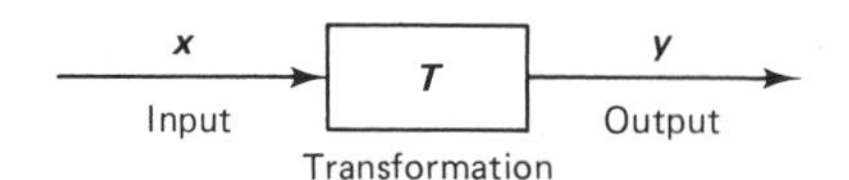

Figure 3.9 General transformation of input signal into output signal.

where $\boldsymbol{x} = \{x(t)\}$ is the input signal being transformed, $\boldsymbol{y} = \{y(t)\}$ is the transformed response signal, and $\boldsymbol{T}$ is the rule (operation) by which signal $\boldsymbol{y}$ is generated from signal $\boldsymbol{x}$. In the typical application, we wish to select the rule $\boldsymbol{T}$ so that the signal $\boldsymbol{x}$ is changed in some desirable manner into signal $\boldsymbol{y}$. It is beneficial to envision this transformation operation as shown in Fig. 3.9.

The transform operation as depicted in expression (3.14) is quite general in the sense that *any* conceivable rule for changing one continuous-time signal into another is allowable. If we do not in some way restrict the types (classes) of operations to be considered, however, any hope of developing a tractable mathematical characterization of continuous-time signal operators is lost. On the other hand, any restricted class of operations must possess enough operational flexibility in order to be capable of achieving a variety of useful signal processing objectives. Moreover, such a class of operations must be readily implemented using available instrumentation if the resultant mathematical characterization is to have any practical meaning. As we have previously seen in the context of discrete-time operations, only a few basic operations are necessary. Fortunately, it is possible to attain relatively sophisticated forms of continuous-time signal processing by means of the three elementary operations of (1) differentiation, (2) multiplication by a constant, and (3) summations of signals. A brief description of these operations follows.

Differentiation Operator

In the differentiation operator, the input signal $\boldsymbol{x}$ is transformed into the response signal $\boldsymbol{y}$ by means of the following rule

$$y(t) = \frac{dx(t)}{dt} \tag{3.15}$$

which holds for all t. Thus, the response signal $\boldsymbol{y}$ is said to be equal to the derivative of the input signal $\boldsymbol{x}$. In using relationship (3.15), we are in effect assuming that the signal $x(t)$ is differentiable for all values of time t. There exist many important signals for which this assumption does not hold. For example, any signal that undergoes an abrupt change in level at a given time instant t_0 (that is, the signal is discontinuous at t_0) does not possess a derivative at t_0 for reasons elaborated upon in Section 3.3. Since such signals are of much theoretical importance, we must in some way resolve this dilemma. This necessitates the introduction of a class of pseudo functions (e.g., the impulse function) that enable us to carry out the operation of differentiation on signals that possess such discontinuities. More will be said about this class of pseudo functions in the next section.

The derivative operator is denoted by the symbol $\boldsymbol{D}$ (that is, $\boldsymbol{D}$ for differentiation) so that continuous-time signals related by operator rule (3.15) may be equivalently represented as

$$\left.\begin{aligned} &y = \boldsymbol{D}\boldsymbol{x} \\ &\text{where} \\ &y(t) = \frac{dx(t)}{dt} \qquad \text{for all } t \end{aligned}\right\} \tag{3.16}$$

The differential operator is shown in Fig. 3.10a. We may readily extend the concept of the differential operator. Specifically, the mth derivative of a continuous-time signal $\boldsymbol{x}$ is operationally denoted by

$$\left.\begin{aligned} &y = \boldsymbol{D}^m\boldsymbol{x} \\ &\text{where} \\ &y(t) = \frac{d^m x(t)}{dt^m} \qquad \text{for all } t \end{aligned}\right\} \tag{3.17}$$

with m a nonnegative integer. The operation $\boldsymbol{D}^m$ is to be interpreted as that of applying operator $\boldsymbol{D}$ successively m times to the signal that is being operated upon.

It is well known that the operations of integration and differentiation are the inverses of each other. Thus, if a continuous-time signal is differentiated (integrated), the process of integration (differentiation) restores the original signal. With this in mind, we now define the integration operator as

$$\left.\begin{aligned} &y = \boldsymbol{D}^{-1}\boldsymbol{x} \\ &\text{where} \\ &y(t) = \int_{-\infty}^{t} x(\tau)\, dt \qquad \text{for all } t \end{aligned}\right\} \tag{3.18}$$

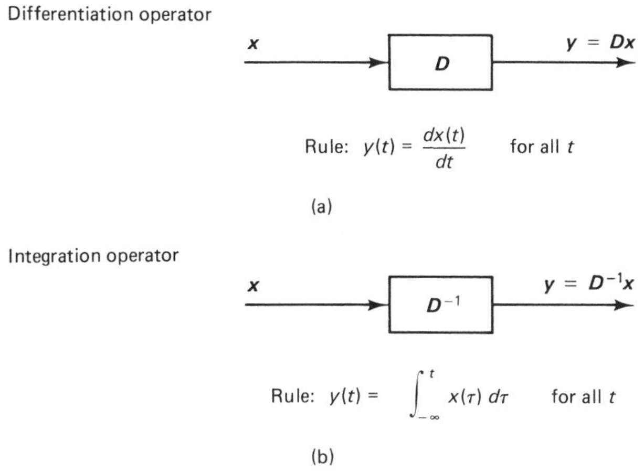

Differentiation operator

$x \rightarrow$ [D] $\rightarrow y = Dx$

Rule: $y(t) = \dfrac{dx(t)}{dt}$ for all t

(a)

Integration operator

$x \rightarrow$ [D^{-1}] $\rightarrow y = D^{-1}x$

Rule: $y(t) = \displaystyle\int_{-\infty}^{t} x(\tau)\, d\tau$ for all t

(b)

Figure 3.10 Elementary operations of differentiation and integration.

with the operator notation $\boldsymbol{D}^{-1}$ being used to designate the integration operator. We have used this notation to point out the fact that $\boldsymbol{D}$ and $\boldsymbol{D}^{-1}$ are inverse operators of each other, that is,

$$\boldsymbol{D}\,\boldsymbol{D}^{-1} = \boldsymbol{D}^{-1}\,\boldsymbol{D} = \boldsymbol{I} \tag{3.19}$$

in which $\boldsymbol{I}$ is the *identity operator* that leaves continuous-time signals unchanged, so that

$$\boldsymbol{I}\boldsymbol{x} = \boldsymbol{x} \tag{3.20}$$

The integration operator $\boldsymbol{D}^{-1}$ is shown in Fig. 3.10b.

Example 3.3

Determine the derivative of the continuous-time signals $\boldsymbol{x}_1$ and $\boldsymbol{x}_2$, where

$$x_1(t) = \cos t \qquad \text{for all } t$$

$$x_2(t) = \begin{cases} \cos t & \text{for } t \geq 0 \\ 0 & \text{for } t < 0 \end{cases}$$

Using standard differentiation, one obtains

$$y_1(t) = \frac{dx_1(t)}{dt} = -\sin t \qquad \text{for all } t$$

$$y_2(t) = \begin{cases} -\sin t & \text{for } t > 0 \\ \text{undefined} & \text{for } t = 0 \\ 0 & \text{for } t < 0 \end{cases}$$

It is noted that the signal $\boldsymbol{x}_2$ does not have a classical derivative at the time instant $t = 0$, since the signal itself has a discontinuity at $t = 0$. It is shown in the next section that it is possible to introduce a pseudo unit-impulse signal in order to effectively allow $\boldsymbol{x}_2$ to possess a derivative at $t = 0$.

Multiplication by a Constant

The operator rule for multiplication by a constant is a straightforward procedure by which the response signal $\boldsymbol{y}$ is related to the input signal $\boldsymbol{x}$ by a scalar multiplier α, that is,

$$\left.\begin{aligned} &\boldsymbol{y} = \alpha\boldsymbol{x} \\ &\text{where} \\ &y(t) = \alpha x(t) \qquad \text{for all } t \end{aligned}\right\} \tag{3.21}$$

This operation is shown in Fig. 3.11a. In the operational expression $\boldsymbol{y} = \alpha\boldsymbol{x}$, it is important to interpret $\boldsymbol{x}$ and $\boldsymbol{y}$ as being continuous-time signals and α as being a constant multiplier (not a continuous-time signal). Whenever this operation is symbolically used, it is important that this distinction be clearly spelled out. In order to avoid misinterpretation, we will normally use a lower case Greek letter to denote the scalar multiplier.

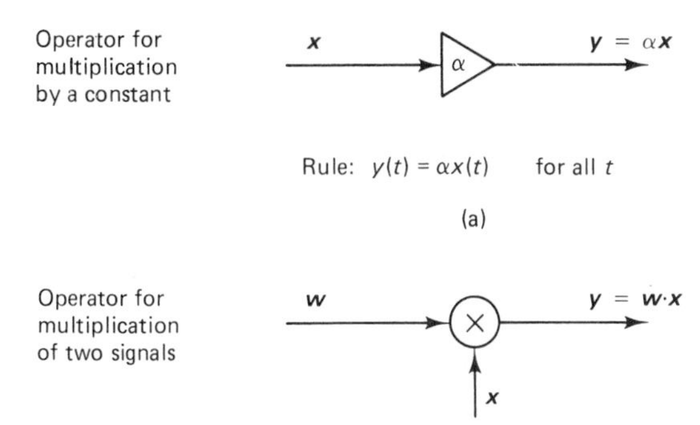

Figure 3.11 Elementary operations of multiplication by a constant and multiplication of two signals.

An associated operation that is used infrequently is that of the product of two continuous-time signals. The continuous-time signal $\boldsymbol{y}$ that results when the two continuous-time signals $\boldsymbol{w}$ and $\boldsymbol{x}$ are multiplied is denoted by

$$\left.\begin{aligned} &\boldsymbol{y} = \boldsymbol{w} \cdot \boldsymbol{x} \\ \text{where} \\ &y(t) = w(t) \cdot x(t) \qquad \text{for all } t \end{aligned}\right\} \tag{3.22}$$

This operation is depicted in Fig. 3.16b. The alert reader will note that this operation simplifies to the multiplication by a constant operation when the signal $\boldsymbol{w}$ is a constant, i.e., $w(t) = \alpha$ for all t.

Summation Operation

The final elementary operation is that in which two continuous-time signals are added to generate the so-called *sum signal*, which is itself a continuous-time signal. This operation is denoted by

$$\left.\begin{aligned} &\boldsymbol{y} = \boldsymbol{w} + \boldsymbol{x} \\ \text{where} \\ &y(t) = w(t) + x(t) \qquad \text{for all } t \end{aligned}\right\} \tag{3.23}$$

Thus, $\boldsymbol{w}$ and $\boldsymbol{x}$ are two continuous-time signals that are combined to form the sum signal $\boldsymbol{y}$. The summation operator is depicted in Fig. 3.12a. It is possible to extend this procedure in an obvious manner to sum the m continuous-time signals $\boldsymbol{x}_1$, $\boldsymbol{x}_2$, $\ldots$, $\boldsymbol{x}_m$, that is,

$$\left.\begin{aligned} &\boldsymbol{y} = \boldsymbol{x}_1 + \boldsymbol{x}_2 + \cdots + \boldsymbol{x}_m \\ \text{where} \\ &y(t) = x_1(t) + x_2(t) + \cdots + x_m(t) \qquad \text{for all } t \end{aligned}\right\} \tag{3.24}$$

Operation for summation of signals

w + y = w + x
+
x

Rule: $y(t) = w(t) + x(t)$ for all t

(a)

Operations for difference of signals

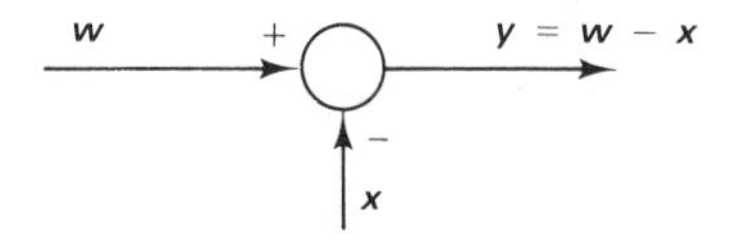

Rule: $y(t) = w(t) - x(t)$ for all t

(b)

Figure 3.12 Elementary operations of summation and difference of two signals.

This operation may be depicted by a summation node with m input terminals instead of the two shown in Fig. 3.12a.

We will also have occasion to generate the difference of two continuous-time signals as defined by

$$\left.\begin{aligned} &y = w - x \\ \text{where} \quad & \\ &y(t) = w(t) - x(t) \qquad \text{for all } t \end{aligned}\right\} \qquad (3.25)$$

This operation is illustrated in Fig. 3.12b where it is observed that a negative sign appears at the x terminal.

3.6 FUNDAMENTAL SIGNALS

As in the case of sequences, there are three fundamental continuous-time signals that are used over and over again in our studies. They are the unit-step, the unit-impulse, and the exponential signals.

Unit-Step Signal

In modeling different phenomena, it frequently happens that a relevant signal undergoes an abrupt change in value (i.e., it has a discontinuity). This is a most natural consequence of the fact that all real-world processes have an associated time instant when a dynamical action is begun. For example, this occurs when a d.c. voltage is suddenly applied to an electrical network, when one abruptly steps on the acceler-

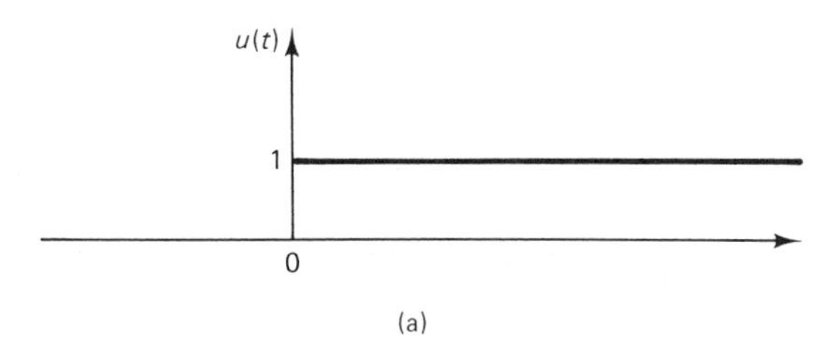

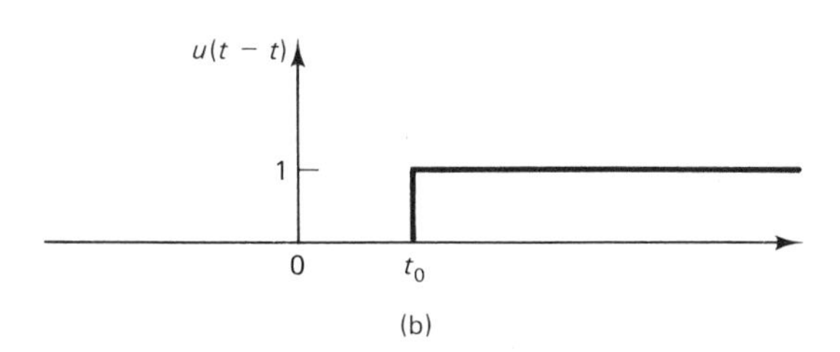

Figure 3.13 (a) Unit-step signal and (b) time-shifted unit-step signal.

ator pedal of an automobile, or when drugs are suddenly administered to the blood system of a hospital patient. In each of these cases, the relevant signal is seen to undergo an abrupt change in level. Since such situations arise so often in the real world, any attempt at generating realistic models must certainly include signals that reflect this characteristic.

With this in mind, we now introduce the unit-step signal, which has the vector notation $\boldsymbol{u} = \{u(t)\}$ and is defined by[3]

$$u(t) = \begin{cases} 1 & \text{for } t > 0 \\ 0 & \text{for } t < 0 \end{cases} \tag{3.26}$$

The unit-step signal has the graphical representation shown in Fig. 3.13a. It is seen to be a signal that is one for positive values of its argument and is zero when its argument is negative. Furthermore, the unit-step signal is seen to have a discontinuity of value one at the origin, since $u(0^+) = 1$ and $u(0^-) = 0$. The unit-step can therefore be used in those frequently occurring situations when a relevant signal suddenly becomes active in a constant manner.

Let us now consider a time-shifted version of the unit-step signal as denoted by $\{u(t - t_0)\}$. From the definition of the unit-step signal (3.26), it follows that

$$u(t - t_0) = \begin{cases} 1 & \text{for } t > t_0 \\ 0 & \text{for } t < t_0 \end{cases} \tag{3.27}$$

where it is observed that the argument $t - t_0$ is positive (negative) whenever $t > t_0$ ($t < t_0$). Thus, the signal $\{u(t - t_0)\}$ is generated by simply shifting the unit-step

[3]The unit-step signal is undefined at $t = 0$, but $u(0)$ may be arbitrarily set to $\frac{1}{2}$ so as to equal $[u(0^+) + u(0^-)]/2$. Unless otherwise stated, we shall hereafter take $u(0)$ to be $\frac{1}{2}$.

signal to the right by t_0 seconds as shown in Fig. 3.13b. The shifted unit-step gives us the flexibility of now suddenly applying a constant driving signal at any time instant t_0.

As implied at the beginning of this section, the unit-step signal may be used to represent signals that have discontinuities. To see how, let us consider the signal $x(t)$ shown in Fig. 3.14a, which has a discontinuity of value α at $t = t_0$ and is everywhere else continuous. It is apparent that this signal can be represented by the decomposition

$$x(t) = x_1(t) + \alpha u(t - t_0) \tag{3.28}$$

where $x_1(t)$ is given by the expression

$$x_1(t) = \begin{cases} x(t) - \alpha & \text{for } t > t_0 \\ x(t) & \text{for } t < t_0 \end{cases}$$

and appears as shown in Fig. 3.14b. Clearly, the signal $x_1(t)$ is everywhere continuous. The signal $x(t)$ has then been represented by the sum of an everywhere continuous signal $x_1(t)$ and a weighted, unit-step–type signal $\alpha u(t - t_0)$. The real scalar α corresponds to the value of the discontinuity of signal $x(t)$ at time t_0.

It should be apparent that any signal that has a finite number of discontinuities of finite value may always be represented by the sum of an everywhere continuous signal and a number of appropriately weighted and shifted unit-step signals. That this observation is important follows from the fact that idealized signals with discontinuities tend to arise frequently in practical applications. If we are to then perform appropriate operations on such signals (e.g., differentiation), it is essential that these operations be well-defined for unit-step–type signals.

To demonstrate the analytical difficulties that can arise when treating dis-

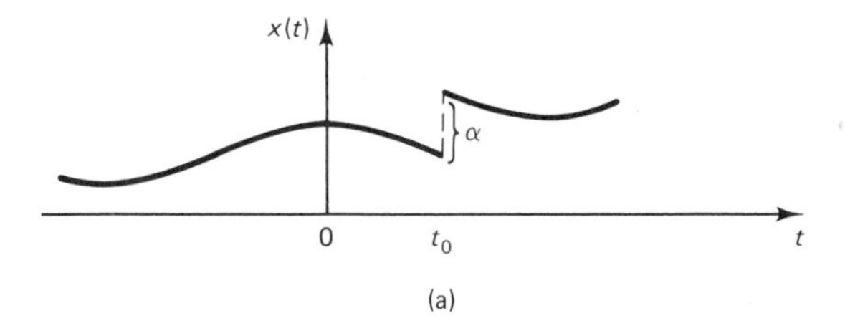

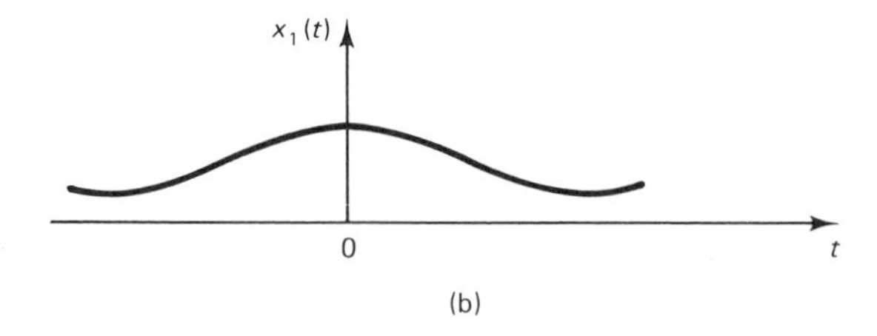

Figure 3.14 Representation of a discontinuous signal using unit-step signals.

continuous signals, let us now find the derivative of the unit-step signal. Applying the basic definition of differentiation as given by relationship (3.4), it follows that

$$\frac{du(t)}{dt} = 0 \qquad \text{for all } t \neq 0 \tag{3.29a}$$

On the other hand, we have at $t = 0$

$$\frac{du(0)}{dt} = \lim_{\varepsilon \to 0} \frac{u(\varepsilon) - u(0)}{\varepsilon}$$

If ε approaches zero from the left, it follows that $u(\varepsilon^-) = 0$, whereas $u(\varepsilon^+) = 1$ if ε approaches zero from the right. In each case, we have

$$\frac{du(0)}{dt} = \lim_{\varepsilon \to 0} \frac{1}{2|\varepsilon|} = \infty \tag{3.29b}$$

where $u(0)$ has been taken to be equal to $\frac{1}{2}$ as suggested in footnote 3 in this chapter. Thus, the unit-step signal's derivative is everywhere well-behaved except at the origin, where it fails to exist (it is unbounded).

As we have just shown, any signal with a discontinuity may be expressed as a sum of a continuous signal and a unit-step–type signal. If we are to then differentiate signals with discontinuities, the problem of differentiating the unit-step signal must be resolved. Unfortunately, if we appeal to classical differential calculus methods, a problem resolution will never be attained. It is possible to achieve a satisfactory solution, however, by introducing the concept of the unit-impulse signal.

Unit-Impulse Signal

The salient property which the unit-step's derivative must possess is expressible in the following integral identity

$$\int_{-\infty}^{\infty} \phi(t) \frac{du(t)}{dt} \, dt = \phi(0) \tag{3.30}$$

where $\phi(t)$ is any signal that is required to be continuous at the origin. This expression is actually not an integral operation in the classical sense, since its value is seen to be dependent on the behavior of the integrand term $\phi(t)$ at the single time instant zero and not at all on its behavior for $t \neq 0$. It is beneficial to envision this procedure of mapping the signal $\phi(t)$ into the number $\phi(0)$ as an integral operation, however, since most of the fundamental properties satisfied by the signal $du(t)/dt$ are obtained by using standard integration methods. This approach invariably leads to correct results and significantly simplifies the resultant analysis. A completely rigorous approach, however, requires the introduction of the mathematical concept of *generalized functions.* We will forgo this more rigorous development comforted by the foreknowledge that an integral interpretation of operator (3.30) yields correct results.

Let us demonstrate this last assertion by giving a mathematically nonrigorous proof of property (3.30) in which the function $\phi(t)$ is further constrained to possess a first derivative and to equal zero for time approaching plus infinity. Specifically, one obtains, after integrating the right side of integral (3.30) by parts,

$$\int_{-\infty}^{\infty} \phi(t) \frac{du(t)}{dt} dt = \phi(t)u(t)\Big|_{-\infty}^{\infty} - \int_{-\infty}^{\infty} \frac{d\phi(t)}{dt} u(t)\, dt$$
$$= 0 - \int_{0}^{\infty} \frac{d\phi(t)}{dt} dt$$
$$= \phi(t)\Big|_{0}^{\infty} = -\phi(0)$$

where we have used the fact that $\phi(\infty) = 0$. It must be mentioned that property (3.30) holds for the larger class of functions which are continuous at the origin and that the further assumption of differentiability is not required in general.

Since the derivative of the unit-step signal arises so often in applications, it is expedient to introduce the following special notation for it:

$$\delta(t) = \frac{du(t)}{dt} \tag{3.31}$$

The signal $\boldsymbol{\delta}$ is called the unit-impulse (or Dirac delta) function.[4] We may rewrite the basic property of the unit-step's derivative as given by expression (3.30) in terms of this new notation, that is,

$$\int_{-\infty}^{\infty} \phi(t)\delta(t)\, dt = \phi(0) \tag{3.32}$$

where the signal $\phi(t)$ must be continuous at the origin.

It is important to note that the unit-impulse signal, $\boldsymbol{\delta}$, is not defined in the classical manner of assigning numerical values to $\delta(t)$ for each value of t, as for example, the unit-step signal (3.26). Instead, its essential features are expressed by the defining integral relationship (3.32). This is not unexpected since we have just seen that $du(t)/dt$ is not defined at the origin (it is unbounded), and so a classical description is not possible.

There is much to be gained in terms of insight by constructing standard signals that have properties approximating the integral relationship (3.32). With this in mind, let us consider the signal $u_\varepsilon(t)$ and its derivative as depicted in Fig. 3.15.[5] Clearly, as the positive parameter ε approaches zero, the signal $u_\varepsilon(t)$ approaches the

[4]Recall from Chapter 2 the notation for the Kronecker delta signal (2.26). Distinction between the two is made by considering the argument: discrete-time, Kronecker; continuous-time, Dirac.

[5]Integration of the signal $du_\varepsilon(t)/dt$ is easily shown to result in the signal $u_\varepsilon(t)$.

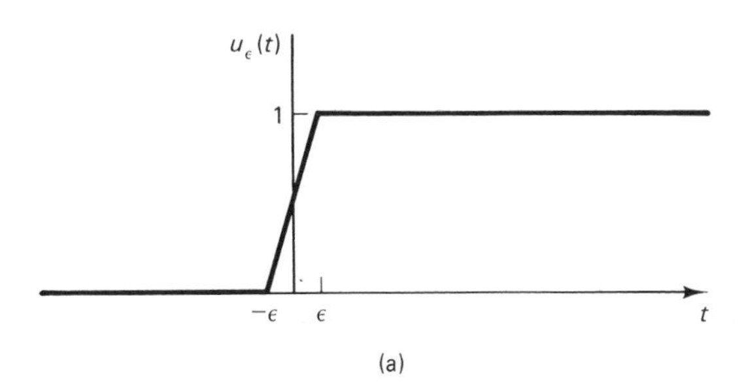

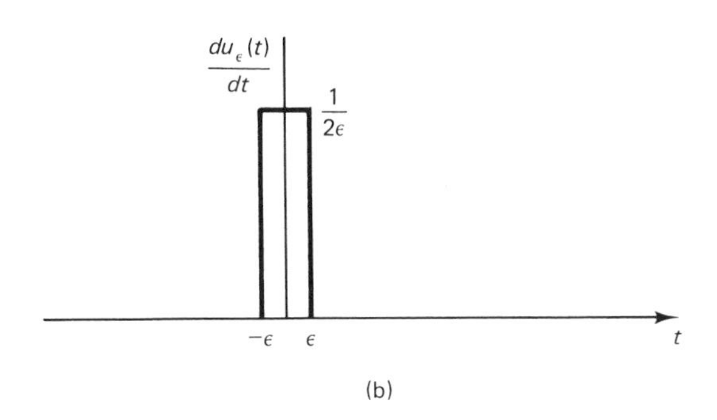

Figure 3.15 Approximation of unit-step signal (a) and its derivative (b).

unit-step signal. It is then intuitively appealing to equate the unit-step signal's derivative with $du_\varepsilon(t)/dt$ as ε approaches zero. To further substantiate this feeling, let us substitute $du_\varepsilon(t)/dt$ for $\delta(t)$ in relationship (3.32) to obtain

$$\int_{-\infty}^{\infty} \phi(t) \frac{du_\varepsilon(t)}{dt}\, dt = \int_{-\varepsilon}^{\varepsilon} \phi(t) \frac{1}{2\varepsilon}\, dt$$

If we now let ε approach 0^+, the integrand can be approximated by $\phi(0)/2\varepsilon$, since $\phi(t)$ is taken to be continuous at the origin. It is then concluded that

$$\int_{-\infty}^{\infty} \phi(t) \frac{du_\varepsilon(t)}{dt} \approx \phi(0)$$

and this approximation can be made as accurate as desired by choosing ε to be sufficiently small but positive.

The signal $du_\varepsilon(t)/dt$ is then seen to satisfy the integration property characterizing the unit-impulse signal (3.32) as $\varepsilon \to 0^+$. It is therefore only natural that we ascribe to the signal $\boldsymbol{\delta}$ the basic features embodied in $du_\varepsilon(t)/dt$ as ε approaches zero. Thus, we can envision the unit-impulse signal as being a *pulse* centered at the origin that has zero width, infinite height, and unit area.[6] With this in

[6]There are many standard signals that, in a limiting sense, satisfy integral relationship (3.32). For example, see Problem 3.27.

mind, one often encounters the following *improper* definition of the unit impulse:

$$\left.\begin{aligned} &\delta(t) = 0 \qquad \text{for } t \neq 0 \\ &\text{and} \\ &\int_{-\infty}^{\infty} \delta(t)\, dt = 1 \end{aligned}\right\} \tag{3.33}$$

The physical interpretation of the unit-impulse signal as being a pulse of infinite height and zero width and having unit area is of immense intuitive value, and its use generally yields correct solutions to most practical problems. It must be pointed out, however, that one is on much firmer ground (mathematically) when using the defining integral relationship (3.32) in proving the many properties satisfied by the unit-impulse signal.[7] We take this latter approach in what is to follow.

In using the integral relationship (3.32) to define the unit-impulse signal, the many properties which it satisfies are readily proven by using a standard change of variables operation. To illustrate this, let us consider the so-called *sampling* (or *sifting*) property, which states that

$$\int_{-\infty}^{\infty} \phi(t)\delta(t - t_0)\, dt = \phi(t_0) \tag{3.34}$$

where $\phi(t)$ is assumed to be continuous at the time instant t_0. This property is proved by making the change of variables $\tau = t - t_0$ and then using the defining integral relationship (3.32), that is,

$$\int_{-\infty}^{\infty} \phi(t)\delta(t - t_0)\, dt = \int_{-\infty}^{\infty} \phi(\tau + t_0)\delta(\tau)\, d\tau \equiv \phi(t_0)$$

Thus, the time-shifted unit-impulse $\delta(t - t_0)$ is said to sample the signal $\phi(t)$ at its point of application (i.e., at $t = t_0$).

In order to aid our intuitive feel for the unit-impulse signal, its unshifted and shifted versions are shown in Fig. 3.16. The vertical arrow notation shown in the figure represents a pulse of zero width, infinite height, and unit area as suggested by properties (3.33). With this interpretation in mind, it follows that

$$u(t) = \int_{-\infty}^{t} \delta(\tau)\, d\tau = \begin{cases} 1 & \text{for } t > 0 \\ 0 & \text{for } t < 0 \end{cases} \tag{3.35}$$

since when t is less than zero, the integrand is identically zero, and when t is greater than zero, the integrand has unity area. This result is not unexpected since the unit-impulse signal was originally conceived of as being the derivative of the unit-step signal.

[7]A completely mathematical, rigorous treatment of the unit-impulse signal requires the theory of generalized functions (distributions). See M.J. Lighthill, *An Introduction to Fourier Analysis and Generalized Functions* (New York: Cambridge University Press, 1959).

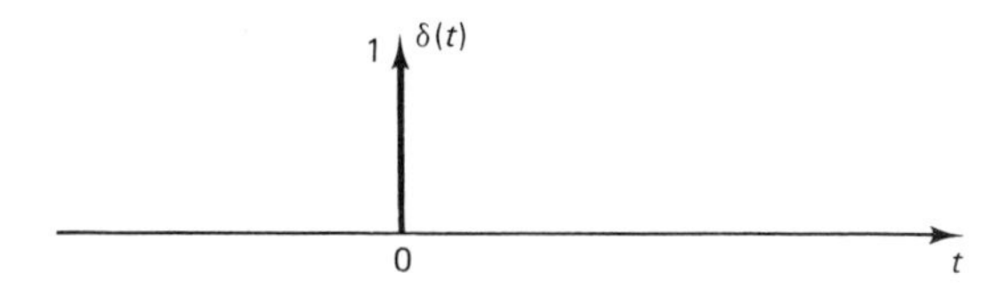

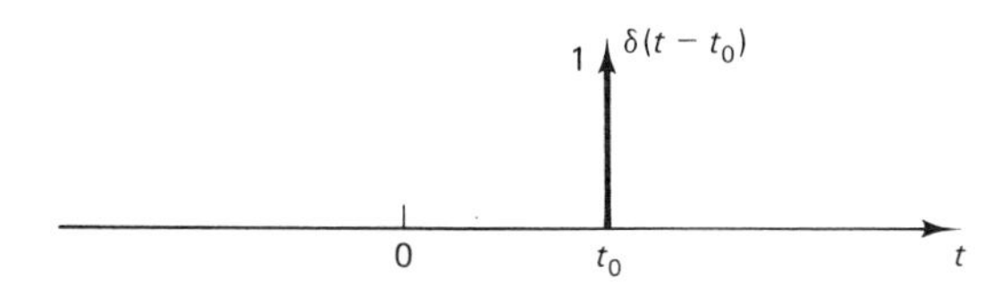

Figure 3.16 Depiction of unit impulse (a) and its shifted version (b).

Another of the more fundamental properties that characterize the unit-impulse signal is given by

$$\delta(at) = \frac{1}{|a|}\,\delta(t) \tag{3.36}$$

where a is any real number. If we let $a = -1$ in this relationship, we determine that the unit-impulse signal is an even function of time, that is,

$$\delta(-t) = \delta(t) \tag{3.37}$$

One can also show that the following equivalence holds:

$$f(t)\delta(t - t_0) = f(t_0)\delta(t - t_0) \tag{3.38}$$

for all functions $f(t)$ that are continuous at time t_0. This relationship is extremely useful and significantly shortens what would be an otherwise time-consuming task when differentiating signals. Thus, one should always replace the term $f(t)\delta(t - t_0)$ by its equivalent $f(t_0)\delta(t - t_0)$ whenever $f(t)$ is continuous at t_0. If $f(t)$ is not continuous at t_0, then the signal $f(t)\,\delta(t - t_0)$ is undefined at t_0.

It is possible to extend the unit-impulse signal concept to higher-order derivatives of the unit step. For example, let us consider the second derivative as given by $d\delta(t)/dt = d^2u\,(t)/dt^2$. Using integration by parts, one readily obtains the following integral identity:

$$\int_{-\infty}^{\infty} \frac{d\delta(t)}{dt}\,\phi(t)\,dt = -\int_{-\infty}^{\infty} \delta(t)\,\frac{d\phi(t)}{dt}\,dt = \frac{-d\phi(0)}{dt} \tag{3.39}$$

for all functions $\phi(t)$ that have a continuous derivative at the origin and go to zero at $t = \infty$. The second derivative of the unit-step signal is commonly referred to as the *unit-doublet* and is expressed as follows:

$$\frac{d\delta(t)}{dt} = \frac{d^2u(t)}{dt^2} \tag{3.40}$$

In a similar manner, by sequentially applying integration by parts, one may show that

$$\int_{-\infty}^{+\infty} \frac{d^n\delta(t)}{dt^n}\,\phi(t)\,dt = (-1)^n \frac{d^n\phi(t)}{dt^n} \tag{3.41}$$

for all test functions $\phi(t)$ that have a continuous nth derivative at the origin and go to zero at $T = \infty$. The nth derivative of the unit-step signal is called the *unit-n tuplet* and is expressed as follows:

$$\frac{d^{n-1}\delta(t)}{dt^{n-1}} = \frac{d^n u(t)}{dt^n} \tag{3.42}$$

The unit-singlet (or unit-impulse), -doublet, and -triplet signals are illustrated in Fig. 3.17.

One can use many of the standard rules of differentiation and integration

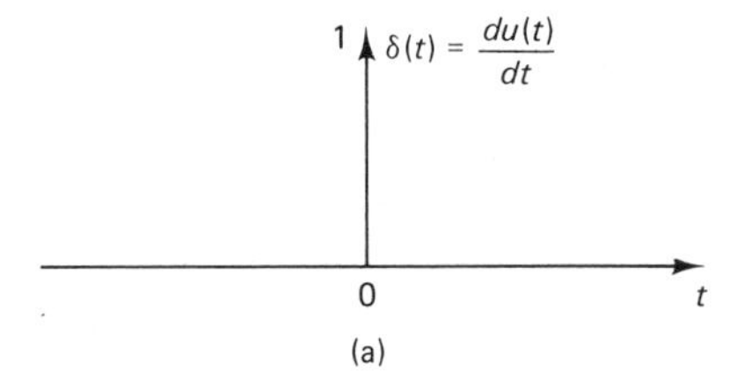

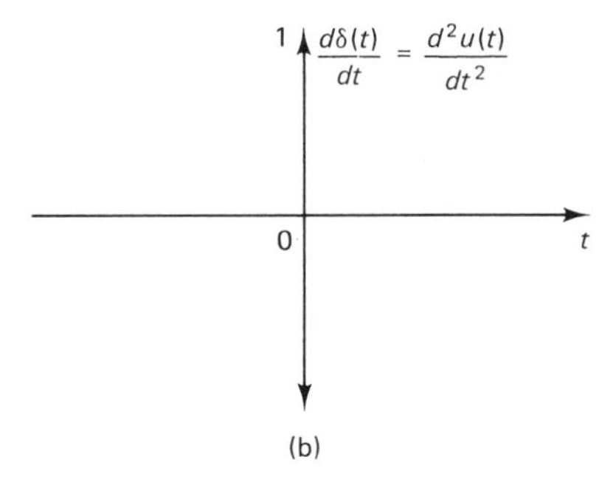

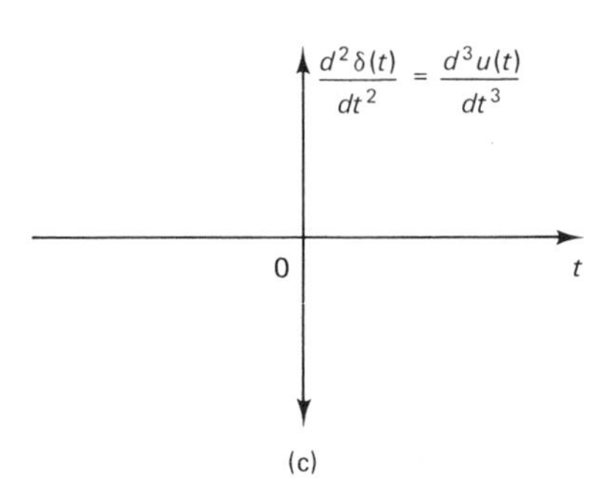

Figure 3.17 Sketches of the (a) unit-singlet, (b) unit-doublet, and (c) unit-triplet.

when derivatives of the unit-step signal arise. For example, let us find the second derivative of the signal.

$$x(t) = w(t)u(t) \tag{3.43}$$

which is seen to be identically zero for negative values of time. Applying the standard rule governing the derivative of a product of time functions, we have

$$\begin{aligned}\frac{dx(t)}{dt} &= w(t)\frac{du(t)}{dt} + \frac{dw(t)}{dt}u(t)\\ &= w(0)\delta(t) + \frac{dw(t)}{dt}u(t)\end{aligned}$$

where we have utilized property (3.38). Continuing in this manner, it follows that

$$\begin{aligned}\frac{d^2x(t)}{dt^2} &= w(0)\frac{d\delta(t)}{dt} + \frac{dw(t)}{dt}\frac{du(t)}{dt} + \frac{d^2w(t)}{dt^2}u(t)\\ &= w(0)\frac{d\delta(t)}{dt} + \frac{dw(0)}{dt}\delta(t) + \frac{d^2w(t)}{dt^2}u(t) \qquad (3.44)\\ &= w(0)\frac{d\delta(t)}{dt} + \frac{dw(0)}{dt}\delta(t) + \frac{d^2w(t)}{dt^2}u(t)\end{aligned}$$

where it has been assumed that the signal $w(t)$ possesses a second derivative that is continuous at the origin.

Example 3.4

Determine the second derivative of the time function

$$x(t) = [1 + \sin \omega t]u(t)$$

It is observed that this signal is in the same format as representation (3.43). Using the procedure that lead to result (3.44), one finds that

$$\frac{dx(t)}{dt} = \delta(t) + \omega \cos \omega t \; u(t)$$

$$\frac{d^2x(t)}{dt^2} = \frac{d\delta(t)}{dt} + \omega\delta(t) - \omega^2 \sin \omega t \; u(t)$$

The signals $x(t)$, $dx(t)/dt$, and $d^2x(t)/dt^2$ are shown in Fig. 3.18.

Complex Exponential Signal

It will soon become apparent that the exponential signal is the most important signal relative to our study of continuous-time signals and linear systems. One can gain an appreciation for why this is so by noting that the derivative and integral of the complex exponential signal as defined by

$$x(t) = e^{st} \qquad \text{for all } t \tag{3.45}$$

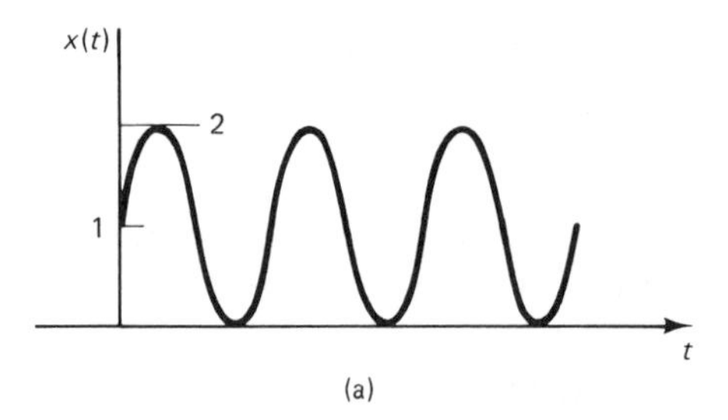

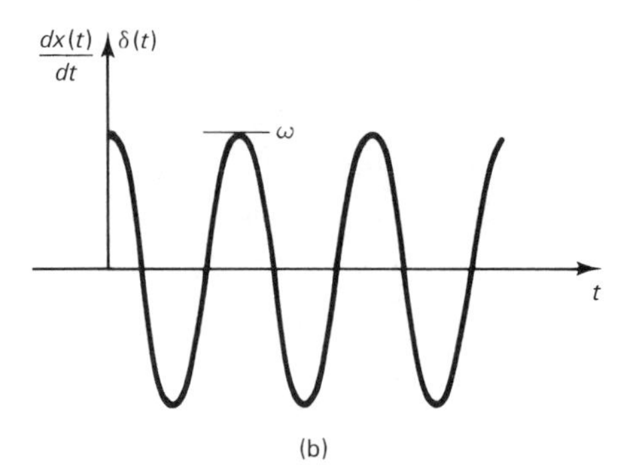

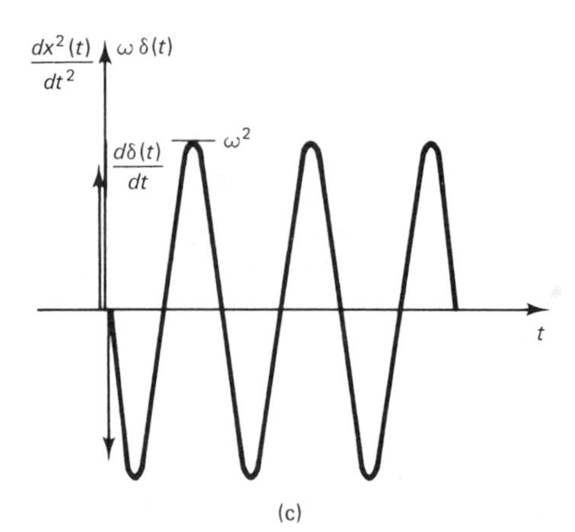

Figure 3.18 Sketches of (a) $x(t)$, (b) dx/dt, and (c) d^2x/dt^2 from Example 3.4.

is given by

$$\frac{dx(t)}{dt} = se^{st} \quad \text{and} \quad \int_{-\infty}^{\tau} e^{s\tau}\, d\tau = \frac{1}{s}\, e^{st} \tag{3.46}$$

where s is a fixed complex number. Thus, the complex exponential signal's derivative and integral are each seen to be a scalar multiplied version of itself. In linear algebra terminology, the exponential signal (3.45) is an *eigenfunction* of the deriva-

algebra terminology, the exponential signal is said to be an *eigenfunction* of the derivative operator $\boldsymbol{D}$ with *eigenvalue* s, since it satisfies the operational relationship[8]

$$\boldsymbol{Dx} = s\boldsymbol{x} \tag{3.47}$$

where $x(t) = e^{st}$. The implications of this observation, although not readily apparent at this time, are certainly thought-provoking when it is realized that only exponential signals satisfy this characterization, i.e., satisfy the differential equation $dx(t)/dt = sx(t)$. This observation in conjunction with the foreknowledge that the differential and integral operators play fundamental roles in linear system theory should serve as convincing evidence of the exponential signal's importance.

The complex exponential signal as represented by expression (3.45) has a time behavior that is completely determined by the selection of the complex number s in the exponent, that is,

$$s = \sigma + j\omega$$

where σ and ω are real numbers and $j = \sqrt{-1}$. It is possible to represent a large class of signals by an appropriate linear combination of exponential signals (i.e., a Fourier representation). To gain an appreciation for why this is so, let us first consider the case where s is restricted to be real (i.e., $\omega = 0$), thereby giving rise to the class of real exponentials:

$$x(t) = e^{\sigma t} \qquad \text{for all } t \tag{3.48}$$

A plot of three possible real exponential signals is given in Fig. 3.19a. The resultant signal characterization is seen to depend on the selection of the so-called *damping* parameter σ. When σ is positive (negative), the real exponential signal is seen to increase (decrease) in value for increasing t, but it remains constant for a selection of $\sigma = 0$.

If s is restricted to be imaginary (i.e., $\sigma = 0$), the class of complex sinusoidal signals is thereby generated:

$$x(t) = e^{j\omega t} = \cos \omega t + j \sin \omega t \tag{3.49}$$

This signal is seen to be a complex function and therefore does not play a direct role in real valued signal applications. It is possible, however, to suitably combine complex sinusoidal signals in order to generate real sinusoidal signals. For example, one may utilize the signals $e^{j\omega t}$ and $e^{-j\omega t}$ to obtain

$$\cos \omega t = \frac{e^{j\omega t} + e^{-j\omega t}}{2} \tag{3.50}$$

A plot of this signal is shown in Fig. 3.18b. The parameter ω is seen to control the frequency of oscillation of the signal, i.e., the rapidity with which the given sinusoid oscillates between its maximum $(+1)$ and minimum (-1) values.

[8]Similarly, the complex exponential signal is seen to be an eigenfunction of the integral operator $\boldsymbol{D}^{-1}$ with corresponding eigenvalue of $1/s$ since $\boldsymbol{D}^{-1}\boldsymbol{x} = (1/s)\boldsymbol{x}$.

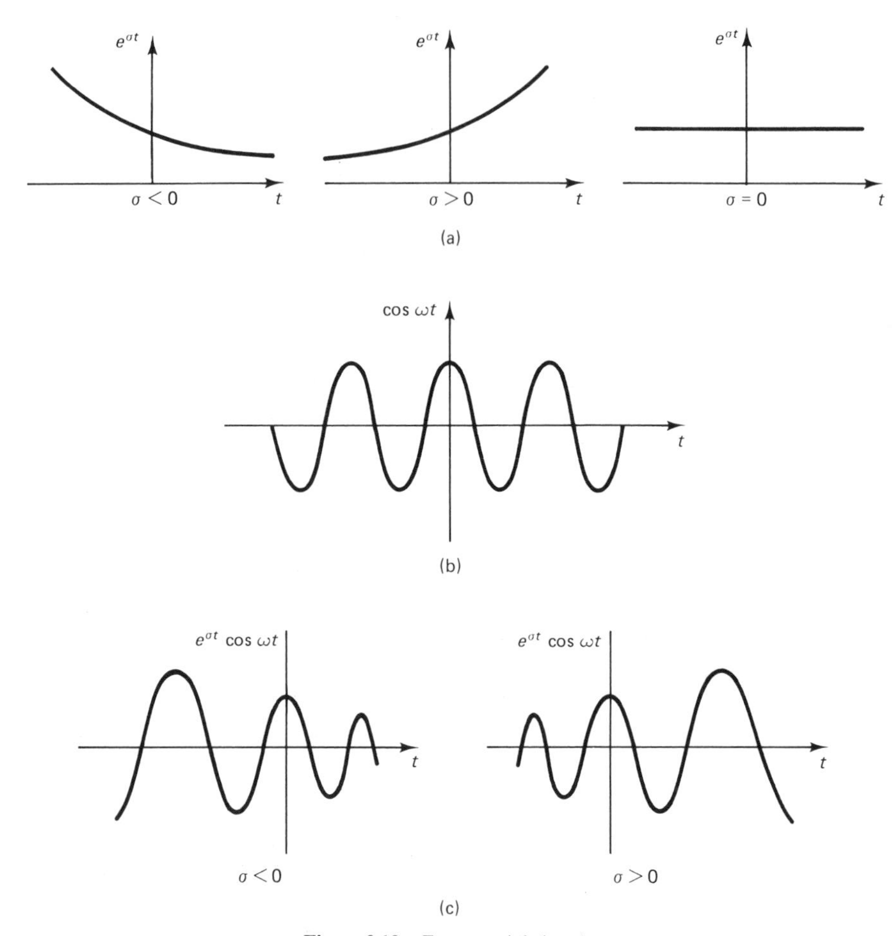

Figure 3.19 Exponential signals.

In a similar manner, one may generate the so-called exponential-varying sinusoid by the following combination of complex exponentials:

$$e^{\sigma t} \cos \omega t = \frac{e^{(\sigma + j\omega)t} + e^{(\sigma - j\omega)t}}{2} \tag{3.51}$$

Two possible exponential-varying sinusoids are shown in Fig. 3.18c.

It should be apparent that a large variety of signals can be generated from the simple complex exponential signal $e^{st} = e^{(\sigma + j\omega)t}$ by suitably selecting the parameters σ and ω. In our treatment of the Fourier representation of signals in Chapters 8 and 9, it is shown that it is possible to approximate, arbitrarily close, virtually any signal of interest by an appropriate linear combination of suitably chosen complex exponential signals. The class of complex exponential signals is indeed a very rich one and has been the center of interest of mathematicians and scientists for the past few centuries.

3.7 MEASURE OF SIGNAL SIZE

As in the discrete-time case, the ability to analytically analyze a continuous-time signal is considerably enhanced by introducing the concept of signal size (or norm). In this text, we are primarily interested in three norms that arise in a natural manner in different contexts. For example, it is shown in Chapter 4 that the $\mathscr{L}_1$ norm of the signal $\boldsymbol{x} = \{x(t)\}$ as defined by

$$\mathscr{L}_1 \text{ norm:} \qquad \|\boldsymbol{x}\|_1 = \int_{-\infty}^{\infty} |x(t)|\, dt \tag{3.52}$$

is useful in characterizing the stability of a linear signal operator. In describing system stability, it is necessary to use the notion of *bounded signals*. The signal $\boldsymbol{x}$ is said to be bounded if its $\mathscr{L}_\infty$ norm as specified by

$$\mathscr{L}_\infty \text{ norm:} \qquad \|\boldsymbol{x}\|_\infty = \sup_t |x(t)| \tag{3.53}$$

is finite. In most situations, one may replace the operator *sup* (supremum, or the least upperbound) by the term maximum. Finally, the energy in the signal $\boldsymbol{x}$ is measured by its $\mathscr{L}_2$ norm as defined by

$$\mathscr{L}_2 \text{ norm:} \qquad \|\boldsymbol{x}\| = \left[\int_{-\infty}^{\infty} |x(t)|^2\, dt\right]^{1/2} \tag{3.54}$$

To illustrate these norm measures, the following example is offered.

Example 3.5

Evaluate the $\mathscr{L}_1$, $\mathscr{L}_2$, and $\mathscr{L}_\infty$ norms associated with the *one-sided* exponential signal $x(t) = 3e^{-4t}u(t)$. Applying definitions (3.52), (3.53), and (3.54), it is found that

$$\|\boldsymbol{x}\|_1 = \int_0^{\infty} 3e^{-4t}\, dt = \frac{3}{4}$$

$$\|\boldsymbol{x}\|_2 = \left[\int_0^{\infty} [3e^{-4t}]^2\, d(t)\right]^{1/2} = \frac{3\sqrt{2}}{4}$$

$$\|\boldsymbol{x}\|_\infty = \max\,[3e^{-4t}u(t)] = 3$$

3.8 PROBLEMS

3.1. Graph the following signals.

(a) $x_a(t) = \begin{cases} 0 & \text{for } t \le 0 \\ t & \text{for } 0 < t \le 1 \\ \cos 2\pi t & \text{for } 1 < t \end{cases}$

(b) $x_b(t) = \begin{cases} 1 & \text{for } 2n \le t < 2n+1 \\ 0 & \text{for other } t \end{cases} \qquad \text{for } n = 0, \pm 1, \pm 2, \ldots$

(c) $x_c(t) = \exp(-|t|)$

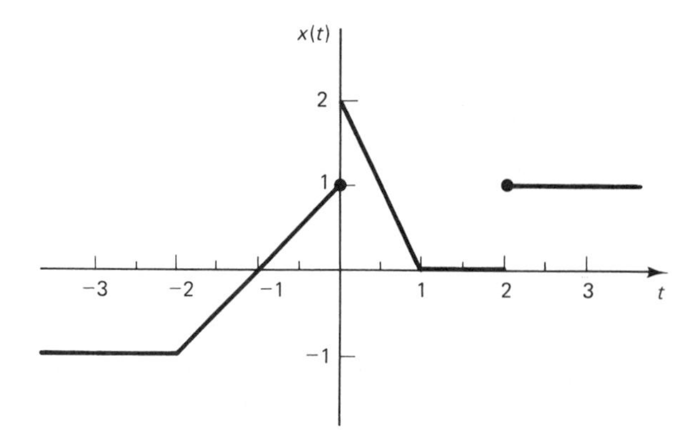

Figure P3.2

3.2. Identify the signal in Fig. P3.2 by the appropriate formulas.

3.3. For the signals in Problem 3.1, determine if they are continuous for all t. If not, indicate those values of t for which they are discontinuous.

3.4. In Fig. P3.2, $x(0) = 1$ as denoted by the dotted value at $t = 0$.
(a) Determine $x(0^+)$, $x(1^+)$, $x(1^-)$, $x(2^-)$, and $x(2^+)$.
(b) For which points t in the set $\{-2, -1, 0, 1, 2\}$ is $\boldsymbol{x}$ continuous?

3.5. From Fig. P3.2 determine
(a) those points for which dx/dt does not exist.
(b) the value of dx/dt at $t = -3, -1, 3$.

3.6. Sketch graphs of the signals $x_a(2t)$ and $x_a(t/2)$ where $x_a(t)$ is given in Problem 3.1a.

3.7. If t is the normal time-axis variable and a signal $x(t) = \cos 2\pi t$ has a period of one second, calculate the period of $x(\tau)$ where $\tau = t \times 10^{-3}$ is in milliseconds.

3.8. What positive shift of time variable is required to convert $\cos 2\pi t$ into $\sin 2\pi t$, i.e., find $t_0 > 0$ such that $\cos 2\pi(t - t_0) = \sin 2\pi t$? Is this value of t_0 unique?

3.9. Graph the following time-shifted signals where $x(t)$ is as shown in Fig. P3.2:
(a) $x(t + 2)$
(b) $x(t - 1)$
(c) $x(t - 2)$

3.10. Graph the following transposed signals where $x(t)$ is as shown in Fig. P3.2:
(a) $x(-t)$
(b) $x(2 - t)$
(c) $x(-t - 1)$

3.11. Find the derivative $\boldsymbol{Dx}$ and graph the result for $x(t) = x_a(t)$ from Problem 3.1.

3.12. Repeat Problem 3.11 for $x(t) = x_c(t)$ from Problem 3.1. Is the derivative signal continuous at $t = 0$?

3.13. Graph the product $x_b(t) \cdot x_c(t)$ where $\boldsymbol{x}_b$ and $\boldsymbol{x}_c$ are taken from Problem 3.1.

3.14. Graph the sum $x_b(t) + x_c(t)$ where $\boldsymbol{x}_b$ and $\boldsymbol{x}_c$ are given in Problem 3.1.

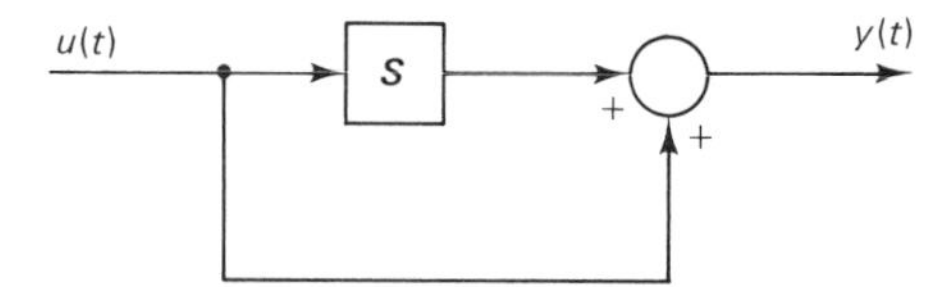

Figure P3.15

3.15. Sketch the output signal y arising from the system shown in Fig. P3.15 when the input is a unit-step signal. Recall that S is a unit-delay operator.

3.16. Sketch the signals given by the following combinations of unit-step functions:

(a) $u(t) - u(t - T)$

(b) $tu(t) - tu(t - T)$

(c) $tu(t) - (t - T)u(t - T)$

3.17. Describe the signal x in Fig. P3.17 in terms of unit-step functions.

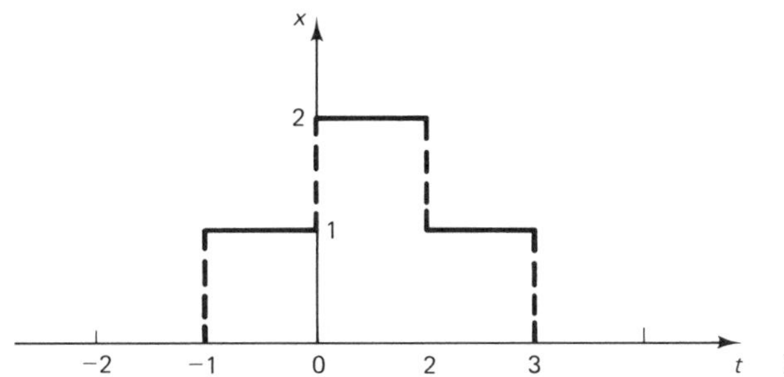

Figure P3.17

3.18. Describe the signal x given as a graph in Fig. P3.18 as a continuous signal y plus any necessary step functions.

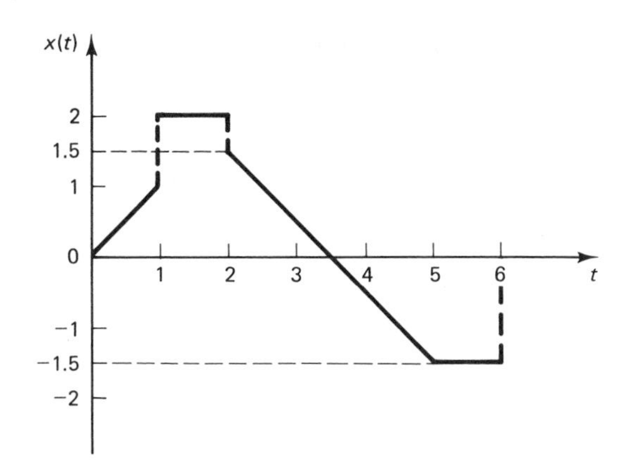

Figure P3.18

3.19. Use the sifting property of δ, given in expression (3.34) to evaluate the following integrals:

(a) $\int_{-\infty}^{\infty} \cos t\, \delta(t)\, dt$

(b) $\int_{0}^{10} t^2 \delta(t - 2)\, dt$

3.20. Evaluate y in Fig. P3.20 using the sifting property of δ.

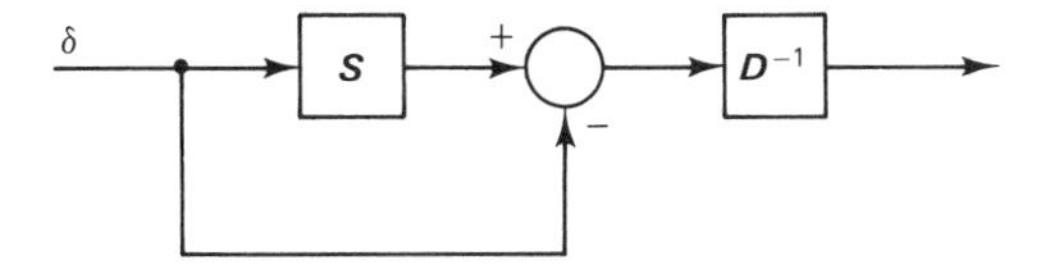

Figure P3.20

3.21. Find the value of y as a function of t if

$$y(t) = \int_0^5 \tau\delta(\tau - t)\, d\tau$$

3.22. **(a)** Sketch the sequence of signals $\{T\ \Delta(t/T)\}$ as T takes on values $1, \frac{1}{2}, \frac{1}{4}, \ldots,$ and $\Delta(t)$ is defined in Fig. P3.22. What is the limiting signal as T approaches zero?

(b) Now sketch the derivative sequence $\{d/dt\ [T\ \Delta(t/T)]\}$ as T takes on values $1, \frac{1}{2}, \frac{1}{4}, \ldots$. Notice that the limiting signal is the unit-doublet of expression (3.40).

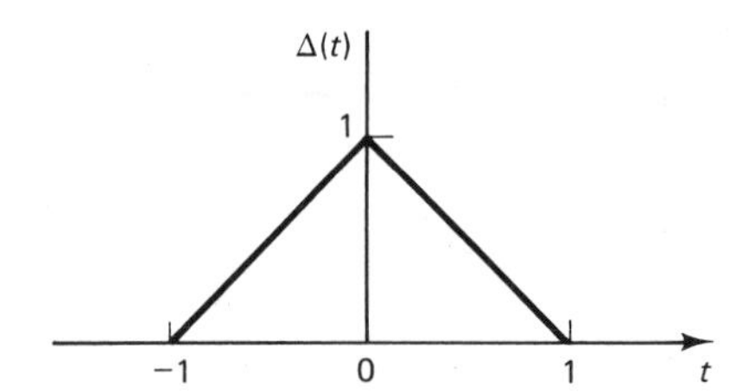

Figure P3.22

3.23. Determine and sketch the derivatives and second derivatives of

(a) $\cos t\ [u(t) - u(t - 2\pi)]$

(b) $t \sin t\ u(t)$

3.24. **(a)** Show that $[\boldsymbol{D}^2 + 2\boldsymbol{D} + 3\boldsymbol{I}]e^{St} = (s^2 + 2s + 3)e^{st}$

(b) Generalize to any polynomial operator $p(\boldsymbol{D})$.

3.25. **(a)** If $Dx(t) = e^{st}$, where s is a constant, show that $x(t) = (1/s)e^{st}$.

(b) If $(D + 2)x(t) = e^{st}$, show that $x(t) = e^{st}/(s + 2)$.

Hint: Assume a solution $x(t) = ce^{st}$.

3.26. Plot the graphs of

(a) Real $[e^{jt}]$

(b) Real $[e^{(-1+j)t}]$

(c) $e^{\sigma t}$ for $\sigma = 1, 2$

(d) $e^{-\sigma t}$ for $\sigma = 1, 2$

3.27. Show that the following two functions have properties associated with the Dirac delta function as the positive parameter ε approaches zero.

(a) Triangular function as defined by

$$\delta_\varepsilon(t) = \begin{cases} \dfrac{1}{\varepsilon} - \dfrac{1}{\varepsilon^2}|t| & \text{for } -\varepsilon \le t \le \varepsilon \\ 0 & \text{otherwise} \end{cases}$$

(b) Gaussian function as defined by

$$\delta_\varepsilon(t) = \frac{1}{\varepsilon\sqrt{2\pi}}\, e^{-(t^2/2\varepsilon^2)}$$

4

Linear Operations on Signals

4.1 INTRODUCTION

In Chapter 1, the concepts of signals and signal operators were introduced. Signals are important since they represent information concerning a dynamic phenomenon of interest to the investigator. On the other hand, a signal operation is important since it is often necessary to alter signals in order either to enhance their information content (e.g., filtering) or to use them to effect an auxiliary objective (e.g., control). With this in mind, we now examine some of the more important features of signal operations as represented by the general operator expression

$$y = Tx \tag{4.1}$$

where x is the excitation (input) signal being operated upon, T is the signal operator, and y is the response (output) of this operation. As previously indicated, T is a well-defined rule that relates the excitation and response signals in a unique manner.

In the operational representation (4.1), the signal operator T is completely general. We must in some way restrict this generality, however, if any hope of obtaining a better understanding of its features is to be had. To see why this is so, let us consider the task of characterizing a completely general signal operator. An intuitive procedure for achieving this would be to excite the signal operator T by a number of different excitations x_k and record the corresponding responses $y_k = Tx_k$ for $k = 1, 2, 3, \ldots$. We could then compile a list of these experimentally determined representative excitation-response pairs (x_k, y_k) for $k = 1, 2, 3, \ldots$. Clearly, this listing would convey some useful information relative to the behavior of the signal operation. We would typically never be in a position, however, to confidently predict a response from an excitation that is not found in this listing. Such is the dilemma when considering a general system operator. Due to this difficulty, we now restrict our attention to the special class of linear operators. Fortunately, linear operators are not only relatively easy to characterize, they are also very important in practical applications.

4.2 LINEAR DISCRETE-TIME SIGNAL OPERATIONS

To begin our analysis of linear signal operators, we first examine the discrete-time case. The discrete-time operator $\boldsymbol{T}$ as characterized by the so-called *convolution summation* expression

$$y(n) = \sum_{k=-\infty}^{\infty} h(k)x(n-k) \tag{4.2a}$$

is said to be *linear*. The *characteristic sequence* $\{h(n)\}$ that appears in this summation is a fixed signal that governs the dynamical behavior of the linear signal operation. To compute the nth element of the response signal $\{y(n)\}$, it is necessary to carry out the indicated multiplications $h(k)x(n-k)$ and then sum these products. Clearly, this operation is well-defined and therefore constitutes a legitimate signal operation. Moreover, this operation is readily implementable on digital computers since it only involves multiplication and summation operations. It is readily shown by making a term-for-term comparison that the convolution summation above may be *equivalently* expressed as

$$y(n) = \sum_{k=-\infty}^{\infty} h(n-k)x(k) \tag{4.2b}$$

By restricting the signal operation to be of the simple product-summation format (4.2a) or (4.2b), a performance tradeoff has been made. Namely, a significant loss in the types of signal operations available has thereby been incurred. Offsetting this negative feature is the welcomed realization that linear signal operations possess salient properties that make their use in signal processing applications extremely valuable. Some of these more important properties are described and developed in the sections that follow.

For the purposes of this text, a linear discrete-time operator is defined to be any well-defined rule (that is, an algorithm) that may be equivalently represented as a convolution summation of format (4.2). In particular, we are particularly concerned with the subset of such operators that are governed by a linear recursive difference equation of order (p, q) as given by

$$y(n) = \sum_{k=0}^{q} b_k x(n-k) - \sum_{k=1}^{p} a_k y(n-k) \tag{4.3}$$

In this expression p and q are nonnegative integers specifying the order of the recursive operator, and a_k and b_k are coefficients that characterize the dynamical behavior of the operator. Upon examination of this algorithm, it is seen that the response element's value at time n is equal to a linear combination of the present and q most recent excitation values and the p most recent response values. It is subsequently shown that the linear recursive operator (4.3) may be equivalently represented in the convolution summation format (4.2). As such, linear recursive operators possess all the nice features associated with general linear operators.

To the novice, signal operation (4.2) may still appear to be rather abstract. In specific applications where numerical values are assigned to the $h(k)$ coefficients, however, an interpretation of the linear signal operation is straightforward. The next two examples illustrate this point.

Example 4.1

In many signal processing applications requiring low-pass filtering, the signal operation as defined by

$$y(n) = \frac{1}{N}[x(n) + x(n-1) + \cdots + x(n-N+1)]$$

is very useful. The filter order integer N (for example, $N = 10$) specifies the filter's memory. For this filter, the characteristic sequence elements are specified by $h(n) = 1/N$ for $0 \leq n \leq N - 1$ and $h(n) = 0$ otherwise. Let us now find this filter's response to the unit-step excitation. Upon setting the excitation $x(n) = u(n)$ in the operation above, it is found that

$$y(n) = \begin{cases} 0 & \text{for } n < 0 \\ \dfrac{n+1}{N} & \text{for } 0 \leq n \leq N-1 \\ 1 & \text{for } N-1 \leq n \end{cases}$$

Example 4.2

As another example, let us consider a bank savings account system in which interest is being paid at the rate of r compounded at N conversion periods per year. If

$x(n)$ = totality of deposits made into the account at the end of the nth conversion period

$y(n)$ = totality of funds in the account at the end of the nth conversion period

it was shown in Section 2.8 that these two signals are related by

$$y(n) = \left(1 + \frac{r}{N}\right)y(n-1) + x(n) \tag{4.4}$$

It is subsequently shown that this signal operation is linear and may be therefore put into the convolution summation format (4.2).

Let us now examine a specific savings account history under a savings plan which pays interest at the rate of 5 percent compounded semiannually. In this case, $r = 0.05$ and $N = 2$. It is assumed that when the account was established there was an initial deposit of \$1000. This corresponds to $y(0) = 1000$ (or equivalently $u(0) = 1000$). Thereafter, the totality of deposits made during successive conversion periods is \$476, \$355, $-$\$217 (a withdrawal from the account), \$727, and so forth. This history of deposits corresponds to the input sequence

$$x(1) = 476, \quad x(2) = 355, \quad x(3) = -217, \quad x(4) = 727, \ldots$$

Using expression (4.4), we are able to calculate the state of the account at the close of successive conversion periods. Specifically, at the end of the first half year ($k = 1$), we have

$$y(1) = (1 + 0.025)y(0) + x(1)$$
$$= 1.025 \times 1000 + 476 = 1051$$

In a similar manner,

$$y(2) = 1.025y(1) + x(2)$$
$$= 1.025 \times 1501 + 355 = 1893.53$$
$$y(3) = 1.025y(2) + x(3)$$
$$= 1.025 \times 1893.53 - 217 = 1723.87$$
$$y(4) = 1.025y(3) + x(4)$$
$$= 1.025 \times 1723.87 + 727 = 2493.96$$

and so forth. The beginning student is cautioned not to fall into the common error of misinterpreting discrete time n. In this example, n stands for the number of half years in time, and *not* the number of years.

4.3 LINEAR CONTINUOUS-TIME SIGNAL OPERATIONS

In the case of continuous-time signal operators, the convolution integral relationship

$$y(t) = \int_{-\infty}^{\infty} h(\tau)x(t - \tau)\, d\tau \tag{4.5a}$$

constitutes a linear operation by which the response $\{y(t)\}$ is readily computed from the excitation $\{x(t)\}$. The *characteristic signal* $\{h(t)\}$, which completely governs the dynamical behavior of this operation, is shortly shown to be equal to the operator's *unit-impulse response*. More is said about this characteristic signal in a later section. To compute the response signal's value at time t, it is then necessary to perform the product $h(\tau)x(t - \tau)$ and then integrate this product over the range $-\infty < \tau < +\infty$. Upon comparison of this continuous-time operation with the corresponding discrete-time operation (4.2), a remarkable similarity is in evidence. The primary difference arises from the interchanging of the summation and integration operations. Since integration is often interpretable as a limit of summation, this difference is not major. Finally, by making a change in the integration variable, it is readily shown that the convolution integral above may be *equivalently* expressed as

$$y(t) = \int_{-\infty}^{\infty} h(t - \tau)x(\tau)\, d\tau \tag{4.5b}$$

It is important to note that many relevant practical phenomena may be characterized by a linear convolution integral operation of form (4.5). Specifically,

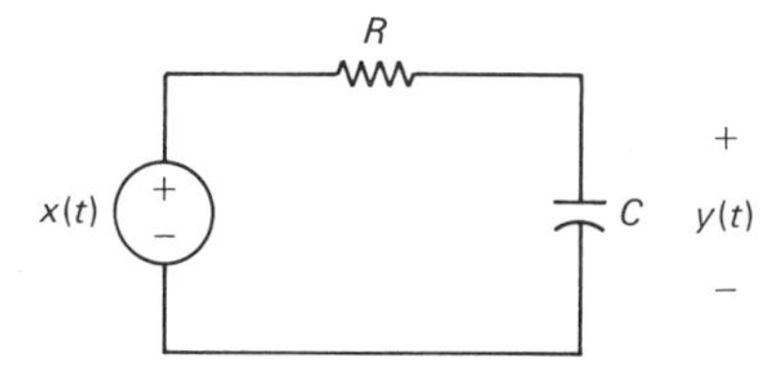

Figure 4.1 Simple *RC* network.

any phenomenon that is governed by a linear differential equation of order (p, q) as specified by

$$\frac{dy^p(t)}{dt^p} + \sum_{k=1}^{p-1} a_k \frac{dy^k(t)}{dt^k} = \sum_{k=0}^{q} b_k \frac{dx^k(t)}{dt^k} \tag{4.6}$$

is equivalently represented by convolution integral (4.5). In this differential equation, p and q are nonnegative integers specifying the operator's order, and a_k and b_k are fixed coefficients that determine the dynamical behavior of the operator. With this in mind, we are then able to apply the concepts developed below to such important areas as circuit theory, mechanical systems, economical models, and any phenomenon that is describable by a constant-coefficient linear differential equation. To illustrate this point, the following simple example is offered.

Example 4.3

In network analysis, we may directly apply the concepts described above to a circuit's descriptive equations that arise from the application of Kirchhoff's laws. To give a simple illustration of this point, let us consider the elementary differentiating network shown in Fig. 4.1. Applying Kirchhoff's voltage law, the following first-order linear differential equation is found to relate the circuit's input and output voltages:

$$\frac{dy(t)}{dt} + \frac{1}{RC} y(t) = \frac{1}{RC} x(t) \tag{4.7a}$$

A procedure for finding the equivalent convolution integral representation is presented in Chapter 6 and results in

$$y(t) = \int_{-\infty}^{t} e^{-[(t-\tau)/RC]} x(\tau)\, d\tau \tag{4.7b}$$

In this case, the operator's characteristic signal is seen to be specified by $h(t) = e^{-t/RC}u(t)$. To establish the equivalencies of expressions (4.7a) and (4.7b), one may substitute integral expression (4.7b) into differential equation (4.7a) and show that this latter equation is satisfied.

4.4 LINEAR OPERATORS

The signal operators as represented by relationships (4.2) and (4.5) possess the important feature of being "linear." To establish the concept of linearity, let us take the general operator approach used previously. This will help the reader to become more familiar with general operator theory as well as to gain an appreciation for the common bonds shared by linear discrete- and linear continuous-time signal operators. We further explore the linearity properties relative to these two specific operators in the next two sections.

Central to the concept of operator linearity are the notions of operator *homogeneity* and *additivity*. A formal definition of these two important algebraic ideas is now given.

Definition 4.1. A signal operator $\boldsymbol{T}$ is *homogeneous* if for an arbitrary constant α and any input-output pair $(\boldsymbol{x}, \boldsymbol{y})$ where $\boldsymbol{y} = \boldsymbol{Tx}$, it follows that

$$\boldsymbol{T}(\alpha \boldsymbol{x}) = \alpha \boldsymbol{Tx} \tag{4.8}$$

In words, if we multiply the excitation signal $\boldsymbol{x}$ by the scalar α, the corresponding response is multiplied by the same scalar factor. A homogeneous operator, therefore, has a response behavior that is insensitive to amplitude levels in the sense of expression (4.8).

Definition 4.2. The signal operator $\boldsymbol{T}$ is said to be *additive* if, for any two input-output pairs $(\boldsymbol{x}_1, \boldsymbol{y}_1)$ and $(\boldsymbol{x}_2, \boldsymbol{y}_2)$ where $\boldsymbol{y}_k = \boldsymbol{Tx}_k$ for $k = 1, 2$, the system response $\boldsymbol{y}$ to the sum excitation $\boldsymbol{x} = \boldsymbol{x}_1 + \boldsymbol{x}_2$ is

$$\boldsymbol{T}(\boldsymbol{x}_1 + \boldsymbol{x}_2) = \boldsymbol{Tx}_1 + \boldsymbol{Tx}_2 \tag{4.9a}$$

or equivalently

$$\boldsymbol{y} = \boldsymbol{y}_1 + \boldsymbol{y}_2 \tag{4.9b}$$

Thus, an additive operator is one for which the operation on the sum of two (or more) signals produces a response that is equal to the sum of the corresponding individual responses.

We are now in a position to prove the most fundamental property associated with linear operators. Undoubtedly, the reader has been previously exposed to this property under the name *superposition principle* as widely used in network theory.

Theorem 4.1. Let $\boldsymbol{L}$ be a signal operator that is both homogeneous and additive. The operator $\boldsymbol{L}$ is then said to be *linear* since it satisfies the *superposition property*

$$\boldsymbol{L}(\alpha_1 \boldsymbol{x}_1 + \alpha_2 \boldsymbol{x}_2) = \alpha_1 \boldsymbol{Lx}_1 + \alpha_2 \boldsymbol{Lx}_2 \tag{4.10}$$

for all possible choices of the scalars α_1 and α_2 and excitation signals $\boldsymbol{x}_1$ and $\boldsymbol{x}_2$.

To prove this linear relationship, it is noted that since $\boldsymbol{L}$ is additive, it immediately follows that $\boldsymbol{L}(\alpha_1 \boldsymbol{x}_1 + \alpha_2 \boldsymbol{x}_2) = \boldsymbol{L}(\alpha_1 \boldsymbol{x}_1) + \boldsymbol{L}(\alpha_1 \boldsymbol{x}_2)$. Finally, the homogeneity property is applied to the two right-side terms to yield $\boldsymbol{L}(\alpha_1 \boldsymbol{x}_1) = \alpha_1 \boldsymbol{L}(\boldsymbol{x}_1)$ and $\boldsymbol{L}(\alpha_2 \boldsymbol{x}_2) = \alpha_2 \boldsymbol{L}(\boldsymbol{x}_2)$, which establishes relationship (4.10). A *linear operator* is therefore a well-defined rule that is said to preserve linear combinations. Namely, if the input signal is expressible as a linear combination of signals, the corresponding output signal is equal to the same linear combination of the individual response signals. It is this very property that makes the use and analysis of linear operators so rich and valuable.

4.5 LINEARITY OF THE DISCRETE- AND CONTINUOUS-TIME CONVOLUTION OPERATORS

We now put to use the ideas just developed to establish the linearity of the discrete-time and continuous-time operators considered in Sections 4.2 and 4.3. This task is indeed straightforward and involves only the simplest of manipulations. Let us begin this investigation by considering the discrete-time case.

Linear Discrete-Time Operator

To establish the linearity of the discrete-time operator as governed by the convolution summation

$$y(n) = \sum_{k=-\infty}^{\infty} h(k)x(n-k) \tag{4.11}$$

it is necessary to find this operator's response to the linear combination excitation

$$x(n) = \alpha_1 x_1(n) + \alpha_2 x_2(n) \tag{4.12}$$

In this determination, it is important that the scalars α_1 and α_2 and the signals $\{x_1(n)\}$ and $\{x_2(n)\}$ be kept arbitrary. The response arising from the given excitation is formally obtained by substituting expression (4.12) into relationship (4.11), yielding

$$\begin{aligned} y(n) &= \sum_{k=-\infty}^{\infty} h(k)[\alpha_1 x_1(n-k) + \alpha_2 x_2(n-k)] \\ &= \alpha_1 \sum_{k=-\infty}^{\infty} h(k)x_1(n-k) + \alpha_2 \sum_{k=-\infty}^{\infty} h(k)x_2(n-k) \end{aligned} \tag{4.13}$$

In this manipulation, the decomposing of the single summation in the first line of (4.13) into two summations corresponds to the additive property, whereas the factoring out of the scalars α_1 and α_2 is associated with the homogeneity property. The terms that are *multiplying scalars* α_1 and α_2 are recognized as being the separate responses of operator (4.11) to the excitations $\{x_1(n)\}$ and $\{x_2(n)\}$, respectively. Thus, operator expression (4.11) satisfies the superposition principle, thereby establishing its linearity.

Linear Continuous-Time Operator

The linearity of the continuous-time convolution integral operator as expressed by

$$y(t) = \int_{-\infty}^{\infty} h(\tau)x(t-\tau)\, d\tau \tag{4.14}$$

is established in exactly the same manner. Namely, the input signal as specified by the linear combination

$$x(t) = \alpha_1 x_1(t) + \alpha_2 x_2(t) \tag{4.15}$$

is first substituted into relationship (4.14). Upon using the additivity and homogeneity properties possessed by the integration operator, the resultant response is found to be

$$y(t) = \alpha_1 \int_{-\infty}^{\infty} h(\tau)x_1(t-\tau)\,d\tau + \alpha_2 \int_{-\infty}^{\infty} h(\tau)x_2(t-\tau)\,d\tau \tag{4.16}$$

The integral terms multiplying the scalars α_1 and α_2 are recognized as being the response of operator (4.14) to the excitation $\{x_1(t)\}$ and $\{x_2(t)\}$, respectively. Thus, the convolution integral operation (4.14) is also seen to be linear.

From the developments noted above, it is important to appreciate the commonality with which the additivity and homogeneity properties were used in establishing linearity for the specific discrete-time (4.11) and continuous-time (4.14) operators as well as the general linear operator of Theorem 4.1. As such, instead of separately developing further properties associated with specific operators, it is more expedient to develop properties corresponding to a general class of operators (for example, linear operators) and then associate these properties to specific operators that happen to belong to that general class.

4.6 UNIT-IMPULSE RESPONSE DETERMINATION

The linear convolution summation and integral expressions play fundamental roles when investigating the important class of linear signal operations. In particular, the dynamical behavior of these operations is totally determined by the fixed *characteristic signals* $\{h(n)\}$ and $\{h(t)\}$ that appear in the associated convolution equations. Unfortunately, these characteristic signals are typically not given in explicit form in many applications. Most often, these signals must be determined from either the linear difference or differential equation that governs the phenomenon under study or from experimentally obtained excitation-response pairs. As we will now show, the unit-impulse signal is a most natural excitation to be used for this purpose.

Discrete-Time Case

To ascertain the characteristic response signal $\{h(n)\}$ associated with a linear discrete-time operator, use is made of the fact that such an operator may be equivalently represented as

$$y(n) = \sum_{k=-\infty}^{\infty} h(k)x(n-k) \tag{4.17}$$

Let us now excite this signal operator with the unit–impulse (or Kronecker delta) input

$$x(n) = \delta(n) \tag{4.18a}$$

Upon making the given excitation substitution $x(n-k) = \delta(n-k)$ in expression (4.17), the corresponding unit–impulse response is given by

$$y_\delta(n) = \sum_{k=-\infty}^{\infty} h(k)\,\delta(n-k) \\ = h(n) \tag{4.18b}$$

In arriving at this result, use has been made of the fact that the summand term $\delta(n-k)$ equals zero for all summation index values of k except at $k = n$, where it equals one. Thus, we reach the remarkable result that a discrete-time linear operator's *characteristic signal* $\{h(n)\}$ is simply equal to that operator's response to the unit–impulse (or Kronecker delta) signal. We may therefore obtain this characteristic signal by either empirically exciting the operator by the unit–impulse and recording the resultant response or by using the z-transform approach to be subsequently developed. The former procedure is of particular usefulness when seeking to effect a linear model of a phenomenon using an empirical approach.

Example 4.4

Determine the convolution summation representation for the simple linear digital filter operator

$$y(n) = x(n) + \alpha y(n-1) \tag{4.19a}$$

where α is a fixed constant. Using the arguments discussed above, the system's characteristic signal is obtained by exciting the operator with the unit–impulse signal. The associated response identified by a δ subscript is

$$y_\delta(n) = \delta(n) + \alpha y_\delta(n-1)$$

Evaluating this expression at $n = 0$, we have

$$y_\delta(0) = \delta(0) + \alpha y_\delta(-1) = 1 + \alpha 0 = 1$$

where the initial condition $y_\delta(-1)$ has been taken to equal zero, since the unit-impulse excitation is identically prior to $n = 0$. Continuing on in this fashion, we have

$$y_\delta(1) = \delta(1) + \alpha y_\delta(0) = 0 + \alpha \cdot 1 = \alpha$$

$$y_\delta(2) = \delta(2) + \alpha y_\delta(1) = 0 + \alpha \cdot \alpha = \alpha^2$$

or in general

$$y_\delta(n) = \alpha^n \qquad \text{for } n \geq 0$$

Since a linear operator's characteristic signal $\{h(n)\}$ was shown to equal its unit–Kronecker delta response, we have therefore established that

$$h(n) = \alpha^n u(n)$$

Moreover, the convolution summation that equivalently represents (4.19a) is therefore

$$y(n) = \sum_{k=0}^{\infty} \alpha^k x(n-k) \tag{4.19b}$$

The two expressions (4.19a) and (4.19b) are said to be equivalent in the sense that each will produce *the same* response to *any* excitation.

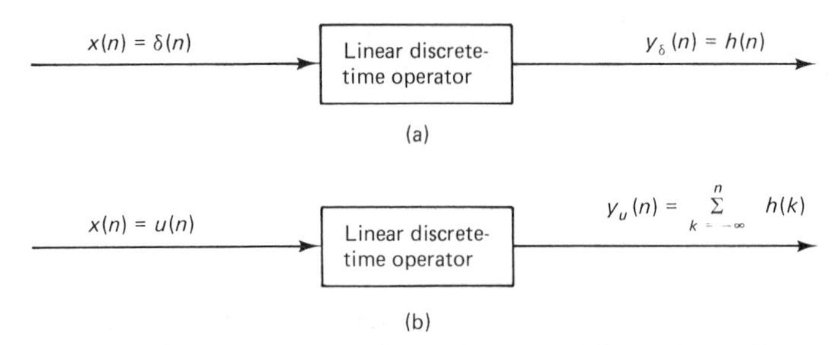

Figure 4.2 Useful excitation-response pairs for a general linear discrete-time operator: (a) unit-Kronecker excitation and (b) unit-step excitation.

It is possible to determine a linear discrete-time operator's characteristic signal by using other excitation-response pairs. To demonstrate this point, let us now evaluate a linear operator's response to the unit-step excitation

$$x(n) = u(n) \tag{4.20a}$$

Insertion of this excitation into the convolution summation representation (4.17) produces the corresponding response

$$y_u(n) = \sum_{k=-\infty}^{\infty} h(k)u(n-k) = \sum_{k=-\infty}^{n} h(k) \tag{4.20b}$$

where use of the fact that $u(n-k) = 0$ for $k > n$ has been made. Next, we form the "first difference" of this unit-step response to conclude that

$$h(n) = y_u(n) - y_u(n-1) \tag{4.21}$$

Thus, the system's characteristic signal is seen to be equal to the first difference of its unit-step response $y_u(n)$. This affords an alternative method for empirically obtaining a linear discrete-time operator's characteristic signal. This result should not be too surprising since the superposition principle could have been directly applied to the excitation $\delta(n) = u(n) - u(n-1)$ to obtain these same results (for example, see Problems 4.1 and 4.2). These two basic excitation-response pairs are depicted in Fig. 4.2.

Continuous-Time Case

We now examine a similar set of excitation-response pairs that govern a linear continuous-time operator as represented by the convolution integral

$$y(t) = \int_{-\infty}^{\infty} h(\tau)x(t-\tau)\, d\tau \tag{4.22}$$

Using the integral properties of the unit–impulse (or Dirac delta) function as described in Chapter 3, it directly follows that this linear operator's response to the unit–impulse excitation

$$u(t) = \delta(t) \tag{4.23a}$$

is given by

$$y_\delta(t) = h(t) \tag{4.23b}$$

This excitation-response pair is completely analogous to that found for the corresponding discrete-time case. To find a linear continuous-time operator's *characteristic signal*, we may then conceptually apply the unit-Dirac excitation and record the corresponding response. Unfortunately, there does not exist a physically generatable signal that possesses the properties associated with a unit-Dirac function (that is, a pulse of zero width, infinite amplitude, and unit area). In its place, however, one may use a high amplitude pulse of unit area whose width is small compared with the time constant of the linear operator. What constitutes a linear operator's time constant is discussed later.

Instead of using the physically unimplementable unit-Dirac signal to obtain the characteristic signal, let us instead use the more practical unit-step excitation. Applying the unit-step signal excitation

$$x(t) = u(t) \tag{4.24a}$$

to linear operator (4.22), the corresponding unit-step response is given by

$$y_u(t) = \int_{-\infty}^{t} h(\tau)\, d\tau \tag{4.24b}$$

If we differentiate this response with respect to time using the Leibnitz differentiating rule, we obtain[1]

$$h(t) = \frac{dy_u(t)}{dt} \tag{4.25}$$

Thus, a linear continuous-time operator's characteristic signal $\{h(t)\}$ is equal to the derivative of that operator's unit-step response. We illustrate these important excitation-response properties in Fig. 4.3. The analogy with the discrete-time case shown in Fig. 4.2 is striking.

Example 4.5

Let us determine the unit-impulse and unit-step responses for the simple RC network shown in Fig. 4.1. In Example 4.3, it was shown that the first-order differential equation governing the excitation-response voltages was

$$\frac{dy(t)}{dt} + \frac{1}{RC}\, y(t) = \frac{1}{RC}\, x(t)$$

Upon setting the excitation signal $x(t) = u(t)$, the corresponding response signal is required to satisfy

$$\frac{dy_u(t)}{dt} + \frac{1}{RC}\, y_u(t) = \frac{1}{RC}\, u(t) \tag{4.27}$$

[1]Leibnitz differentiating rule: Let $f(x, t)$ be a continuous function and have a continuous derivative $\partial f(x, t)/\partial t$ in a domain of the (x, t) plane, which includes the rectangle $t_1 \le t \le t_2$ and $a(t) \le x \le b(t)$ where $a(t)$ and $b(t)$ are defined and have continuous derivatives for $t_1 < t < t_2$. Then for $t_1 < t < t_2$,

$$\frac{d}{dt}\int_{a(t)}^{b(t)} f(x, t)\, dx = f(b(t), t)\,\frac{db(t)}{dt} - f(a(t), t)\,\frac{da(t)}{dt} + \int_{a(t)}^{b(t)} \frac{\partial f(x, t)}{\partial t}\, dx \tag{4.26}$$

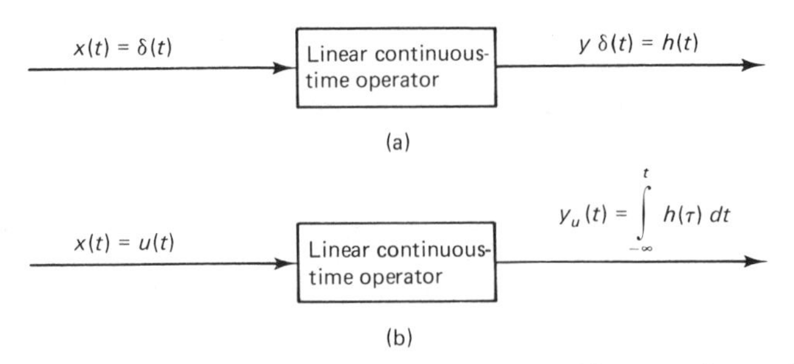

Figure 4.3 Useful excitation-response pairs for a general linear continuous-time operator: (a) unit-Dirac excitation and (b) unit-step excitation.

Using ordinary differential equation theory or the Laplace transform approach to be developed in the next chapter, the desired solution is found to be

$$y_u(t) = (1 - e^{-t/RC})u(t)$$

in which the initial condition $y_u(0^-)$ is taken to be zero. The validity of this solution is readily shown by demonstrating that it satisfies equation (4.27). In accordance with relationship (4.25), the associated unit-impulse response is given by

$$\begin{aligned} h(t) &= \frac{d}{dt}(1 - e^{-t/RC})u(t) \\ &= (1 - e^{-t/RC})\frac{du\,(t)}{dt} + \frac{1}{RC}e^{-t/RC}u(t) \\ &= (1 - e^{-t/RC})\delta(t) + \frac{1}{RC}e^{-t/RC}u(t) \\ &= \frac{1}{RC}e^{-(t/RC)}u(t) \end{aligned}$$

4.7 CAUSAL LINEAR OPERATORS

A linear operator is said to be *causal* if its response at any time is dependent only on the excitation behavior up to that time instant and not on its future behavior. All real-time signal operations must be causal in nature, and the study of such operators is therefore of obvious relevancy. In accordance with this concept, we therefore conclude that the most general causal discrete-time linear operator must be of the form

$$\begin{aligned} y(n) &= \sum_{k=0}^{\infty} h(k)x(n-k) \\ &= \sum_{k=-\infty}^{n} h(n-k)x(k) \end{aligned} \tag{4.28}$$

In each of these summations, it is seen that $y(n)$ is dependent only on the present, $x(n)$, and past, $x(n-1)$, $x(n-2)$, $\ldots$) excitation elements. The characteristic signal $\{h(n)\}$ identifying this causal operator is seen to be identically zero for negative arguments, that is,

$$h(n) = 0 \qquad \text{for all } n < 0 \tag{4.29}$$

In accordance with the developments made in the last section, we see that a causal linear operator's response to the unit-impulse excitation, i.e., the characteristic signal $\{h(n)\}$, must of necessity satisfy (4.29), since the excitation is identically zero for negative time.

In an entirely similar fashion, it is clear that a causal continuous-time linear operator must be governed by a convolution integral of form

$$\begin{aligned} y(t) &= \int_0^{\infty} h(\tau)x(t-\tau)\,d\tau \\ &= \int_{-\infty}^{t} h(t-\tau)x(\tau)\,d\tau \end{aligned} \tag{4.30}$$

From these integral expressions, it is clear that the characteristic signal $\{h(t)\}$ associated with any linear operator is identically zero for negative time arguments, that is,

$$h(t) = 0 \qquad \text{for all } t < 0 \tag{4.31}$$

Since a system's characteristic signal is the response of a linear operator to the unit-impulse excitation, this is not an unexpected result.

4.8 TIME INVARIANCE

One of the more important properties possessed by the linear discrete- and continuous-time operators considered in Sections 4.2 and 4.3 is that of time invariance. A signal operation is said to be *time-invariant* if its input-output relationship does not change as time evolves. For example, operators governed by linear difference or differential equations with constant coefficients are so characterized. Thus, an electronic circuit that is composed of a configuration of resistors, capacitors, and inductors whose values are *fixed* constitutes a time-invariant signal operation.

Discrete-Time Invariance

Mathematically, the discrete-time operator $\boldsymbol{T}$ is said to be time-invariant if it satisfies the relationship

$$\boldsymbol{T}\{x(n-n_0)\} = \{y(n-n_0)\} \tag{4.32}$$

for all excitations $\{x(n)\}$ and time shifts n_0 where $\{y(n)\} = \boldsymbol{T}\{x(n)\}$. Namely, if $\{y(n)\}$

denotes the operator's response to the excitation $\{x(n)\}$, its response to the time-shifted excitation $\{x(n - n_0)\}$ is equal to the time-shifted response $\{y(n - n_0)\}$. It is a simple matter to show that the linear discrete-time operator as governed by

$$y(n) = \sum_{k=-\infty}^{\infty} h(k)x(n - k) \tag{4.33}$$

is time-invariant. The basis for this time invariance arises from the fact that the unit-impulse response of this operator remains the same except for an appropriate time shift independent of when the unit-impulse excitation is applied (that is, $x(n) = \delta(n - n_0)$ implies $y_\delta(n - n_0) = h(n - n_0)$).

Continuous-Time Invariance

The extension of the discrete-time invariance concept to continuous-time invariance is straightforward. In particular, the continuous-time operator $\boldsymbol{T}$ is said to be time-invariant if

$$\boldsymbol{T}\{x(t - t_0)\} = \{y(t - t_0)\} \tag{4.34}$$

holds for all excitations $\{x(t)\}$ and time shifts t_0. In this expression, $\{y(t)\}$ denotes the response of operator $\boldsymbol{T}$ to the excitation $\{x(t)\}$. Using elementary manipulations, one may readily show that the linear continuous-time operator

$$y(t) = \int_{-\infty}^{\infty} h(\tau)x(t - \tau)\, d\tau \tag{4.35}$$

is time-invariant.

4.9 OPERATOR STABILITY

The notion of operator stability is central to any practical application. Implicit in the word *stability* is the thought that a signal processing operation is well-behaved as long as the signals generated in its operation are themselves well-behaved. There are various measures for specifying if a signal is well-behaved. For the purposes of this section, the discrete-time signal $\{x(n)\}$ is said to be well-behaved if its individual elements are bounded in amplitude, that is, if its ℓ_∞ norm (see Section 2.7) as defined by

$$\| \{x(n)\} \|_\infty = \max |x(n)| \tag{4.36}$$

is finite.[2] A discrete-time signal is therefore said to be bounded if each of its elements has a magnitude that does not exceed some finite level. In a similar fashion, the continuous-time signal $\{x(t)\}$ is bounded if its associated $\mathscr{L}_\infty$ norm as measured by

$$\| \{x(t)\} \|_\infty = \max |x(t)| \tag{4.37}$$

[2]From a strictly mathematical point of view, the maximum operation should be replaced by the supremum operation (or the *least upper bound*).

is finite. Thus, if a continuous-time signal has an amplitude whose magnitude never exceeds a finite level, the signal is said to be *bounded*. With these concepts of boundedness in mind, it is possible to introduce the notion of *signal operator stability*.

Definition 4.3. Let T correspond to a discrete- or continuous-time signal operator. This operator is said to be *stable* if its response to any bounded excitation is itself bounded, that is,

$$\| Tx \|_\infty < +\infty \quad \text{if} \quad \| x \|_\infty < +\infty \tag{4.38}$$

The measure $\| \cdot \|_\infty$ is taken to be (4.36) in the discrete-time case or (4.37) in the continuous-time case.

Clearly, the desire to use stable signal operations arises from the natural inclination to work with signals that are reasonably sized. If a signal should become too large due to an unstable signal operation, then either the danger of equipment destruction or data overflow is present. We now show that there exists a straightforward method for determining the stability characterization of the linear convolution operators considered in Sections 4.2 and 4.3.

Discrete-Time Case

Let us now examine the condition under which the linear discrete-time signal operator as governed by

$$y(n) = \sum_{k=-\infty}^{\infty} h(k)x(n-k) \tag{4.39}$$

is stable. Using standard rules obeyed by the magnitude operator, it directly follows that the response element $y(n)$ is bounded above by

$$\begin{aligned} |y(n)| &= \left| \sum_{k=-\infty}^{\infty} h(k)x(n-k) \right| \\ &\le \sum_{k=-\infty}^{\infty} |h(k)| \cdot |x(n-k)| \end{aligned}$$

This inequality becomes an equality if and only if the sign of each of the summand terms $h(k)x(n-k)$ are all positive or all negative for $-\infty < k < +\infty$. Since the excitation is required to be bounded, each element $|x(n-k)|$ must be less than or equal to the excitation signal's norm $\| \{x(n)\} \|_\infty$. We have therefore established the bounding

$$|y(n)| \le \| \{x(n)\} \|_\infty \cdot \sum_{k=-\infty}^{\infty} |h(k)|$$

which holds for all time n. For system stability, it is necessary that $|y(n)|$ be finite for all n as long as $\| \{x(n)\} \|_\infty$ is itself finite. From these arguments, it follows that a necessary and sufficient condition for signal operator (4.39) to be stable is that

$$\| \boldsymbol{h} \|_1 = \sum_{k=-\infty}^{\infty} |h(k)| < +\infty \tag{4.40}$$

Thus, if the operator's characteristic signal has a finite ℓ_1 norm, the stability of the operator is assured.

Continuous-Time Case

Using analogous reasoning, it may be shown that the linear continuous-time operator

$$y(t) = \int_{-\infty}^{\infty} h(\tau)x(t-\tau)\, d\tau \tag{4.41}$$

is stable in the sense of Definition 4.3 if and only if

$$\| \boldsymbol{h} \|_\infty = \int_{-\infty}^{\infty} |h(\tau)|\, d\tau < +\infty \tag{4.42}$$

We are again led to the conclusion that operator stability requires that the operator's characteristic signal have a finite $\mathscr{L}_1$ norm.

Example 4.6

The signal operators considered in Examples 4.1 and 4.3 are each stable since

(a) for Example 4.1, we have

$$\sum_{k=-\infty}^{\infty} |h(k)| = \sum_{k=0}^{N-1} \frac{1}{N} = 1$$

(b) in Example 4.3, it follows that

$$\int_{-\infty}^{\infty} |h(\tau)|\, d\tau = \int_{0}^{\infty} e^{-t/RC}\, dt = RC$$

4.10 RESPONSE TO EXPONENTIAL EXCITATIONS

When examining the exponential response characteristics of a linear operator, an invaluable analysis tool is revealed, namely, the concept of studying the constituent time-domain signals (e.g., the excitation, response, and unit-impulse response) in an alternative transform domain is suggested. To illustrate why this is so, we now give a preliminary introduction to the z-transform and the Laplace transform. A more thorough study is made in the two chapters to follow.

Discrete-Time Case

Let the linear discrete-time operator as specified by

$$y(n) = \sum_{k=-\infty}^{\infty} h(k)x(n-k) \tag{4.43}$$

be excited by the two-sided exponential

$$x(n) = z^n \tag{4.44}$$

where z is a complex-valued scalar. Upon inserting this excitation into operator relationship (4.43), we have

$$y(n) = \sum_{k=-\infty}^{\infty} h(k)z^{n-k} = z^n \sum_{k=-\infty}^{\infty} h(k)z^{-k}$$

where the exponential term z^{n-k} has been decomposed into the equivalent factored form $z^n z^{-k}$ and the common product term z^n is then taken outside the summation on index k. Since the resultant summation is a function of the characteristic signal $\{h(n)\}$ and the complex-valued scalar z, it can be expressed as the function

$$H(z) = \sum_{n=-\infty}^{\infty} h(n)z^{-n} \tag{4.45}$$

In summary, when the linear discrete-time operator (4.43) is excited by the exponential signal (4.44), the corresponding response is equal to the product of the excitation and this function $H(z)$, that is,

$$y(n) = H(z)z^n \tag{4.46}$$

The entity $H(z)$ appearing here is referred to as the *z-transform* of the signal $\{h(n)\}$. As is seen in later chapters, the *z-transform* plays a most vital role in the analysis of discrete-time signals and linear operations on such signals. It is to be noted that the result as shown in expression (4.46) arises due to two factors: (1) the specific linear form of the convolution summation (4.43), and (2) the ability to decompose the exponential excitation signal so that $x(n-k)$ can be represented as $x(n)x(-k)$.

Continuous-Time Case

Let the linear continuous-time operator as governed by the convolution integral

$$y(t) = \int_{-\infty}^{\infty} h(\tau)x(t-\tau)\, d\tau \tag{4.47}$$

be excited by the two-sided exponential

$$x(t) = e^{st} \tag{4.48}$$

The corresponding response is then given by

$$y(t) = \int_{-\infty}^{\infty} h(\tau)e^{s(t-\tau)}\, d\tau = e^{st} \int_{-\infty}^{\infty} h(\tau)e^{-s\tau}\, d\tau$$

where the scalar s is taken to be complex-valued. As in the discrete-time case above, this response is a product of the excitation signal and the function

$$H(s) = \int_{-\infty}^{\infty} h(t)e^{-st}\, dt \tag{4.49}$$

This function of the complex variable s is commonly referred to as the *Laplace transform* of the characteristic signal $\{h(t)\}$. It too plays a most important role in studying continuous-time signals and linear operations on such signals.

Upon examination of the z-transform (4.45) and Laplace transform (4.49) expressions, it is seen that they are each procedures for converting (or transforming) a time-domain signal into a function of a complex variable. As is seen in the next two chapters, these two transforms share many common properties. This is not surprising, given the exponential response characterization taken in this section.

4.11 EVALUATION OF THE CONVOLUTION OPERATION

In this chapter, we have been primarily concerned with convolution-type signal operations. If such operations are to be used in generating useful excitation-response results, it is essential to have a well-based understanding of their evaluation. For both discrete- and continuous-time linear convolution operations, it is possible to provide a common graphical interpretation that yields useful insights. As we will show, the convolution process entails the separate operations of time transposition, time shifting, and summation (or integration).

Discrete-Time Convolution

The operation of discrete-time convolution as governed by the expression

$$\begin{aligned} y(n) &= \sum_{k=-\infty}^{\infty} h(k)x(n-k) \\ &= \cdots + h(-1)x(n+1) + h(0)x(n) + h(1)x(n-1) + \cdots \end{aligned} \tag{4.50}$$

is directly obtained upon making a simple interpretation. Namely, upon examination of the second line of relationship (4.50) it is seen that the elements $h(k)$ in the products $h(k)x(n-k)$ have increasing arguments as one goes from left to right. On the other hand, the elements $x(n-k)$ in the products $h(k)x(n-k)$ have a decreasing argument as one proceeds from left to right. With this thought in mind, let us envision two strips of paper on which the signals $h(k)$ versus k and the time transpose $x(-k)$ versus k are plotted as shown in Fig. 4.4. We note that the signal elements $\{h(k)\}$ and $\{x(-k)\}$ have been entered in the left-to-right fashion as just

$h(k)$ versus k:	$\ldots \; h(-2) \; h(-1) \; h(0) \; h(1) \; h(2) \; \ldots$
$x(-k)$ versus k:	$\ldots \; x(2) \; x(1) \; x(0) \; x(-1) \; x(-2) \ldots$

Figure 4.4 Two paper strips upon which the element values of $h(k)$ versus k and the time transpose $x(-k)$ versus k are entered.

described. To compute the response element $y(n)$, we follow the three-step procedure as described below.

1. Shift the time transpose plot $x(-k)$ versus k by the amount of n time units to the right until the value $x(n)$ is vertically aligned with $h(0)$ at $k = 0$.
2. Multiply each of the signal elements $h(k)$ and $x(n - k)$ which appear one above the other to generate the product signal $h(k)x(n - k)$.
3. Sum the elements of the product signal to form $y(n)$.

With the two strips as positioned in Fig. 4.4, this procedure produces the response element $y(0)$ as is evident from expression (4.50) with n set to zero. To compute $y(1)$, we simply right shift the $\{x(n)\}$ paper strip by one time unit to the right and carry out steps (2) and (3). If we proceed in this orderly fashion for each value of n, the desired response $\{y(n)\}$ would be generated.

Example 4.7

Using the approach above, convolve the two signals

$$h(n) = x(n) = \begin{cases} 1 & 0 \le n \le 2 \\ 0 & \text{otherwise} \end{cases}$$

The paper strips would appear as shown below

$h(k)$ versus k:	$\ldots$ 0 0 0 0 1 1 1 0 0 0 $\ldots$
	↑ (under the first 1)
$x(-k)$ versus k:	$\ldots$ 0 0 1 1 1 0 0 0 0 0 $\ldots$
	↑ (under the third 1)

in which the underarrow indicates the time origin. Using the approach listed above, it is found that

$$y(n) = \delta(n) + 2\delta(n-1) + 3\delta(n-2) + 2\delta(n-3) + \delta(n-4)$$

Continuous-Time Convolution

By this time, it should not be surprising that the ideas central to the discrete-time convolution interpretation can be readily extended to the continuous-time convolution operation as specified by

$$y(t) = \int_{-\infty}^{\infty} h(\tau)x(t - \tau)\, d\tau$$

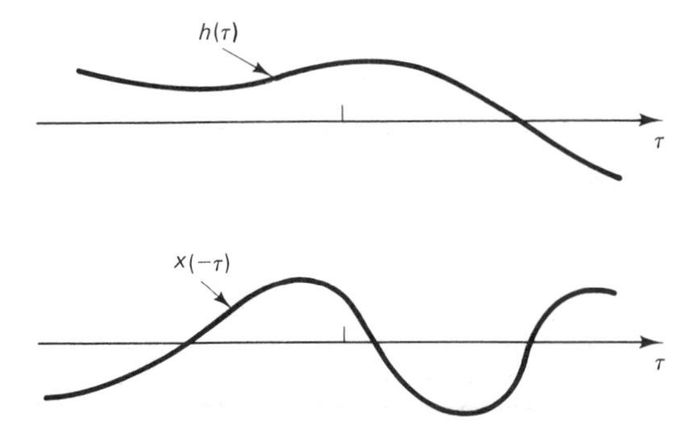

Figure 4.5 Plots of $h(\tau)$ versus τ and the time transpose $h(-\tau)$ versus τ.

We first prepare two paper strips upon which the signals $h(\tau)$ and $x(-\tau)$ are plotted as shown in Fig. 4.5. To compute the response value $y(t)$, we next carry out the three-step procedure outlined below.

1. Shift the time transpose plot $x(-\tau)$ versus τ by the amount of t seconds to the right until the value $x(t)$ is vertically aligned with $h(0)$ at $\tau = 0$.
2. Multiply the two signals $h(\tau)$ and the right-shifted signal $x(t - \tau)$.
3. Integrate the product signal $h(\tau)x(t - \tau)$ to produce the response value $y(t)$.

When evaluating the response of a linear operator to a pulse-like excitation, the convolution interpretation given above is helpful in ascertaining time intervals over which the integrand product $h(\tau)x(t - \tau)$ takes on a different description. The following example illustrates this point.

Example 4.8

Determine the response of the RC circuit considered in Example 4.3 to the pulse excitation

$$x(t) = \begin{cases} 1 & \text{for } 0 \le t \le T \\ 0 & \text{otherwise} \end{cases}$$

The unit-impulse response associated with the RC network was previously shown to be $h(t) = e^{-t/RC}u(t)$. Plots of $h(\tau)$ and $x(-\tau)$ are given in Fig. 4.6. Using the three-step procedure given above, the required response is found to be

$$y(t) = \begin{cases} 0 & \text{for } t < 0 \\ RC(1 - e^{-t/RC}) & \text{for } 0 \le t \le T \\ RCe^{-t/RC}(e^{T/RC} - 1) & \text{for } T < t \end{cases}$$

It is to be noted that a different response formula applies for the time intervals $0 \le t \le T$ and $t > T$. This is readily appreciated since the transposed-shifted pulse, i.e., $x(t - \tau)$, has a different overlap characterization as is appreciated upon examination of Fig. 4.6.

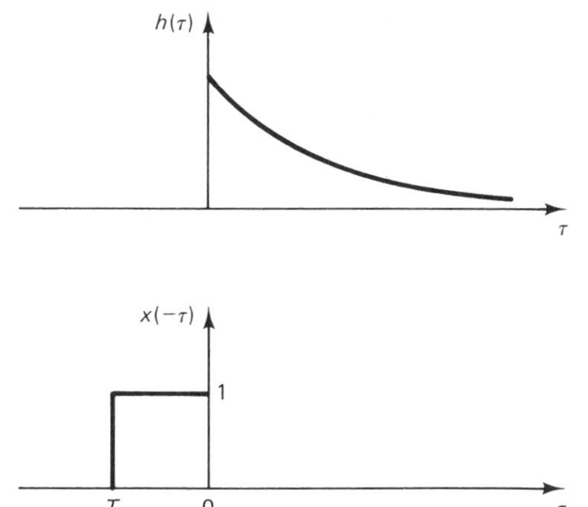

Figure 4.6 Sketch of signals appearing in Example 4.8.

4.12 LINEAR TIME-VARYING SIGNAL OPERATORS

There exists a more general class of linear signal operators than those considered up to this point, namely, operators whose excitation-response dynamics change with time but yet still satisfy the superposition principle. Any signal operation that is characterized by a linear difference or differential equation whose coefficients vary with time would fall into this class of time-varying linear operators. As in the time-invariant case, these operators may also be written in the form of a convolution operation. The most general linear discrete-time signal operation can always be representable in the convolution summation format

$$y(n) = \sum_{k=-\infty}^{\infty} h(n, k)x(k) \tag{4.51}$$

The unit-impulse response $\{h(n, k)\}$ that characterizes this *time-varying* operation is seen to be a function of two variables instead of one, as was the case in the time-invariant operation (4.2). We may interpret the entity $h(k, n)$ as being the response's value at time n caused by a unit–Kronecker delta excitation applied at time k. In a similar manner, the most general linear continuous-time signal operation is always expressible in the convolution integral format

$$y(t) = \int_{-\infty}^{\infty} h(t, \tau)x(\tau)\, d\tau \tag{4.52}$$

The entity $h(t, \tau)$ is interpretable as being the response of the linear operator at time t to the unit-Dirac excitation applied at time τ.

Although time-varying linear operators are important in describing various commonly encountered phenomena, we shall hereafter restrict our attention to their more specialized time-invariant counterparts (for which $h(n, k) = h(n - k)$ or $h(t, \tau) = h(t - \tau)$). This is done primarily for two reasons. First, linear time invariants are useful in themselves for describing many relevant dynamic situations. Second, it is possible to provide a more thorough analysis of time-invariant operators. Unfortunately, one must often use rather restrictive approaches when studying time-variant operators.

4.13 PROBLEMS

4.1. Determine the response of the system governed by the linear operator rule

$$y(n) = 2x(n) - 5x(n-1) + 4x(n-3)$$

to the (a) unit-impulse signal and (b) unit-step signal. What interesting interpretation can be given to the unit-impulse response? If $\{y_\delta(n)\}$ and $\{y_u(n)\}$ denote the unit-impulse and unit-step responses, respectively, show that $y_\delta(n) = y_u(n) - y_u(n-1)$.

4.2. Repeat Problem 4.1 for the system

$$y(n) = b_0 x(n) + b_1 x(n-1) + \cdots + b_q x(n-q)$$

in which the b_k are fixed scalars and q is a nonnegative integer specifying the system order.

4.3. Show that the response of the two systems

(a) $y_1(n) = \sum_{k=0}^{\infty} \alpha^k x(n-k)$

(b) $y_2(n) = x(n) + \alpha y_2(n-1)$

are identical when $x(n) = \delta(n)$ and $x(n) = u(n)$ where α is a fixed scalar. It will be subsequently shown that these two systems are equivalent. In determining the response of system (b), take the initial condition $y(-1)$ to be zero.

4.4. In expression (1.11), a numerical integration algorithm was described by

$$y_a(n) = Tx(n) + y_a(n-1)$$

Determine the unit-impulse and unit-step response of this discrete-time signal operator in which the initial condition $y(-1) = 0$.

4.5. In Section 2.8, a model for the amortization process was described by the difference equation

$$y(n) = (1 + r)y(n-1) - x(n)$$

Determine this system's unit-impulse and unit-step responses.

4.6. Show that the convolution integral (4.7b) satisfies differential equation (4.7a).

4.7. Using ordinary differential equation theory, determine the unit-step response of the RC network considered in Example 4.3. Show that the same result will arise when using the equivalent convolution integral (4.7b). If $\{y_u(t)\}$ denotes the unit-step response, show that $dy_u(t)/dt = h(t) = e^{-t/RC}u(t)$.

4.8. Determine if the following continuous-time operations on $\boldsymbol{x}$ are linear:

(a) $y_a(t) = a + x(t)$

(b) $y_b(t) = bx(t)$

(c) $y_c(t) = cx^2(t)$

(d) $y_d(t) = \sin[x(t)]$

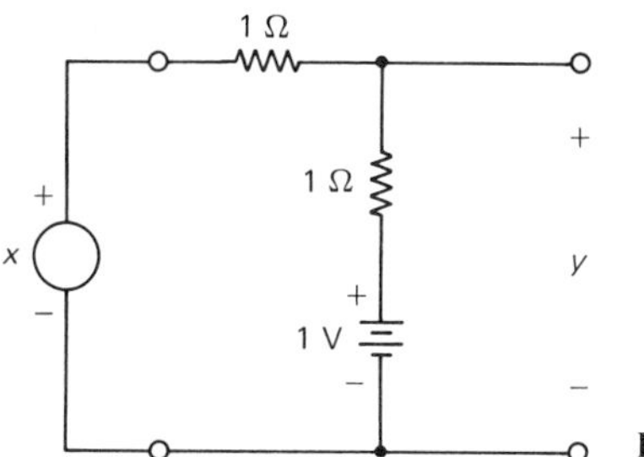

Figure P4.9

4.9. Determine whether the network in Fig. P4.9 viewed as an operation, $\boldsymbol{y} = \boldsymbol{T}[\boldsymbol{x}]$, is both additive and homogeneous.

4.10. Demonstrate that the continuous-time operator $\boldsymbol{T} = \boldsymbol{D} + \boldsymbol{I}$, where $\boldsymbol{D}$ is the deriv' operation and $\boldsymbol{I}$ is the identity operator (i.e., $Tx = x$), is linear.

4.11. **(a)** For the discrete-time operator $\boldsymbol{T} = \boldsymbol{S}^2 + 2\boldsymbol{S} + 3\boldsymbol{I}$, show that $\boldsymbol{T}$ is linear where $\boldsymbol{S}$ denotes the unit-shift operator.

(b) Assuming that T used on continuous-time signals is interpreted as if $\boldsymbol{S}$ is an ideal unit-time delay, show that $\boldsymbol{T}$ is linear.

4.12. Determine the response to a unit-pulse for each of the following discrete-time operators:

(a) $\boldsymbol{T}_a = \boldsymbol{S} + \boldsymbol{S}^{-1}$

(b) $\boldsymbol{T}_b = \boldsymbol{S}^2 + 5\boldsymbol{S} + 6\boldsymbol{I}$

(c) The operator $\boldsymbol{T}_c$ described in Fig. P2.10 where $\boldsymbol{y} = \boldsymbol{T}_c[\boldsymbol{x}]$.

4.13. Find the responses of the operators $\boldsymbol{T}_a$, $\boldsymbol{T}_b$, and $\boldsymbol{T}_c$ described in Problem 4.12 to the excitation $x = \{\ldots 0, 0, \underset{\uparrow}{1}, 2, 3, 0, 0, \ldots\}$.

4.14. Determine the unit-impulse and unit-step responses of the discrete-time operators

(a) $y_1(n) = 3x(n + 1) + 7x(n - 1) + 4x(n - 2)$

(b) $y_2(n) = 4x(n - 1) - 0.5y_2(n - 1)$

(c) $y_3(n) = \sum_{k=0}^{N-1} (-1)^k x(n - k)$

If $\{y_u(n)\}$ and $\{y_\delta(n)\}$ denote the associated unit-step and unit-impulse responses, respectively, show that $y_\delta(n) = y_u(n) - y_u(n - 1)$.

4.15. Given that a system's unit-impulse response function is $h(t) = \exp\{-t\}\, u(t)$, determine and sketch the system response $y(t)$ when the input is

(a) $x_a(t) = u(t) - u(t - 10)$

(b) $x_b(t) = u(t) - u(t - 1)$

(c) $x_c(t) = u(t) - u(t - 0.1)$

(d) Compare $x_c(t)$ with $h(t)$.

4.16. Determine whether the signal operators described as follows are (a) causal, (b) linear, (c) time-invariant:

(a) $\boldsymbol{T}\{x(n)\} = \{h(n)x(n)\}$ and $\boldsymbol{T}\{x(t)\} = \{h(t)x(t)\}$

(b) $\boldsymbol{T}\{x(n)\} = \{3x(n - 2)\}$ and $\boldsymbol{T}\{x(t)\} = \{3x(t - 2)\}$

(c) $\boldsymbol{T}\{x(n)\} = \{6x(n - 1) + 7\}$ and $\boldsymbol{T}\{x(t)\} = \{6x(t - 1) + 7\}$

(d) $\boldsymbol{T}\{x(n)\} = \{x^2(n)\}$ and $\boldsymbol{T}\{x(t)\} = \{x^2(t)\}$

(e) $\boldsymbol{T}\{x(n)\} = \{nx(n)\}$ and $\boldsymbol{T}\{x(t)\} = \{tx(t)\}$

4.17. A discrete-time system or operator is said to be causal if its characteristic unit-impulse response $h(n)$ is zero for $n < 0$. Determine if the operators $\boldsymbol{T}_a$, $\boldsymbol{T}_b$, and $\boldsymbol{T}_c$ of Problem 4.12 are causal.

4.18. Determine if the following discrete-time systems characterized by their unit-impulse sequences are stable:

(a) $h_a(n) = u(n)$

(b) $h_b(n) = \begin{cases} 2^{-n} & \text{for } n \geq 0 \\ 0 & \text{otherwise} \end{cases}$

(c) If your calculations indicate instability, find a bounded input sequence for which the output is unbounded.

4.19. Find the response of the discrete-time operators

(a) $y(n) = 2x(n-1) - 5x(n-3)$

(b) $y(n) = x(n) + x(n-1) + \cdots + x(n-N)$

(c) $y(n) = x(n) + \dfrac{1}{2}\, y(n-1)$

to the exponential excitation $x(n) = z^n$. (Hint: Let $y(n) = H(z)z^n$.) Express your answer as $y(n) = H(z)z^n$.

4.20. For the operation described by the network of Fig. P4.20

(a) Describe the operator $\boldsymbol{y} = \boldsymbol{Tx}$ in terms of the derivative $\boldsymbol{D}$, or integral $\boldsymbol{D}^{-1}$, operator.

(b) Calculate the unit-impulse response $h(t)$ of operator $\boldsymbol{T}$.

(c) Determine $\boldsymbol{y}$ by convolution if $\boldsymbol{x}$ is given by $x(t) = u(t) - u(t-1)$.

(d) Using the ideas of Chapter 3 concerning complex exponential signals, calculate $y(t)$ if $x(t) = \cos t$.

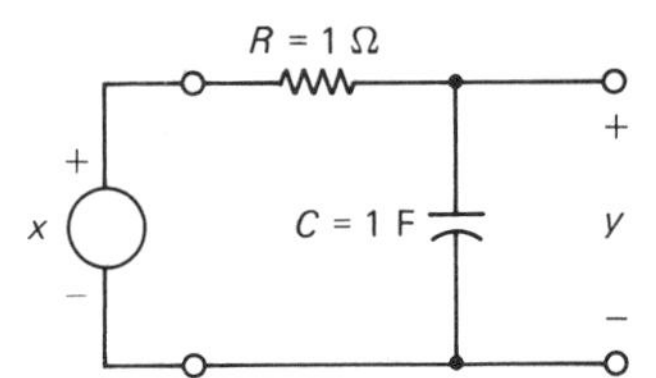

Figure P4.20

4.21. Given the system of Problem 4.20, find the system's exponential transfer function $H(s)$ and determine the output $\boldsymbol{y}$ for the following inputs:

(a) $x_a(t) = e^{-2t}$

(b) $x_b(t) = e^{jt}$

(c) $x_c(t) = 2\cos t$

4.22. Determine if the following continuous-time systems described by their unit-impulse response functions are stable:

(a) $h_a(t) = u(t)$

(b) $h_b(t) = e^{-t}u(t)$

If your calculations indicate instability, find a bounded input signal for which the output is unbounded.

4.23. Convolve the following sequences:

(a) $\boldsymbol{h}_a = \{\ldots, 0, 0, \underset{\uparrow}{1}, 1, \ldots, 1, \ldots\}$

(b) $h_b = \{\ldots, 0, 0, \underset{\uparrow}{1}, 2, 3, 2, 1, 0, 0, \ldots\}$

with $x = \{\ldots, 0, 0, \underset{\uparrow}{1}, 1, 1, -1, -1, 1, -1, 0, 0, \ldots\}$

4.24. Calculate the convolution of the signal x with itself where x is the sequence given in Problem 4.23. What property of this sequence might be useful?

4.25. Two sequences x and y are said to be inverse sequences if $x * y = \delta$ where $*$ denotes the convolution summation operator. Show that

$$x = \{\ldots, 0, 0, \underset{\uparrow}{1}, 1, 1, 0, 0, \ldots\}$$

and

$$y = \{\ldots, 0, 0, \underset{\uparrow}{1}, -1, 0, 1, -1, 0, 0, \ldots\}$$

are inverse sequences.

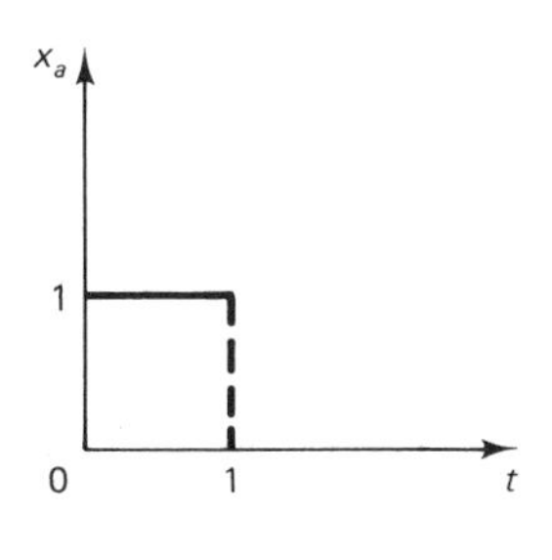

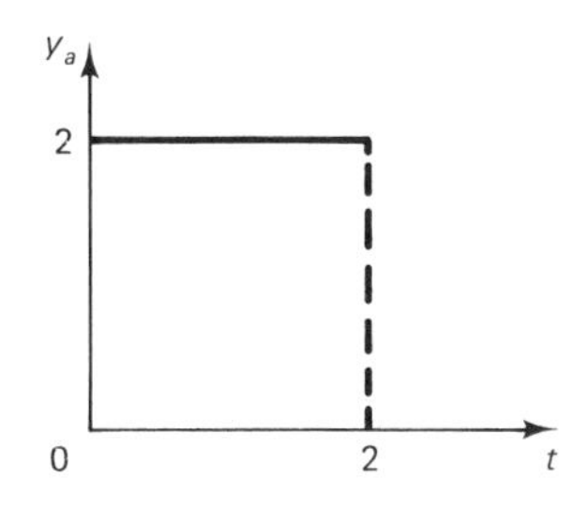

(a)

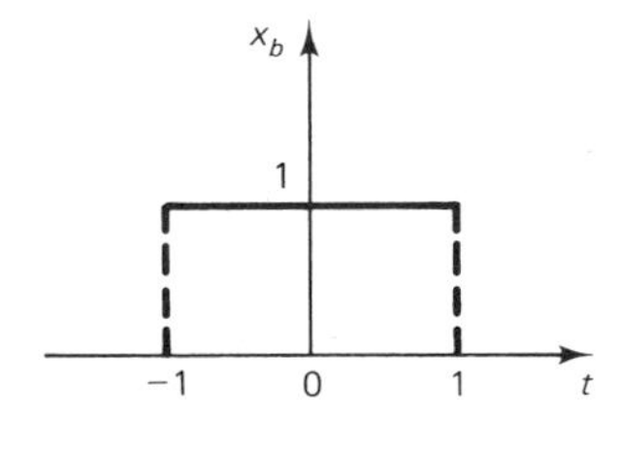

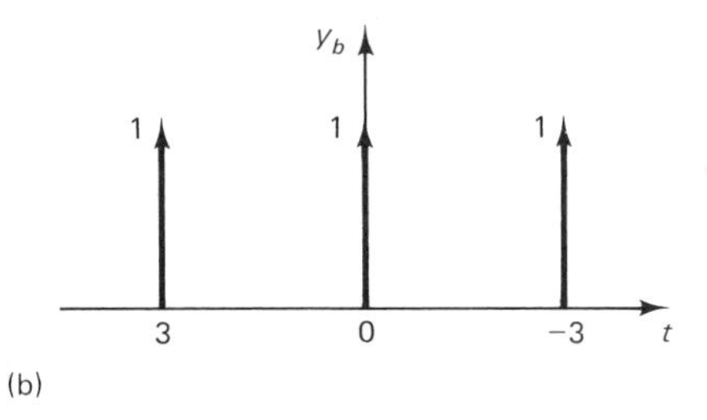

(b)

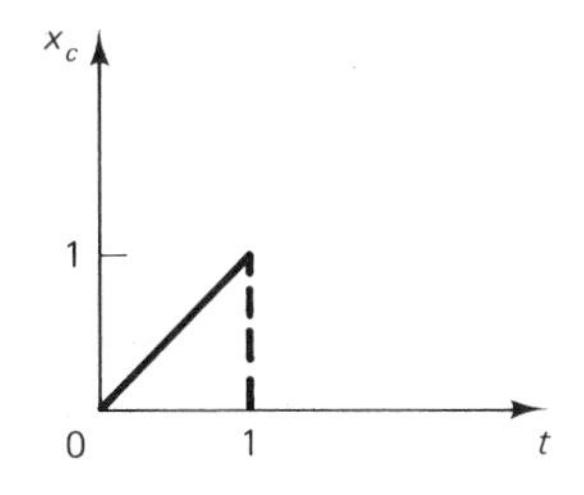

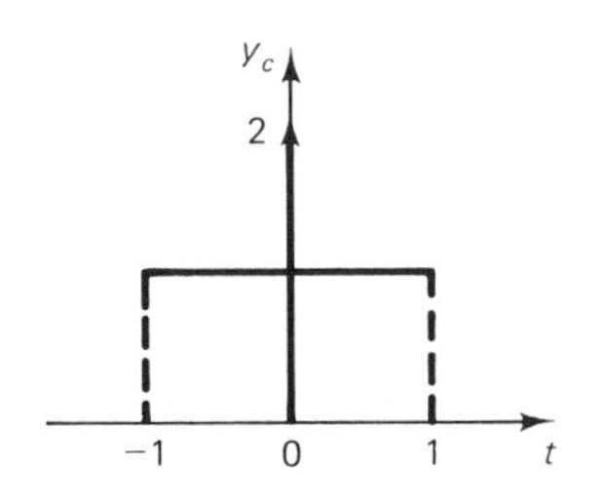

(c)

Figure P4.27

4.26. To eliminate random measurement errors, a measured sequence is sometimes averaged over several values. Show that this "sliding averager" can be described by a convolution with $h(n) = u(n) - u(n - N + 1)$ with a scaling factor of $1/N$ where N is the number of sequence values being averaged.

4.27. Perform the continuous-time convolution between the pairs of signals $\boldsymbol{x}$ and $\boldsymbol{y}$ shown in Fig. P4.27. Sketch the results. In parts (b) and (c) the $\boldsymbol{y}$ signal contains impulse functions with amplitudes as shown.

4.28. **(a)** Show that two continuous-time signals $\boldsymbol{x}$ and $\boldsymbol{y}$ that are nonzero only over intervals of lengths M and N, respectively, have a convolutional interval of length $(M + N)$.

(b) Determine the length of the convolution between two discrete-time sequences of length M and N.

4.29. A useful operation in signal processing of noisy signals is one for which the unit-impulse response is $u(t) - u(t - T)$. This operation is sometimes called a *sliding integrator*. Investigate the effect of this operation on the following input signals; use $T = 1$ second:

(a) $x_a(t) = tu(t)$

(b) $x_b(t) = 10 + \sin 2\pi t$

4.30. In Fig. P4.30 is described a mechanical vibrational system. If M is the mass of the cart and k is the linear spring constant, write the differential equation relating the input force $f(t)$ and the output position of the mass measured from equilibrium relative to the frame. Translate your results to the operator notation $\boldsymbol{x} = \boldsymbol{T}[\boldsymbol{f}]$ and specify $\boldsymbol{T}$ in terms of $\boldsymbol{D}$.

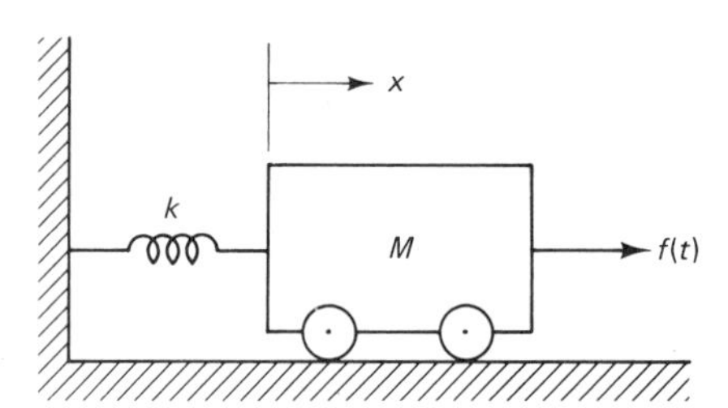

Figure P4.30

5

Laplace Transform

5.1 INTRODUCTION

In the previous chapter, the basic continuous-time system operation of convolution was presented. The integration involved often makes the calculation of the system response very tedious, if not impossible, to obtain in a closed-form expression. In this chapter we present an alternate method of solving for the system response which makes use of the Laplace transform an extremely useful tool.

The Laplace transform converts time-domain signal descriptions into functions of a complex variable. This *complex domain* description of a signal provides new insight into the analysis of signals and systems. In addition, the Laplace transform method often simplifies the calculations involved in obtaining system response signals. The concept of transfer function, introduced briefly at the end of the previous chapter, is developed. In working with transfer functions, linear differential equations describing system operations are transformed into algebraic relations, thus eliminating both the necessity of solving the differential equations using classical techniques and the tedium of convolution integration.

5.2 LAPLACE TRANSFORM INTEGRAL

As shown in the previous chapter, the transfer function completely characterizes the exponential response of a time-invariant linear operator. Recall that this transfer function is formally generated through the process of multiplying the linear operator's characteristic signal $h(t)$ by the signal e^{-st} and then integrating that product over the time interval $(-\infty, +\infty)$. This systematic procedure is more generally known as *taking the Laplace transform* of the signal $h(t)$. It will shortly become apparent that the Laplace transform operation has a much broader setting than merely characterizing a linear operator's exponential response. Undoubtedly, it constitutes one of the more powerful and useful analytical tools that contemporary scientists have at their disposal. The Laplace transform is particularly valuable when determining the solution of a constant coefficient linear differential equation. This is indeed significant when it is realized that many real-world phenomena are modeled by such operators.

The Laplace transform of the continuous-time signal $x(t)$ is designated hereafter by the symbol $X(s)$ and is formally defined by the integration operation

$$X(s) = \int_{-\infty}^{+\infty} x(t)e^{-st}\, dt \tag{5.1}$$

The variable s that appears in this integrand exponential is generally complex-valued and is therefore often expressed in terms of its rectangular coordinates

$$s = \sigma + j\omega \tag{5.2}$$

where $\sigma = \text{Re}\,(s)$ and $\omega = \text{Im}\,(s)$ are referred to as the *real* and *imaginary components* of s, respectively. An examination of integral (5.1) indicates that the Laplace transform is totally dependent on the signal $x(t)$ being transformed and the value assigned to the complex variable s. It is precisely because of this dependency that the natural notation $X(s)$ is incorported.

The signal $x(t)$ and its associated Laplace transform $X(s)$ is said to form a *Laplace transform pair*. This reflects a form of equivalency between the two apparently different entities $x(t)$ and $X(s)$. We may symbolize this interrelationship in the following suggestive manner:

$$X(s) = \mathscr{L}[x(t)]$$

or

$$x(t) \leftrightarrow X(s)$$

where the operator notation $\mathscr{L}$ means to "multiply the signal $x(t)$ being operated upon by the complex exponential e^{-st} and then to integrate that product over the time interval $(-\infty, +\infty)$."

Before embarking on a general discussion of the Laplace transform and its properties, it is perhaps advisable to first demonstrate some typical mechanics one normally employs when evaluating the Laplace transform integral. With this in mind, let us reconsider the specific causal exponential signal

$$x(t) = e^{-t}u(t) \tag{5.3}$$

The Laplace transform of this signal is formally obtained by substituting this signal expression into the defining transform integral (5.1). This yields the following relationship:

$$X(s) = \int_{-\infty}^{+\infty} e^{-t}u(t)e^{-st}\,dt$$

To evaluate this integral, it is first observed that the integrand term $u(t)$ is one for positive t and zero for negative t. Thus, the integral expression for $X(s)$ simplifies to

$$X(s) = \int_{0}^{+\infty} e^{-t}e^{-st}\,dt = \int_{0}^{+\infty} e^{-(s+1)t}\,dt$$

$$= \left. \frac{e^{-(s+1)t}}{-(s+1)} \right|_{0}^{\infty}$$

Unless the complex variable s is judiciously chosen, the upper-limit evaluation (i.e., at $t = +\infty$) for $X(s)$ yields an unbounded result. Since our interest in Laplace transforms is confined to those sets of s that render $X(s)$ a finite value, it is necessary to further investigate this situation. This is readily accomplished by equivalently expressing s as $\sigma + j\omega$ in the exponential term so that

$$X(s) = \left. \frac{e^{-(\sigma+1)t}e^{-j\omega t}}{-(s+1)} \right|_{0}^{+\infty}$$

It is now observed that the upper-limit evaluation is zero if the real part of s is selected so that $\sigma + 1 > 0$ (or equivalently $\sigma > -1$). On the other hand, the upper-limit evaluation is unbounded if $\sigma + 1 < 0$ and is undefined if $\sigma + 1 = 0$ since the meaning of $e^{-j\omega\infty}$ is unknown. At the lower limit (i.e., $t = 0$), the evaluation of $e^{-(s+1)t}$ is one for all selections of s.

The Laplace transform of the elementary signal $e^{-t}u(t)$ has therefore been completed and is given by

$$X(s) = \begin{cases} \dfrac{1}{s+1} & \text{for Re } (s) > -1 \\ \text{undefined} & \text{for Re } (s) = -1 \\ \text{unbounded} & \text{for Re } (s) < -1 \end{cases}$$

where we have chosen to denote σ as Re (s). This information may be conveniently displayed as shown in Fig. 5.1, in which the vertical line specified by Re $(s) = -1$ separates the regions Re $(s) > -1$ (section with cross-hatched lines) and Re $(s) < -1$ (section without cross-hatched lines). Thus, for any value of s whose real component exceeds minus one, the corresponding Laplace transform integral simplifies to $1/(s + 1)$. This result may be concisely expressed as

$$\mathscr{L}[e^{-t}u(t)] = \frac{1}{s+1} \qquad \text{for Re } (s) > -1$$

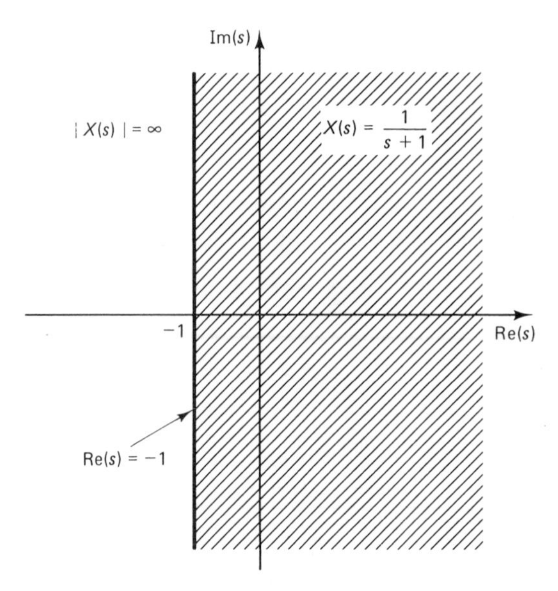

Figure 5.1 Depiction of the Laplace transform of signal $e^{-t}u(t)$.

5.3 REGION OF ABSOLUTE CONVERGENCE

In evaluating the Laplace transform integral that corresponds to a given signal, it is generally found that this integral will exist (that is, the integral has finite magnitude) for only a restricted set of s values. This was illustrated in the previous section when finding the Laplace transform of the signal $e^{-t}u(t)$. As might be anticipated, we are primarily interested in the behavior of $X(s)$ over that set of s which causes its magnitude to be finite in value. With this in mind, the following definition of region of *absolute convergence* is offered.

Definition 5.1. The set of complex numbers s for which the magnitude of the Laplace transform integral is finite is said to constitute the region of absolute convergence for that integral transform.

It is shown in Appendix 5A (Section 5.12) that this region of convergence is always expressible as

$$\sigma_+ < \text{Re}\,(s) < \sigma_- \tag{5.4}$$

where σ_+ and σ_- denote real parameters that are related to the causal and anticausal components, respectively, of the signal whose Laplace transform is being sought. The procedure for determining these convergent region parameters is made evident in the next section and in Appendix 5. From expression (5.4), it is apparent that a function has a Laplace transform only if $\sigma_+ < \sigma_-$.

5.4 LAPLACE TRANSFORMS OF BASIC SIGNALS

In the next chapter, it is shown that Laplace transform theory can be effectively used for determining the response of a constant coefficient linear differential equation operator to many inputs. This solution process entails the determination of the Laplace transform of the input signal. Clearly, the practicality of this approach is, to a large extent, then dependent on one's ability to evaluate the Laplace transform integrals of various signals that are encountered in routine applications. The objective of this section is that of familiarizing the reader with some of the mechanics used in evaluating the Laplace transform integral.

Causal Exponential Signals

The exponential signal is the most fundamental of all signals in characterizing linear operators. As such, it is apparent that the Laplace transform of this signal will play an important role in much of what is to follow. With this in mind, let us now find the Laplace transform of the first-order causal exponential signal

$$x_1(t) = e^{-at}u(t) \tag{5.5}$$

where the constant "a" can in general be a complex number. This exponential signal then, in fact, represents the entire class of exponential signals (e.g., $e^{-t}u(t)$ is obtained by letting $a = 1$). The Laplace transform of this general exponential signal is determined upon evaluating the associated Laplace transform integral

$$\begin{aligned} X_1(s) &= \int_{-\infty}^{+\infty} e^{-at}u(t)e^{-st}\,dt = \int_0^{+\infty} e^{-(s+a)t}\,dt \\ &= \left.\frac{e^{-(s+a)t}}{-(s+a)}\right|_0^{+\infty} \end{aligned} \tag{5.6}$$

In order for $X_1(s)$ to exist, it must follow that the real part of the exponential argument be positive, that is,

$$\text{Re}\,(s + a) = \text{Re}\,(s) + \text{Re}\,(a) > 0$$

If this was not the case, the evaluation of expression (5.6) at the upper limit $t = +\infty$ would either be unbounded if $\text{Re}\,(s) + \text{Re}\,(a) < 0$ or undefined when $\text{Re}\,(s) + \text{Re}\,(a) = 0$. On the other hand, the upper-limit evaluation is zero when $\text{Re}\,(s) + \text{Re}\,(a) > 0$, as is already apparent. The lower limit evaluation at $t = 0$ is equal to $1/(s + a)$ for all choices of the variable s.

The Laplace transform of exponential signal $e^{-at}u(t)$ has therefore been found and is given by

$$\mathcal{L}[e^{-at}u(t)] = \frac{1}{s + a} \qquad \text{for Re}\,(s) > -\text{Re}\,(a) \tag{5.7}$$

In arriving at this result, it is comforting to note that the region of absolute convergence obtained by directly evaluating the Laplace transform integral is in fact identical to that which would have been obtained using the approach of Appendix 5 (i.e., $\sigma_+ = -\text{Re}\,(a)$ and $\sigma_- = +\infty$). With this in mind, we shall hereafter determine the Laplace transform of any signal by directly evaluating the transform integral with the foreknowledge that the associated region of absolute convergence will be a byproduct of this integral evaluation.

Let us now continue our study by finding the Laplace transforms of higher-order causal exponential signals. In particular, let us begin by considering

$$x_2(t) = te^{-at}u(t) \tag{5.8}$$

where again "a" is taken to be generally a complex-valued constant. The Laplace transform of this signal is formally obtained by evaluating the following integral expression:

$$X_2(s) = \int_0^{+\infty} te^{-(s+a)t}\,dt$$

This evaluation may be achieved by using the standard integration-by-parts routine with $u(t) = t$ and $dv(t) = e^{-(s+a)t}\,dt$. This results in

$$X_2(s) = \left.\frac{te^{-(s+a)t}}{-(s+a)}\right|_0^{+\infty} + \frac{1}{s+a}\int_0^{+\infty} e^{-(s+a)t}\,dt \tag{5.9}$$

It is now observed that the entity $te^{-(s+a)t}$ when evaluated at the upper limit $t = +\infty$ is bounded and defined only if the variable s is selected so that

$$\text{Re}\,(s) + \text{Re}\,(a) > 0 \tag{5.10}$$

in which case the upper-limit evaluation yields zero. Furthermore, the integral in relationship (5.9) is recognized as being equal to the Laplace transform of the signal $e^{-at}u(t)$ which was found to be $1/(s + a)$. This integral was previously found to exist only if the complex variable s is selected in order to satisfy inequality (5.10). Thus, the Laplace transform of the second-order exponential signal has been obtained and is given by

$$\mathscr{L}[te^{-at}u(t)] = \frac{1}{(s+a)^2} \qquad \text{for Re}\,(s) > -\text{Re}\,(a) \tag{5.11}$$

It is possible to continue this integration-by-parts routine to obtain the Laplace transform of higher-order causal exponential signals. This is found to result in the following Laplace transform pair:

$$\mathscr{L}[t^k e^{-at}u(t)] = \frac{k!}{(s+a)^{k+1}} \qquad \text{for Re}\,(s) > -\text{Re}\,(a) \tag{5.12}$$

where "a" is a complex constant and "k" is a nonnegative integer. From this transform pair, it is observed that the region of absolute convergence is totally dependent on the parameter a and not at all on the order of the exponential k.

Anticausal Exponential Signals

It may appear to many that the region of absolute convergence is of minor importance in characterizing the Laplace transform of a given signal. The temptation to concentrate our interest on the analytical expression for $X(s)$ and ignore the associated region of absolute convergence is indeed great. However, as we now demonstrate, this can often lead to erroneous conclusions since the region of absolute convergence will in fact give valuable information relative to the causal and anticausal nature of the signal being transformed.

With this in mind, let us now determine the Laplace transform of the first-order anticausal signal as given by

$$x_3(t) = -e^{-at}u(-t) \tag{5.13}$$

where a is generally a complex-valued constant. This transform is given by

$$X_3(s) = \int_{-\infty}^{+\infty} -e^{-at}u(-t)e^{-st}\,dt = \frac{e^{-(s+a)t}}{s+a}\Bigg|_{-\infty}^{0}$$

In order for the lower-limit evaluation to exist, it is necessary that the complex variable s be selected so that $\mathrm{Re}\,(s + a) < 0$. If this is the case, then the lower-limit evaluation is zero and we have established the following Laplace transform pair:

$$\mathscr{L}[-e^{-at}u(-t)] = \frac{1}{s+a} \qquad \text{for } \mathrm{Re}\,(s) < -\mathrm{Re}\,(a) \tag{5.14}$$

It is interesting to note that the analytical expressions for the Laplace transforms of the two different signals $e^{-at}u(t)$ and $-e^{-at}u(-t)$ are each given by $1/(s + a)$. One might then erroneously conclude that it is impossible to differentiate between these seemingly identical Laplace transforms. This is, however, far from the truth when the corresponding regions of convergence are taken into account. Namely, the transform $1/(s + a)$ holds in the region $\mathrm{Re}\,(s) > -\mathrm{Re}\,(a)$ for the causal signal $e^{-at}u(t)$ and in the region $\mathrm{Re}\,(s) < -\mathrm{Re}\,(a)$ for the anticausal signal $-e^{-at}u(-t)$. These regions share no points in common as is indicated in Fig. 5.2 and the two seemingly identical transforms are seen to be, in fact, different. This is an extremely important point which will be utilized when discussing the topic of inverse Laplace transform.

Using the integration-by-parts routine, it is possible to generate further Laplace transform pairs relative to higher-order anticausal exponential signals. The following general transform pair may be readily established:

$$\mathscr{L}[-(-t)^k e^{-at}u(-t)] = \frac{k!}{(s+a)^{k+1}} \qquad \text{for } \mathrm{Re}\,(s) < -\mathrm{Re}\,(a) \tag{5.15}$$

where "a" is a complex constant and "k" is a nonnegative integer.

Example 5.1

We may use the previously derived Laplace transform pairs to determine the transforms of signals that are of a disguised exponential nature. For example, it is readily seen that

$$\mathscr{L}[u(t)] = \frac{1}{s} \qquad \text{for } \mathrm{Re}\,(s) > 0$$

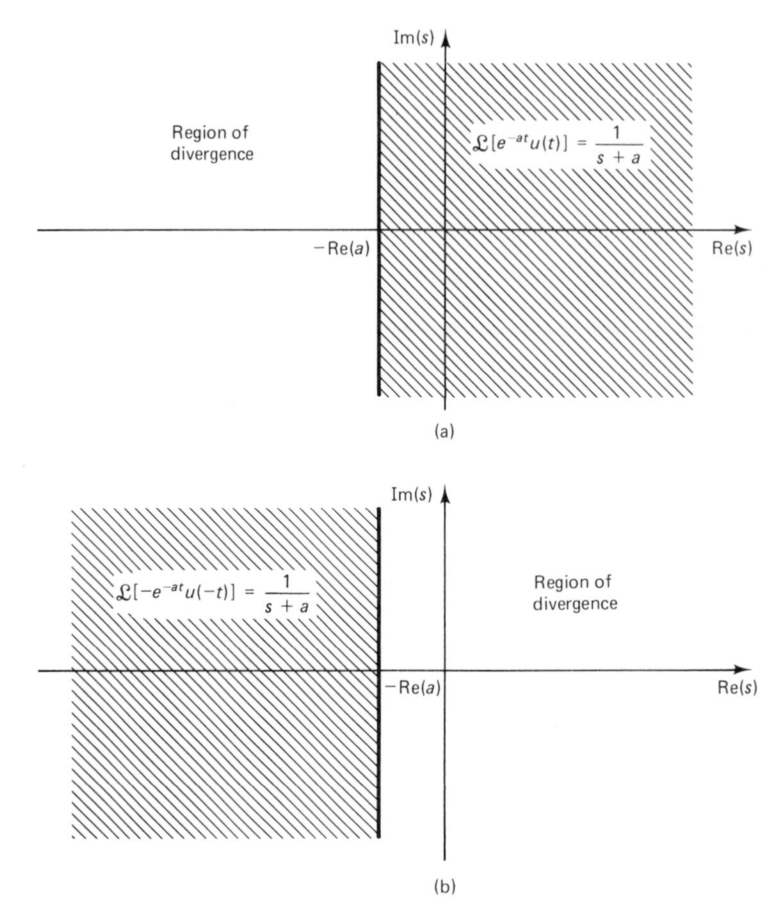

Figure 5.2 Laplace transforms and the associated regions of absolute convergence for (a) $e^{-at}u(t)$ and (b) $-e^{-at}u(-t)$.

and more generally

$$\mathscr{L}[t^k u(t)] = \frac{k!}{s^{k+1}} \qquad \text{for Re } (s) > 0$$

which are each derivable from relationship (5.12) by simply setting $a = 0$.

Impulse Signals

We conclude this section by determining the Laplace transforms of unit-impulse-type signals that arise when differentiating discontinuous continuous-time signals. To begin this investigation, it is readily seen that the Laplace transform of the unit-impulse (or Dirac delta) signal is given by

$$\mathscr{L}[\delta(t)] = \int_{-\infty}^{\infty} \delta(t)e^{-st}\, dt = 1 \qquad \text{for all } s \tag{5.16}$$

where we have made use of the integration property possessed by unit-impulse signals [see relationship (3.32)]. Continuing on in this manner, the following Laplace transform pair can be established:

$$\mathscr{L}\frac{d^k\delta(t)}{dt^k} = s^k \qquad \text{for all } s \tag{5.17}$$

by appealing to the integration-by-parts process.

Laplace Transform Pair Tables

It is possible to continue this direct integration procedure to determine the Laplace transforms of a variety of signals that arise frequently in applications. In order to avoid any repetitions of Laplace transform evaluations, it is convenient to display

TABLE 5.1 TABLE OF LAPLACE TRANSFORM PAIRS

	Time Signal $x(t)$	Laplace Transform $X(s)$	Region of Absolute Convergence
1.	$e^{-at}u(t)$	$\dfrac{1}{s+a}$	$\text{Re}(s) > -\text{Re}(a)$
2.	$t^k e^{-at}u(t)$	$\dfrac{k!}{(s+a)^{k+1}}$	$\text{Re}(s) > -\text{Re}(a)$
3.	$-e^{-at}u(-t)$	$\dfrac{1}{(s+a)}$	$\text{Re}(s) < -\text{Re}(a)$
4.	$(-t)^k e^{-at}u(-t)$	$\dfrac{k!}{(s+a)^{k+1}}$	$\text{Re}(s) < -\text{Re}(a)$
5.	$u(t)$	$\dfrac{1}{s}$	$\text{Re}(s) > 0$
6.	$\delta(t)$	1	all s
7.	$\dfrac{d^k\,\delta(t)}{dt^k}$	s^k	all s
8.	$t^k u(t)$	$\dfrac{k!}{s^k}$	$\text{Re}(s) > 0$
9.	$\text{sgn}\, t = \begin{cases} 1, & t \geq 0 \\ -1, & t < 0 \end{cases}$	$\dfrac{2}{s}$	$\text{Re}(s) = 0$
10.	$\sin \omega_0 t\; u(t)$	$\dfrac{\omega_0}{s^2+\omega_0^2}$	$\text{Re}(s) > 0$
11.	$\cos \omega_0 t\; u(t)$	$\dfrac{s}{s^2+\omega_0^2}$	$\text{Re}(s) > 0$
12.	$e^{-at} \sin \omega_0 t\; u(t)$	$\dfrac{\omega}{(s+a)^2+\omega_0^2}$	$\text{Re}(s) > -\text{Re}(a)$
13.	$e^{-at} \cos \omega_0 t\; u(t)$	$\dfrac{s+a}{(s+a)^2+\omega_0^2}$	$\text{Re}(s) > -\text{Re}(a)$

the results of our efforts in a so-called Laplace transform pair table. In this manner, instead of having to continuously reevaluate the Laplace transforms of standard signals, we could then simply refer to such a table and read out the desired transform. Table 5.1, which gives some Laplace transform pairs, is sufficient for our needs. It displays the time signal $x(t)$ and its corresponding Laplace transform and region of absolute convergence.

5.5 PROPERTIES OF LAPLACE TRANSFORM

The Laplace transform provides a systematic procedure for converting a function of a real variable "t" into a function of a complex variable "s." The functions $x(t)$ and $X(s)$ are then said to be *equivalent representations*, in that one may be obtained from the other and vice versa. In any investigation related to continuous-time signals, one may then conduct his or her study in either the natural time domain t or, if desired, in the complex domain s. As we will see, in some very important application areas (for example, solving linear differential equations), the complex-domain representation often yields desired results in a more expedient manner and also provides a great deal of useful insight into the underlying phenomena being studied. With this in mind, we now present properties that indicate how certain operations effected in one domain of the equivalent representation manifest themselves in the other domain.

Linearity

One of the most fundamental properties of the Laplace transform is revealed when finding the Laplace transform of a signal that is composed of a linear combination of two other signals, that is,

$$x(t) = \alpha_1 x_1(t) + \alpha_2 x_2(t) \tag{5.18}$$

where α_1 and α_2 are constants. The Laplace transform of $x(t)$ is formally given by substituting this representation of $x(t)$ into the defining relationship (5.1):

$$X(s) = \int_{-\infty}^{\infty} [\alpha_1 x_1(t) + \alpha_2 x_2(t)]e^{-st}\, dt$$

in which the term within the braces is recognized as being $x(t)$. Using standard integration operations, this integral may be expressed as

$$X(s) = \alpha_1 \int_{-\infty}^{\infty} x_1(t)e^{-st}\, dt + \alpha_2 \int_{-\infty}^{\infty} x_2(t)e^{-st}\, dt$$

The first and second integrals in this expression are seen to be the Laplace transforms of the signals $x_1(t)$ and $x_2(t)$, respectively, thereby yielding

$$X(s) = \alpha_1 X_1(s) + \alpha_2 X_2(s) \tag{5.19}$$

Thus, the Laplace transform is seen to preserve linear combination of time signals in the sense that $X(s)$ is the same linear combination of $X_1(s)$ and $X_2(s)$ as was $x(t)$ of the signals $x_1(t)$ and $x_2(t)$.

To determine the region of absolute convergence for this linear combination signal, it is generally necessary to evaluate the exponential bounding of $x(t)$ along the lines of Appendix 5. In most practical applications, however, this objective may be achieved by simply ascertaining those values of s for which the two integrals leading to expressions in relationship (5.19) are simultaneously bounded. It is therefore apparent that the desired convergence region is at least equal to the set of s which simultaneously lies in the regions of absolute convergence for $X_1(s)$ and $X_2(s)$, that is,

$$\max(\sigma_+^1, \sigma_+^2) < \text{Re}(s) < \min(\sigma_-^1, \sigma_-^2) \tag{5.20}$$

where the pairs (α_+^1, α_-^1) and (α_+^2, α_-^2) identify the regions of convergence for the Laplace transforms $X_1(s)$ and $X_2(s)$, respectively. As the next example illustrates, however, the region of absolute convergence for $X(s)$ can sometimes be larger. In summary, the linearity property indicates that

$$\boxed{\mathscr{L}[\alpha_1 x_1(t) + \alpha_2 x_2(t)] = \alpha_1 X_1(s) + \alpha_2 X_2(s)} \tag{5.21}$$

and the region of absolute convergence is *at least as large* as that given by expression (5.20).

Example 5.2

Determine the Laplace transform of the following pulse-type signal:

$$x(t) = \begin{cases} 2, & 0 \le t < 3 \\ 0, & \text{otherwise} \end{cases}$$

A direct application of the defining Laplace transform integral is seen to give

$$X(s) = \int_0^3 2e^{-st}\, dt = \left. \frac{2e^{-st}}{-s} \right|_0^3$$

$$= \frac{2(1 - e^{-3s})}{s} \qquad \text{for all } s$$

The transform's region of absolute convergence is seen to consist of all finite s due to the fact that the governing integral is seen to exist (i.e., be bounded) for any choice of s (i.e., an exponential bounding of signal $x(t)$ is found to give $\sigma_- = +\infty$ and $\sigma_+ = -\infty$).

An alternate method for determining $X(s)$ is to utilize the linearity property given in which the signal under study can be expressed in the linear combination format

$$x(t) = 2u(t) - 2u(t - 3)$$

According to linearity property (5.21), we have

$$X(s) = 2\mathscr{L}[u(t)] - 2\mathscr{L}[u(t - 3)] \tag{5.22}$$

The Laplace transform for $u(t)$ has already been evaluated and is given by $1/s$ with the associated region of absolute convergence being $\text{Re}(s) > 0$. A simple integration operation indicates that $\mathscr{L}[u(t-3)] = e^{-3s}/s$ for all s such that Re $(s) > 0$. It then follows from expression (5.22) that

$$X(s) = 2\frac{1}{s} - 2\frac{e^{-3s}}{s} = 2\frac{1 - e^{-3s}}{s}$$

which is in agreement with the result obtained by using the direct approach. It is to be noted, however, that the region of absolute convergence is not equal to the intersection of the regions of absolute convergence for $\mathscr{L}[u(t)]$ and $\mathscr{L}[u(t-3)]$, that is, Re $(s) > 0$.

Example 5.3

Let us now demonstrate the utility of the linearity property by finding the Laplace transform of the causal sinusoidal signal

$$x(t) = A \sin(\omega_0 t + \theta)\, u(t) \tag{5.23}$$

where A, ω_0, and θ are the sinusoid's amplitude, radian frequency, and phase angle constants, respectively. Application of Euler's identity to this sinusoid signal is found to yield

$$\begin{aligned} x(t) &= A\,\frac{e^{j(\omega_0 t + \theta)} - e^{-j(\omega_0 t + \theta)}}{2j}\, u(t) \\ &= \left(\frac{Ae^{j\theta}}{2j}\right) e^{j\omega_0 t} u(t) - \left(\frac{-Ae^{-j\theta}}{2j}\right) e^{-j\omega_0 t} u(t) \end{aligned}$$

It is now observed that $x(t)$ is here expressed as a linear combination of the two exponential signals $x_1(t) = e^{j\omega_0 t}u(t)$ and $x_2(t) = e^{-j\omega_0 t}u(t)$. With this in mind, the linearity property is applicable where use of the Laplace transform pair (5.7) with $a = -j\omega_0$ and $a = j\omega_0$ is made. This yields

$$\begin{aligned} X(s) &= \frac{Ae^{j\theta}}{2j} \cdot \frac{1}{s - j\omega_0} - \frac{Ae^{-j\theta}}{2j} \cdot \frac{1}{s + j\omega_0} \\ &= \frac{A}{2j} \cdot \frac{s(e^{j\theta} - e^{-j\theta}) + j\omega_0(e^{j\theta} + e^{-j\theta})}{(s - j\omega_0)(s + j\omega_0)} \\ &= A\,\frac{s\,(\sin\theta) + j\omega_0(\cos\theta)}{s^2 + \omega_0^2} \qquad \text{for Re } (s) > 0 \end{aligned} \tag{5.24}$$

in which the region of absolute convergence is obtained by exponentially bounding $x(t)$ (this results in $\sigma_- = +\infty$, and $\sigma_+ = 0$). In this case, the region of absolute convergence is seen to be equal to the intersect of the regions of absolute convergence for $x_1(t)$ and $x_2(t)$.

Example 5.4

Determine the Laplace transform of the signal

$$x(t) = \frac{d\,\delta(t)}{dt} + \frac{3}{2}e^{-t}u(t) - \frac{2}{3}e^{-2t}u(t) - \frac{1}{6}e^{t}u(-t)$$

Inserting this signal expression into the relationship defining the Laplace transform, it is found that

$$X(s) = \int_{-\infty}^{\infty} \frac{d\,\delta(t)}{dt} e^{-st}\,dt + \frac{3}{2}\int_{-\infty}^{\infty} [e^{-t}u(t)]e^{-st}\,dt$$

$$- \frac{2}{3}\int_{-\infty}^{\infty} [e^{-2t}u(t)]e^{-st}\,dt - \frac{1}{6}\int_{-\infty}^{\infty} [e^{t}u(-t)]e^{-st}\,dt$$

These four integrals are recognized as being the Laplace transforms of the four signals $d\delta(t)/dt$, $e^{-t}u(t)$, $e^{-2t}u(t)$, and $e^{t}u(-t)$ that constitute the signal being transformed. Using entries 1, 3, and 7 of Laplace transform Table 5.1, it follows that

$$X(s) = s + \frac{3}{2}\left(\frac{1}{s+1}\right) - \frac{2}{3}\left(\frac{1}{s+2}\right) - \frac{1}{6}\left(\frac{-1}{s-1}\right)$$

We could have obtained this result by directly appealing to an extended version of the linearity property.

To determine the region of convergence for $X(s)$, it is necessary to find the set of s that will simultaneously cause each of the four integrals above to be bounded. The first integral converges for all values of s, whereas the second, third, and fourth integrals are found to be individually bounded when s is such that

$$-1 < \text{Re}\,(s) < +\infty, \quad -2 < \text{Re}\,(s) < +\infty, \quad -\infty < \text{Re}\,(s) < 1$$

respectively. The set of s which is common to each of these vertical strips is seen to be

$$-1 < \text{Re}\,(s) < 1$$

where $-1 = \max\,(-1, -2, -\infty)$ and $1 = \min\,(+\infty, +\infty, 1)$. Thus, the desired Laplace transform is given by

$$X(s) = s + \frac{3}{2}\left(\frac{1}{s+1}\right) - \frac{2}{3}\left(\frac{1}{s+2}\right) - \frac{1}{6}\left(\frac{-1}{s-1}\right)$$

$$= \frac{s^4 + 23^3 + 2}{s^3 + 23^2 - s - 2} \qquad \text{for } -1 < \text{Re}\,(s) < 1$$

Time-Domain Differentiation

An operation that is used extensively in continuous-time signal processing is that of *differentiation*. With this in mind, it is logical that we would be interested in determining the Laplace transforms of differentiated signals. Since the signals $x(t)$ and $dx(t)/dt$ are associated with each other in an obvious manner, one might anticipate that their corresponding Laplace transforms would be in some way related. This is, in fact, the case as may be established by formally finding the Laplace transform of $dx(t)/dt$, that is,

$$\mathscr{L}\left[\frac{dx(t)}{dt}\right] = \int_{-\infty}^{+\infty} \frac{dx(t)}{dt} e^{-st}\,dt$$

To evaluate this integral expression, the integration-by-parts routine is used with $dv(t) = [dx(t)/dt]\,dt$ and $u(t) = e^{-st}$, so that $v(t) = x(t)$ and $du(t) = -se^{-st}\,dt$. This results in

$$\mathscr{L}\left[\frac{dx(t)}{dt}\right] = x(t)e^{-st}\Big|_{-\infty}^{+\infty} - \int_{-\infty}^{+\infty} x(t)[-se^{-st}]\,dt$$

For the time being, let us assume that the complex variable s lies in the region of absolute convergence of $x(t)$. It then follows that the term $x(t)e^{-st}$ is zero when evaluated at either of the two limits plus and minus infinity. This expression then simplifies to

$$\mathscr{L}\left[\frac{dx(t)}{dt}\right] = s\int_{-\infty}^{+\infty} x(t)e^{-st}\,dt$$

where the term s has been taken outside the integral since it is not a function of time. The integral appearing in this expression is recognized as being the Laplace transform of the signal $x(t)$, which we shall denote by $X(s)$. We have therefore established the property

$$\boxed{\mathscr{L}\left[\frac{dx(t)}{dt}\right] = sX(s)} \tag{5.25}$$

Furthermore, it is clear that the region of absolute convergence of $dx(t)/dt$ is at least as large as that of $x(t)$. In some cases, it is larger as the example to follow shows.

The operation of time-domain differentiation has then been found to correspond to a multiplication by s in the Laplace variable s domain. To emphasize this association, this property may be envisioned as shown in Fig. 5.3. Since the operation of scalar multiplication is generally viewed as being *simpler* than the process of differentiation, it may have occurred to the reader that continuous-time operations involving differentiation might be better examined in the s-domain. It is precisely because of this observation that the analysis of constant-coefficient linear differential equation operators is so readily carried out in the s-domain. More will be said about this most important observation in the next chapter.

It is possible to continue the procedure which led to the above differentiation

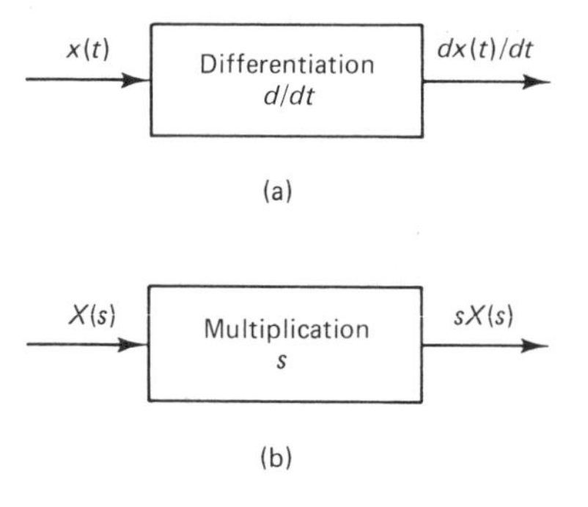

Figure 5.3 Equivalent operations in the (a) time-domain operation and (b) the Laplace transform–domain operation.

property in order to establish higher-order derivative properties. As an example, it immediately follows from relationship (5.25) that

$$\mathscr{L}\left[\frac{d^2x(t)}{dt^2}\right] = s\mathscr{L}\left[\frac{dx(t)}{dt}\right] = s^2X(s)$$

where again $X(s)$ denotes the Laplace transform of the signal $x(t)$, and use of the fact that $d^2x(t)/dt^2$ is the first derivative of $dx(t)/dt$ has been made. More generally, an inductive argument can be applied to establish the relationship

$$\mathscr{L}\left[\frac{d^kx(t)}{dt^k}\right] = s^kX(s) \tag{5.26}$$

which holds for all nonnegative integer values of k. A literal interpretation of this property then indicates that each time-domain differentiation gives rise to a multiplication by s operation in the Laplace transform domain.

Example 5.5

Determine the Laplace transform of the signal $\delta(t)$ using the time-domain differentiating property. In this case, it is seen that $\delta(t)$ is in fact identical to the derivative of the unit-step signal, so that

$$\mathscr{L}[\delta(t)] = \mathscr{L}\left[\frac{du(t)}{dt}\right] = s\mathscr{L}[u(t)]$$

It has been previously found that $\mathscr{L}[u(t)] = 1/s$, so that

$$\mathscr{L}[\delta(t)] = 1 \qquad \text{for all } s$$

The region of absolute convergence for this signal is seen to consist of all s and is therefore larger than the region of convergence for $\mathscr{L}[u(t)]$, i.e., Re $(s) > 0$. It will subsequently be made apparent that this lack of correspondence between the regions of absolute convergence for $\mathscr{L}[du(t)/dt]$ and $\mathscr{L}[u(t)]$ arises due to the appearance of the factor s in the denominator of $\mathscr{L}[u(t)] = 1/s$.

In a similar manner, one may inductively establish the property that

$$\mathscr{L}\left[\frac{d^k\,\delta(t)}{dt^k}\right] = s^k \qquad \text{for all } s \tag{5.27}$$

where k is any nonnegative integer.

Time Shift

The operation of time shifting a signal plays an essential role in describing a pulse-like signal as well as being used in many practical forms of signal processing. In particular, the signal $x(t - t_0)$ is said to be a version of the signal $x(t)$ right-shifted (or delayed) by t_0 seconds. Its Laplace transform is formally given by

$$\mathscr{L}[x(t - t_0)] = \int_{-\infty}^{+\infty} x(t - t_0)e^{-st}\,dt$$

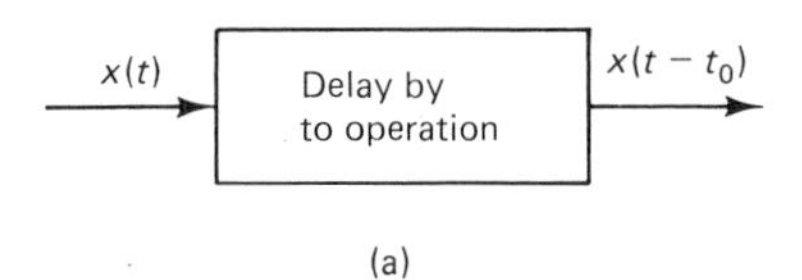

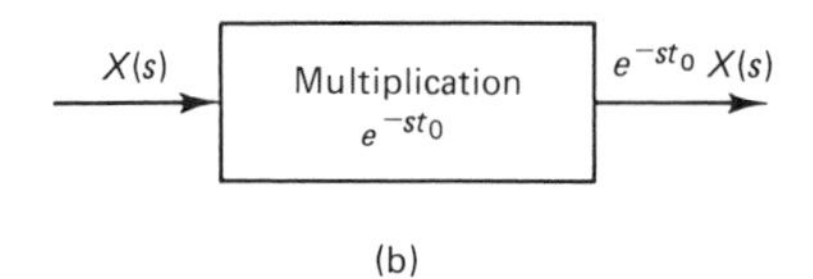

Figure 5.4 Equivalent operations in (a) the time domain and (b) the Laplace transform domain.

A convenient closed-form expression for this integral may be readily obtained by making the standard change of integration variables $\tau = t - t_0$. This is seen to yield

$$\begin{aligned}\mathscr{L}[x(t - t_0)] &= \int_{-\infty}^{+\infty} x(\tau)e^{-s(\tau + t_0)}\, d\tau \\ &= e^{-st_0}\int_{-\infty}^{+\infty} x(\tau)e^{-s\tau}\, d\tau\end{aligned}$$

The integral multiplying the quantity e^{-st_0} is recognized as being the Laplace transform of the signal $x(t)$ with the integration variable being τ instead of t. The desired Laplace transform relationship

$$\boxed{\mathscr{L}[x(t - t_0)] = e^{-st_0}X(s)} \tag{5.28}$$

has therefore been established, where $X(s)$ denotes the Laplace transform of the unshifted signal $x(t)$. Thus, right shifting (delaying) a signal by a t_0-second duration in the time domain is seen to correspond to a multiplication by e^{-st_0} in the Laplace transform domain. As a general rule, anytime a term of the form e^{-st_0} appears in $X(s)$, this implies some form of time shift in the time domain. This most important property is depicted in Fig. 5.4. It should be further noted that the regions of absolute convergence for the signals $x(t)$ and $x(t - t_0)$ are identical.

Example 5.6

Determine the Laplace transform of the pulse-like signal

$$x(t) = \begin{cases} 2t & \text{for } 0 \le t < 1 \\ 0 & \text{otherwise} \end{cases}$$

This triangle pulse signal may be equivalently expressed as

$$x(t) = 2t[u(t) - u(t - 1)]$$

In order to put this signal into a format whereby the time-shifting property can be used, equivalently express $x(t)$ as

$$x(t) = 2tu(t) - 2(t - 1)u(t - 1) - 2u(t - 1)$$

which is observed to be composed of shifted versions of the signals $u(t)$ and $tu(t)$ whose Laplace transforms we already know. From the linearity property (5.21), it follows that

$$X(s) = 2\mathscr{L}[tu(t)] - 2\mathscr{L}[(t-1)u(t-1)] - 2\mathscr{L}[u(t-1)]$$

We next apply the time-shifting property (5.28) to the last two terms on the right side of this relationship with $t_0 = 1$, to yield

$$X(s) = 2\mathscr{L}[tu(t)] - 2e^{-s}\mathscr{L}[tu(t)] - 2e^{-s}\mathscr{L}[u(t)]$$

Finally, the Laplace transforms for $u(t)$ and $tu(t)$ are obtained from Table 5.1 to give the desired result

$$\begin{aligned} X(s) &= 2\left(\frac{1}{s^2}\right) - 2e^{-s}\left(\frac{1}{s^2}\right) - 2e^{-s}\left(\frac{1}{s}\right) \\ &= 2\,\frac{(1 - e^{-s} - se^{-s})}{s^2} \end{aligned}$$

The region of absolute convergence is the entire s-plane as is readily determined by exponentially bounding the given signal (i.e., $\sigma_+ = -\infty$ and $\sigma_- = +\infty$).

It should be noted that this same transform could have been obtained by directly using the Laplace transform integral definition. This latter approach would involve a fair amount of integration and is generally a more time-consuming procedure.

Other Properties

A number of other properties that characterize the Laplace transform may be similarly established. Without going into the details of proof, some of the more important properties are listed in Table 5.2. Application of these properties often enables one to efficiently determine the Laplace transform of seemingly complex time functions as the next example illustrates.

Example 5.7

In standard radio communications, an audio signal $x(t)$ is transmitted through the atmosphere after being amplitude modulated (AM) by carrier signal $A \cos[\omega_0 t]\, u(t)$, where ω_0 denotes the radian frequency identified with the transmitting station. The amplitude-modulated signal is specified as the product of the audio signal and the carrier signal, that is,

$$y(t) = Ax(t)\cos(\omega_0 t)\, u(t)$$

The Laplace transforms of the audio signal $x(t)$ and its amplitude-modulated version are readily obtained by using the Euler identity to give

$$y(t) = \frac{A}{2}\, x(t)e^{j\omega_0 t}u(t) + \frac{A}{2}\, x(t)e^{-j\omega_0 t}u(t)$$

Application of the linearity and frequency-shifting property is seen to yield

$$Y(s) = \frac{A}{2}\,[X(s - j\omega_0) + X(s + j\omega_0)]$$

Thus, the Laplace transform of the amplitude-modulated signal is seen to be simply

TABLE 5.2 LAPLACE TRANSFORM PROPERTIES

Property	Signal $x(t)$ Time Domain	Laplace Transform $X(s)$ s domain	Region of Convergence of $X(s)$ $\sigma_+ < \text{Re}\,(s) < \sigma_-$
Linearity	$\alpha_1 x_1(t) + \alpha_2 x_2(t)$	$\alpha_1 X_1(s) + \alpha_2 X_2(s)$	At least the intersection of the regions of convergence of $X_1(s)$ and $X_2(s)$
Time differentiation	$\dfrac{dx(t)}{dt}$	$sX(s)$	At least $\sigma_+ < \text{Re}\,(s) < \sigma_-$
Time shift	$x(t - t_0)$	$e^{-st_0}X(s)$	$\sigma_+ < \text{Re}\,(s) < \sigma_-$
Time convolution	$\int_{-\infty}^{\infty} h(\tau)x(t-\tau)\,d\tau$	$H(s)X(s)$	At least the intersection of the regions of convergence of $H(s)$ and $X(s)$
Time scaling	$x(at)$	$\dfrac{1}{\lvert a\rvert} X\left(\dfrac{s}{a}\right)$	$\sigma_+ < \text{Re}\left(\dfrac{s}{a}\right) < \sigma_-$
Frequency shift	$e^{-at}x(t)$	$X(s + a)$	$\sigma_+ - \text{Re}\,(a) < \text{Re}\,(s) < \sigma_- - \text{Re}\,(a)$
Multiplication (frequency convolution)	$x_1(t)x_2(t)$	$\dfrac{1}{2\pi j}\int_{c-j\infty}^{c+j\infty} X_1(u)X_2(s-u)\,du$	$\sigma_+^{(1)} + \sigma_+^{(2)} < \text{Re}\,(s) < \sigma_-^{(1)} + \sigma_-^{(2)}$ $\sigma_+^{(1)} + \sigma_+^{(2)} < c < \sigma_-^{(1)} + \sigma_-^{(2)}$
Time integration	$\int_{-\infty}^{t} x(\tau)\,d\tau$	$\dfrac{1}{s}X(s)$ for $X(0) = 0$	At least $\sigma_+ < \text{Re}\,(s) < \sigma_-$
Frequency differentiation	$(-t)^k x(t)$	$\dfrac{d^k X(s)}{ds^k}$	At least $\sigma_+ < \text{Re}\,(s) < \sigma_-$

composed of a linear combination of two frequency-shifted versions of the Laplace transform of the audio signal.

5.6 TIME-CONVOLUTION PROPERTY

We have reserved a separate section for the time-convolution property in order to reflect upon its fundamental importance relative to time-invariant linear operations on signals. Recall that such an operation was governed by the convolution integral

$$y(t) = \int_{-\infty}^{\infty} h(\tau)x(t-\tau)\,d\tau \tag{5.29}$$

where $x(t)$ denotes the input signal; the $h(t)$ characteristic signal identifying the operation process; and $y(t)$ the corresponding response signal. As we now show, there exists a particularly simple relationship between the Laplace transforms of these three continuous-time signals.

The Laplace transform of the signal $y(t)$ is formally given by

$$Y(s) = \int_{-\infty}^{\infty} y(t)e^{-st}\,dt$$

It is now observed that for any value of time t, the signal $y(t)$ is given by the convolution integral relationship (5.29). Thus, substituting this equivalent expression for $y(t)$ into the Laplace transform integral, we have

$$Y(s) = \int_{-\infty}^{\infty} \left[\int_{-\infty}^{\infty} h(\tau)x(t-\tau)\,d\tau\right] e^{-st}\,dt$$

This integral expression may be evaluated by first interchanging the order of the integration variables, that is,

$$Y(s) = \int_{-\infty}^{\infty} \left[\int_{-\infty}^{\infty} x(t-\tau)e^{-st}\,dt\right] h(\tau)\,d\tau$$

Next, the change of integration variables $\alpha = t - \tau$ is made on the inner integral where τ is, of course, a constant relative to the integration variable t. This results in

$$Y(s) = \int_{-\infty}^{\infty} \left[\int_{-\infty}^{\infty} x(\alpha)e^{-s(\alpha+\tau)}\,d\alpha\right] h(\tau)\,d\tau$$

Finally, we express $e^{-s(\alpha+\tau)}$ as $e^{-s\alpha}e^{-s\tau}$ and observe that the multiplicative term $e^{-s\tau}$ is a constant relative to the integration variable α. It may therefore be taken outside the inner integral. The double integration then decomposes into the following product of two single variable integrations:

$$Y(s) = \int_{-\infty}^{\infty} h(\tau)e^{-s\tau}\,d\tau \int_{-\infty}^{\infty} x(\alpha)e^{-s\alpha}\,d\alpha$$

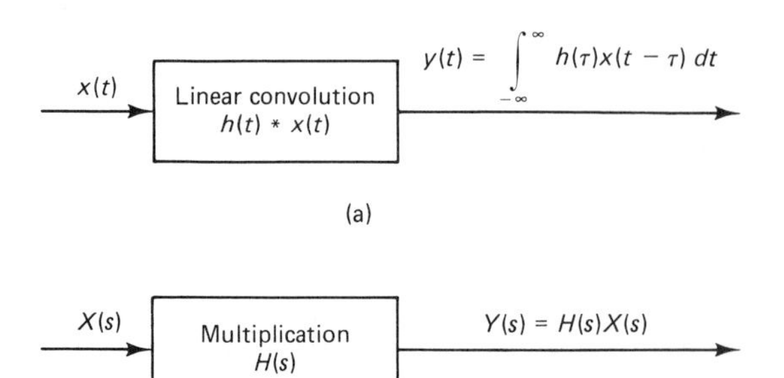

Figure 5.5 Representation of a time-invariant linear operator in (a) the time domain and (b) the s-domain.

The first and second integrals are recognized as being the Laplace transforms of the signals $h(t)$ and $x(t)$, respectively. Thus, the Laplace transform of the response signal is simply given by

$$Y(s) = H(s)X(s) \tag{5.30}$$

where $H(s) = \mathscr{L}[h(t)]$ and $X(s) = \mathscr{L}[x(t)]$. Thus, the convolution of two time domain signals is seen to correspond to the multiplication of their respective Laplace transforms in the s-domain. This is an exceptionally revealing relationship that ultimately enables us to obtain a desirable characterization of the dynamical properties of time-invariant linear operators. One should always keep in mind, however, that the convolution property holds whenever two signals are convolved. The linear operator application here employed serves mainly as a motivation to reenforce the importance of the convolution integral property.

The quantity $H(s)$, which is the Laplace transform of the linear operator's characterizing signal $h(t)$, is referred to as the *transfer function* that corresponds to the underlying linear operator. This notation is indeed a natural one since, by relationship (5.30), the input signal is *transferred* into the response signal by means of a simple multiplicative operation in the s-domain. It is convenient to envision this process in the light of Fig. 5.5.

To determine the response of a time-invariant linear operator to a given input signal, one can proceed in at least two ways: either directly use the convolution integral expression (5.29) or, alternately, use the Laplace transform relationship (5.30). The latter approach is usually the most expedient and requires the two-stage process of determining the Laplace transform functions $H(s)$ and $X(s)$ and then finding the continuous-time signal whose Laplace transform is equal to $H(s)X(s)$. Methods for accomplishing this indirect approach are given in this and the next chapter.

Example 5.8

Find the response of the linear operator whose characteristic signal is given by

$$h(t) = e^{-t}u(t)$$

when the input signal is a step of amplitude 3, that is,

$$x(t) = 3u(t)$$

The Laplace transforms of these signals are obtained from Table 5.1 and are given by

$$H(s) = \frac{1}{s+1} \qquad \text{for Re}(s) > -1$$

and

$$X(s) = \frac{3}{s} \qquad \text{for Re}(s) > 0$$

Using the time convolution property (5.30), the Laplace transform of the resultant response signal is

$$Y(s) = \frac{3}{s(s+1)}$$
$$= \frac{3}{s} - \frac{3}{s+1}$$

The continuous-time signal that has this transform is readily found to be

$$y(t) = 3(1 - e^{-t})u(t)$$

As a check, the reader is requested to use the convolution operation (5.29) with the given signals $h(t)$ and $x(t)$ to obtain this same result.

5.7 TIME-CORRELATION PROPERTY

In a large variety of signal processing applications, the concept of correlating two signals is oftentimes employed. This operation is basically used for determining the structural similarities between any two signals $x(t)$ and $y(t)$ and is formally defined by the integral relationship

$$\phi_{xy}(\tau) = \int_{-\infty}^{\infty} x(t)y(t+\tau)\,dt \tag{5.31}$$

The entity $\phi_{xy}(\tau)$ is referred to as the *cross-correlation function* and measures the relationship between the signal $x(t)$ and a version of signal $y(t)$ shifted to the left by τ seconds. Since this integral expression yields a numerical value for each value of the shift variable τ, it is possible to think of $\phi_{xy}(\tau)$ as being a signal whose independent variable is τ.

The Laplace transform of the correlation function $\phi_{xy}(\tau)$ is obtained by using the basic transform definition. This results in

$$\Phi_{xy}(s) = \int_{-\infty}^{\infty} \phi_{xy}(\tau)e^{-s\tau}\,d\tau$$

where it is noted that the variable of integration is now τ instead of the more standard variable t. Let us now substitute the operational expression (5.31) for $\phi_{xy}(\tau)$

into this integral transform relationship, so that

$$\Phi_{xy}(s) = \int_{-\infty}^{\infty} \left[\int_{-\infty}^{\infty} x(t)y(t+\tau)\, dt \right] e^{-s\tau}\, d\tau$$

To evaluate this two-variable integral, it is beneficial to first interchange the order of integration and then make the change of integration variables $\alpha = t + \tau$ (with t being considered fixed) to eventually obtain

$$\Phi_{xy}(s) = \int_{-\infty}^{\infty} x(t)e^{st}\, dt \int_{-\infty}^{\infty} y(\alpha)e^{-s\alpha}\, d\alpha \tag{5.32}$$

This procedure has then resulted in the equating of a double variable integral with that of the product of two single-variable integrals.

The Laplace transform of the autocorrelation signal is then given by expression (5.32); it exists only if the two involved integrals themselves exist, that is, have finite value. The integral on the right is immediately recognized as being the Laplace transform of the signal $y(t)$, i.e., $Y(s)$, which has a region of absolute convergence of the form

$$\sigma_{y+} < \text{Re}\,(s) < \sigma_{y-} \tag{5.33}$$

On the other hand, a simple change of integration variable $\beta = -t$ indicates that the integral on the left is equal to the Laplace transform of the transposed signal $x(-t)$. Using the scaling property of the Laplace transform with $a = -1$, this integral is then given by $X(-s)$ and its region of absolute convergence is specified by $\sigma_{x+} < \text{Re}\,(-s) < \sigma_{x-}$ or

$$-\sigma_{x-} < \text{Re}\,(s) < -\sigma_{x+} \tag{5.34}$$

where σ_{x-} and σ_{x+} denote the parameters that characterize the region of convergence of the original signal $x(t)$. The transform function $\Phi_{xy}(s)$ therefore equals the product $X(-s)Y(s)$ for those values of s that simultaneously lie in the two regions (5.33) and (5.34). We have therefore proved the following important correlation Laplace transform property

$$\boxed{\Phi_{xy}(s) = X(-s)Y(s)} \tag{5.35a}$$

in which the region of absolute convergence is given by

$$\max\,(-\sigma_{x-}, \sigma_{y+}) < \text{Re}\,(s) < \min\,(-\sigma_{x+}, \sigma_{y-}) \tag{5.35b}$$

Although the operator relationship (5.31) gives the direct procedure for calculating the cross-correlation function that corresponds to two signals, it is often more expedient to use the property noted above to achieve the same objective, as the next example illustrates.

Example 5.9

Determine the cross-correlation function that corresponds to the two signals

$$x(t) = u(t) - u(t-1), \qquad y(t) = u(t) - u(t-2)$$

using transform property 5.35a. The Laplace transforms of these two pulselike signals are found to be

$$X(s) = \frac{1 - e^{-s}}{s} \quad \text{and} \quad Y(s) = \frac{1 - e^{-2s}}{s}$$

and each has a region of absolute convergence consisting of the entire s-plane, that is, $\sigma_+ = -\infty$, $\sigma_- = +\infty$. The Laplace transform of the cross-correlation function in accordance with property (5.35a) is given by

$$\Phi_{xy}(s) = \left(\frac{1 - e^{s}}{-s}\right)\left(\frac{1 - e^{-2s}}{s}\right)$$

$$= -\frac{1}{s^2} + \frac{1}{s^2} e^{-2s} + \frac{1}{s^2} e^{s} - \frac{1}{s^2} e^{-s}$$

Using the time-shifting property of the Laplace transform, it is noted that $\mathscr{L}[(t - t_0)u(t - t_0)] = e^{-t_0 s}/s^2$ and therefore the required correlation function must be given by

$$\phi_{xy}(t) = -tu(t) + (t - 2)u(t - 2) + (t + 1)u(t + 1) - (t - 1)u(t - 1)$$

as can be readily demonstrated by taking the Laplace transform of this postulated correlation function.

The correctness of this cross-correlation function may be verified by directly inserting the given signals into the defining expression (5.31). In using this direct approach, it is found that a great deal of extra computational effort is involved in arriving at the same result. This serves as another illustration of the desirability of using the Laplace transform for analyzing signals as well as standard signal operators.

Autocorrelation Function

An important special case of the cross-correlation function arises when the two signals being correlated are equal to each other, that is, $y(t) = x(t)$. This situation occurs so frequently in relevant applications that the special notation of *autocorrelation* is reserved for it. The autocorrelation function of the signal $x(t)$ is formally defined by

$$\phi_{xx}(\tau) = \int_{-\infty}^{\infty} x(t)x(t + \tau)\, d\tau$$

Using the fact that $y(t) = x(t)$ in the correlation expression, it immediately follows from property (5.35a) that the Laplace transform of the autocorrelation function is specified by

$$\Phi_{xx}(s) = X(-s)X(s)$$

and the corresponding region of absolute convergence is

$$\max\,(-\sigma_{x^-}, \sigma_{x^+}) < \text{Re}\,(s) < \min\,(-\sigma_{x^+}, \sigma_{x^-})$$

It is not difficult to show that this region is nonempty only if the parameter σ_{x^+} is negative and the parameter σ_{x^-} is positive. Thus, the autocorrelation function $\phi_{xx}(t)$

has a Laplace transform only if the region of convergence of $X(s)$ includes the imaginary axis in the s-plane.

5.8 POLES AND ZEROS

In determining the Laplace transform of signals, it often happens that the resultant transform will be equal to a ratio of polynomials in the variable s. A transform that is so characterized is said to be a *rational function* of s and is expressible in the format

$$X(s) = \frac{b_m s^m + b_{m-1}s^{m-1} + \cdots + b_0}{a_n s^n + a_{n-1}s^{n-1} + \cdots + a_0} \tag{5.36}$$

A glance at Table 5.1 should give convincing evidence concerning the importance of rational Laplace transforms. Furthermore, it will be shown in the next chapter that any operator that is governed by a constant-coefficient linear differential equation has a rational transfer function. With these thoughts in mind, it is apparent that a characterization of rational Laplace transforms will play an essential role in our study of signals and operators.

When a Laplace transform is rational, it is possible to factor its constituent numerator and denominator polynomials using standard techniques. To demonstrate this concept, let us consider the rational transform (5.36), which can be factored as

$$X(s) = \frac{b_m}{a_n} \cdot \frac{(s - z_1)(s - z_2) \cdots (s - z_m)}{(s - p_1)(s - p_2) \cdots (s - p_n)} \tag{5.37}$$

where the parameters z_k and p_k denote the roots of the numerator and denominator polynomials, respectively. The elements z_k, which in part characterize this factorization, are referred to as the *zeros* of transform $X(s)$, for the obvious reason that $X(s)$ is itself zero when evaluated at $s = z_1, z_2, \ldots, z_n$. On the other hand, if $X(s)$ is evaluated as $s = p_k$, it is clear that an unbounded result occurs (i.e., a division by zero). The roots of the denominator polynomial $p_1, p_2, \ldots, p_n$ are then referred to as the *poles* or *singularities* of the transform $X(s)$.

The values of the zeros and poles that characterize a rational Laplace transform $X(s)$ characterize in turn the signal $x(t)$ that generated the transform. It is then advantageous to plot the location of these zeros and poles in the complex s-plane in order to emphasize this correspondence. The importance of the pole locations is, in part, made apparent by the following lemma, which is not proved at this time.

Lemma 5.1. If a signal $x(t)$ has a rational Laplace transform $X(s)$, then each boundary that separates a region of convergence from a region of divergence will contain at least one pole of $X(s)$. Furthermore, poles not lying on these boundaries will all be located in the interior of the region of divergence of $X(s)$.

This lemma yields a great deal of insight concerning the interrelationship between the time behavior of a signal and the pole locations of its associated Laplace transform. We make full use of this fact when discussing the concept of inverse Laplace transform in the next section.

Example 5.10

Make a pole-zero plot of the Laplace transform that corresponds to the signal

$$x(t) = -\frac{1}{2}e^{-t}u(t) + \frac{2}{3}e^{-2t}u(t) - \frac{5}{6}e^{t}u(-t)$$

It is readily shown that this signal has the Laplace transform

$$\begin{aligned} X(s) &= -\frac{1}{2}\left(\frac{1}{s+1}\right) + \frac{2}{3}\left(\frac{1}{s+2}\right) + \frac{5}{6}\left(\frac{1}{s-1}\right) \\ &= \frac{s^2+2s+2}{(s+1)(s+2)(s-1)} \\ &= \frac{(s+1+j)(s+1-j)}{(s+1)(s+2)(s-1)} \qquad \text{for } -1 < \text{Re}(s) < 1 \end{aligned}$$

The zero-pole map of this transform is shown in Fig. 5.6 where it is observed that the poles located at $s = -1$ and $s = +1$ lie on the boundaries of the region of convergence. In what is to follow, a pole location is always depicted by a boldface cross (**X**) and a zero location by a boldface circle (**O**).

The dynamic response of a system is directly dependent on its transfer function's pole and zero locations. The poles of a transfer function, for instance, govern the transient behavior of the system in response to either initial energy in the system

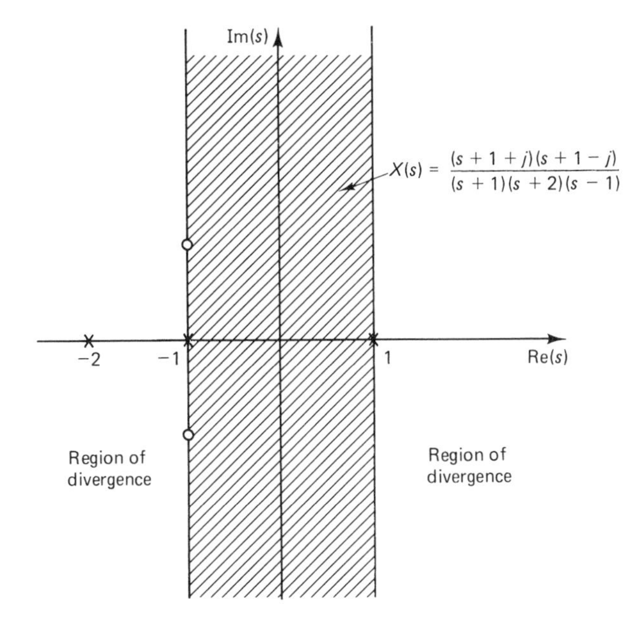

Figure 5.6 Pole-zero plot for Example 5.10.

or external excitation. There are certainly many engineering systems for which it is important to have a particular type of dynamic response. One area of great interest is in the design of feedback control systems. A system is said to contain feedback if the control input is dependent on the output measurements. Typically, the driving force for the system is determined by the difference between a desired (or reference) input and the actual (measured) output. The system is then structured to reduce this error signal to zero in time, thereby making the response conform to the desired input. As mentioned earlier, however, the speed of response to changing commands may be important for a given system design. The following example illustrates how feedback can alter the system poles and therefore the quality of its response.

Example 5.11

The idea of feedback control is employed so as to alter the system performance in order to achieve an improved response behavior. Typically, the poles of the system transfer function are relocated to provide a different transient behavior.

Consider the system in Fig. 5.7a. The voltage $x(t)$ can be considered the system input to drive a d.c. motor positioning system whose response (output) variable is taken to be the shaft angle displacement $\theta(t)$. With the armature current I_a maintained constant (by external circuitry), the mechanical torque $T(t)$ applied to the rotational load is given by

$$T(t) = Ki_f(t) \tag{5.38}$$

where K is a constant parameter of the motor and $i_f(t)$ is the field current. The various electrical and mechanical relationships are

1. $v_f(t) = K_a x(t)$ where K_a = constant amplifier gain
2. $v_f(t) = R_f i_f(t) + L_f(di_f(t)/dt)$ where R_f = field winding resistance and L_f = field winding inductance
3. $T(t) = Ki_f(t)$ where K = motor torque constant

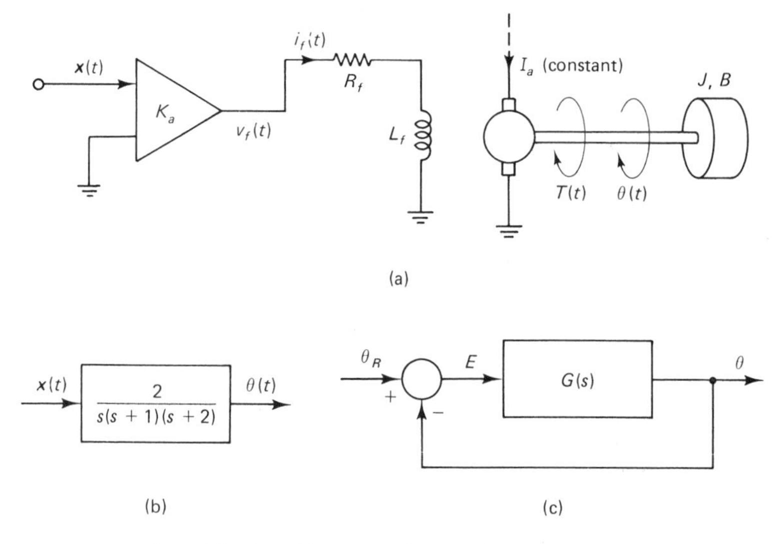

Figure 5.7 A position-control system for Example 5.11.

4. $T(t) = J\ddot{\theta}(t) + B\dot{\theta}(t)$ where J = load moment of inertia and B = coefficient of viscous friction

We now assume that the parameter values satisfy the relations

$$K_a K = 2JL_f, \quad J = B \quad \text{and} \quad R_f = 2L_f$$

Upon transforming the time-domain relationships above and eliminating the intermediate variables $V_f(s)$, $I_f(s)$, and $T(s)$, the transfer function for the system is found to be

$$G(s) = \frac{\theta(s)}{X(s)} = \frac{2}{s(s+1)(s+2)} \tag{5.39}$$

as shown in Fig. 5.7b. This system, unfortunately, does not have a good response for controlling the position θ due to the pole locations.

To improve the performance quality of this system, let us employ the concept of feedback. Namely, let us generate an error signal between a desired reference position θ_R (which might be an input at some remote location) and the actual shaft position θ so that $E(s) = \theta_R(s) - \theta(s)$. This configuration is illustrated in Fig. 5.7c and is referred to as a *unity feedback system.* The use of feedback permits one to achieve accurate control because of the direct comparison between the desired value and the current value of the controlled variable. However, for this example we want to focus on the change in the pole locations of the system caused by feedback.

From Fig. 5.7b we see that

$$\text{open-loop poles} = \{s = 0, \ -1, \ -2\} \tag{5.40}$$

And from expression (5.39) we can write that

$$\theta(s) = G(s)E(s) \tag{5.41}$$

and

$$E(s) = \theta_R(s) - \theta(s) \tag{5.42}$$

where $G(s)$ is the open-loop transfer function given in expression (5.39). Introducing (5.41) into (5.42) and collecting terms, we have the basic feedback relationship that

$$\frac{\theta(s)}{\theta_R(s)} = \frac{G(s)}{1 + G(s)} \tag{5.43}$$

Expression (5.43) is the *closed-loop transfer function*, and for our example

$$\frac{G(s)}{1 + G(s)} = \frac{2}{s^3 + 3s^2 + 2s + 2} \tag{5.44}$$

The closed-loop pole locations are the roots of the denominator polynomial in expression (5.44). Solving for these numerically, we find that the

$$\text{closed-loop poles} = \{s = -2.52138, \ -0.23931 \pm j0.85787\} \tag{5.45}$$

Without actually solving for a particular response, we can imagine that the closed-loop (step) response is quite different from the corresponding open-loop response.

Remark. Although it is not obvious, the closed-loop response matches the level of the input step in steady state; that is to say, the error $e(t)$ between θ_R and $\theta(t)$ goes to zero with time.

5.9 INVERSE LAPLACE TRANSFORM

The Laplace transform provides an extremely convenient vehicle for studying the dynamical behavior of linear systems governed by constant-coefficient linear differential equations. Since such equations are often used in modeling real-world phenomena, the significance of this observation cannot be overstated. Before we can put to full use the true value of the Laplace transform, however, it is necessary to provide a procedure for recovering the signal $x(t)$ that generates a given Laplace transform $X(s)$. Namely, given a transform function $X(s)$ and its region of convergence, how does one go about finding the signal $x(t)$ that generated that transform? This process is called *finding the inverse Laplace transform* and is symbolically denoted as

$$x(t) = \mathscr{L}^{-1}[X(s)]$$

There exists a procedure that is applicable to all classes of transform functions that involves the evaluation of a line integral in the complex s-plane. In particular, the signal $x(t)$ that generates the Laplace transform $X(s)$ can be recovered by means of the relationship

$$x(t) = \frac{1}{2\pi j} \int_{c-j\infty}^{c+j\infty} X(s)e^{st}\, ds \tag{5.46}$$

In this integral, the real number c is to be selected so that the complex number $c + j\omega$ lies entirely within the region of convergence of $X(s)$ for all values of the imaginary component ω. The evaluation of this inverse Laplace transform integral requires an understanding of complex variable theory. Fortunately, for the important class of rational Laplace transform functions, there exists an effective alternate procedure that does not necessitate directly evaluating this integral. This procedure is generally known as the *partial-fraction expansion method.*

Partial-Fraction Expansion Method

As just indicated, the partial-fraction expansion method provides a convenient technique for reacquiring the signal that generates a given rational Laplace transform. Recall that a transform function is said to be rational if it is expressible as a ratio of polynomial in s, that is,

$$X(s) = \frac{B(s)}{A(s)} = \frac{b_m s^m + b_{m-1}s^{m-1} + \cdots + b_1 s + b_0}{s^n + a_{n-1}s^{n-1} + \cdots + a_1 s + a_0} \tag{5.47}$$

The partial-fraction expansion method is based on the appealing notion of equivalently expressing this rational transform as a sum of n elementary transforms whose corresponding inverse Laplace transforms (i.e., generating signals) are readily found in standard Laplace transform pair tables. This method entails the simple five-step process as outlined in Table 5.3. A description of each of these steps and their implementation is now given.

TABLE 5.3 PARTIAL-FRACTION EXPANSION METHOD FOR DETERMINING THE INVERSE LAPLACE TRANSFORM

I. Put rational transform into proper form whereby the degree of the numerator polynomial is less than or equal to that of the denominator polynomial.
II. Factor the denominator polynomial.
III. Perform a partial-fraction expansion.
IV. Separate partial-fraction expansion terms into causal and anticausal components using the associated region of absolute convergence for this purpose.
V. Using a Laplace transform pair table, obtain the inverse Laplace transform.

I. Proper Form for Rational Transform. To accomplish a partial-fraction expansion of any rational function, it is first necessary to put that function into so-called proper form. The rational transform $X(s)$ as specified by (5.47) is said to be in proper form if the degree of its numerator polynomial $B(s)$ does not exceed the degree of its denominator polynomial $A(s)$ (i.e., $m \leq n$). If this is the case, then one may procede directly to Step II of the partial-fraction expansion method. On the other hand, in those very rare situations in which m exceeds n, it is necessary to first perform a long-hand division operation before continuing on with the partial-fraction expansion: one simply divides $A(s)$ into $B(s)$ until the remainder polynomial is less than or equal to n. This division process yields an expression in the proper form as given by

$$X(s) = \frac{B(s)}{A(s)}$$

$$= Q(s) + \frac{R(s)}{A(s)}$$

in which $Q(s)$ and $R(s)$ are the quotient and remainder polynomials, respectively, with the division made so that the degree of $R(s)$ is less than or equal to that of $A(s)$. Let us illustrate this procedure by means of a specific example.

Example 5.12

Put the Laplace transform function

$$X(s) = \frac{s^4 + 2s^3 + 2}{s^3 + 2s^2 - s - 2} \qquad \text{for } -1 < \text{Re}(s) < 1$$

into proper form. Clearly, this rational function is not now in proper form since $m > n$ (i.e., $4 > 3$). A simple division operation, however, rectifies this situation:

$$\begin{array}{r} s \qquad\qquad\qquad \\ s^3 + 2s^2 - s - 2 \;\big)\; s^4 + 2s^3 \qquad\qquad + 2 \\ \underline{s^4 + 2s^3 - s^2 - 2s \quad} \\ s^2 + 2s + 2 \end{array}$$

One division step is seen to result in the remainder polynomial $R(s) = s^2 + 2s + 2$

being of a degree (i.e., 2) less than or equal to that of $A(s) = s^3 + 2s^2 - s - 2$. A proper form for the partial-fraction expansion has thus been obtained and is given by

$$X(s) = s + \frac{s^2 + 2s + 2}{s^3 + 2s^2 - s - 2} \qquad \text{for } -1 < \text{Re}(s) < 1$$

In the great majority of practical applications of the Laplace transform, it is found that the rational transform $X(s)$ is inherently in proper form. In those rare situations in which this is not so, however, one may simply employ this division method to generate a proper form. Whatever the case may be, it is then necessary to carry out a partial-fraction expansion upon a proper-form rational function, that is, either $B(s)/A(s)$ or $R(s)/A(s)$. Therefore, without loss of generality, it is hereafter assumed that the rational Laplace transform function $X(s)$ to be inverted is in the proper format

$$X(s) = \frac{B(s)}{A(s)} = \frac{b_n s^n + b_{n-1}s^{n-1} + \cdots + b_0}{s^n + a_{n-1}s^{n-1} + \cdots + a_1} \tag{5.48}$$

where some of the leading b_k coefficients may be zero in any specific application (e.g., b_n).

II. Factorization of Denominator Polynomial. The next step of the partial-fraction expansion method entails the factorizing of the nth-order denominator polynomial $A(s)$ into a product of n first-order factors. This factorization is always possible and results in the equivalent representation of $A(s)$ as given by

$$A(s) = (s - p_1)(s - p_2) \cdots (s - p_n) \tag{5.49}$$

For moderately large values of n (for example, $n > 3$), one generally has to appeal to numerical algorithms for determining the p_k parameter values that characterize this representation of $A(s)$. The terms $p_1, p_2, \ldots, p_n$ constituting this factorization are called the *roots of polynomial* $A(s)$, or the *poles of* $X(s)$. These roots can be real or complex numbers, or a mixture of both. Furthermore, for the important special case in which the a_k coefficients that characterize polynomial $A(s)$ are all real, it is well-known that complex roots must occur in complex conjugate pairs. Thus, if p happens to be a complex root of $A(s)$, then so will its complex conjugate p^*, that is, if $A(p) = 0$, then $A(p^*) = 0$.

III. Partial-Fraction Expansion. With this factorization of the denominator polynomial accomplished, the rational Laplace transform $X(s)$ can be expressed as

$$X(s) = \frac{B(s)}{A(s)} = \frac{b_n s^n + b_{n-1}s^{n-1} + \cdots + b_0}{(s - p_1)(s - p_2) \cdots (s - p_n)} \tag{5.50}$$

We shall now *equivalently represent* this transform function as a linear combination of elementary transform functions. These elementary transform functions are selected on the basis that each has a readily identifiable inverse. In obtaining such a representation, let us consider separately those cases in which the roots of $A(s)$ are distinct (i.e., all different) from those in which some multiple roots appear.

Case 1: $A(s)$ Has Distinct Roots. When the denominator polynomial $A(s)$ has distinct roots, it is a well-known fact that the rational Laplace transform (5.50) can be equivalently represented by the partial-fraction expansion

$$X(s) = \frac{B(s)}{A(s)} = \alpha_0 + \frac{\alpha_1}{s - p_1} + \frac{\alpha_2}{s - p_2} + \cdots + \frac{\alpha_n}{s - p_n} \tag{5.51}$$

where the α_k are constants that identify the expansion and must be properly chosen for a valid representation. To demonstrate the validity of this partial-fraction expansion and at the same time give a method for evaluating the α_k parameters, let us multiply each side of expansion (5.51) by the polynomial $A(s)$ to give

$$B(s) = \alpha_0 A(s) + \frac{\alpha_1 A(s)}{s - p_1} + \frac{\alpha_2 A(s)}{s - p_2} + \cdots + \frac{\alpha_n A(s)}{s - p_n} \tag{5.52}$$

For the partial-fraction expansion to be valid, it must then be shown that values can be assigned to the α_k parameters so that relationship (5.52) holds. This is readily accomplished by first observing that each of the right-side terms

$$\frac{\alpha_k A(s)}{s - p_k} = \alpha_k(s - p_1)(s - p_2) \cdots (s - p_{k-1})(s - p_{k+1}) \cdots (s - p_n)$$

is in fact equal to an $(n - 1)^{\text{st}}$-degree polynomial due to the cancelation of the $(s - p_k)$ factor in $A(s)$. On the other hand, the term $\alpha_0 A(s)$ is a nth-degree polynomial. Thus, the right-hand side of relationship (5.52) is seen to be equal to an nth-degree polynomial whose coefficients are explicitly dependent on the α_k parameters.

With the information given above in mind, it is then necessary to assign values to the $n + 1$ parameters α_k so that the coefficients of the right-side polynomial (5.52) exactly agree with the coefficients of the left-side polynomial $B(s)$. Namely, the coefficients of the $n + 1$ entities $\{s^n, s^{n-1}, \ldots, s, 1\}$ on each side must be identical. To effect this equivalency of coefficients, it is necessary to solve a system of $n + 1$ linear equations (obtained from equating coefficients) in the $n + 1$ unknown parameters α_k. This always results in a *unique* assignment of values to the α_k parameters. Let us now illustrate this approach.

Example 5.13

Using the method of coefficient equality, make a partial-fraction expansion of the proper-form rational function

$$X_1(s) = \frac{s^2 + 2s - 2}{s^3 + 2s^2 - s - 2}$$

This expansion is begun by first factoring the denominator polynomial into a product of the first-order factors, that is, $s^3 + 2s^2 - s - 2 = (s + 1)(s + 2)(s - 1)$, and then seting up the expansion format, as follows:

$$\frac{s^2 + 2s - 2}{(s+1)(s+2)(s-1)}$$

$$= \alpha_0 + \frac{\alpha_1}{s+1} + \frac{\alpha_2}{s+2} + \frac{\alpha_3}{s-1}$$

$$= \frac{\alpha_0(s+1)(s+2)(s-1) + \alpha_1(s+2)(s-1) + \alpha_2(s+1)(s-1) + \alpha_3(s+1)(s+2)}{(s+1)(s+2)(s-1)}$$

$$= \frac{\alpha_0 s^3 + (2\alpha_0 + \alpha_1 + \alpha_2 + \alpha_3)s^2 + (-\alpha_0 + \alpha_1 + 3\alpha_3)s + (-2\alpha_0 - 2\alpha_1 - \alpha_2 + 2\alpha_3)}{(s+1)(s+2)(s-1)}$$

In order that this expansion be valid for all s, it is necessary to equate the coefficients of the right- and left-side numerator polynomials. This results in the following system of four linear equations in the four α_k unknowns:

$$\begin{aligned}
s^3:&\quad \alpha_0 = 0\\
s^2:&\quad 2\alpha_0 + \alpha_1 + \alpha_2 + \alpha_3 = 1\\
s:&\quad -\alpha_0 + \alpha_1 + 3\alpha_3 = 2\\
1:&\quad -2\alpha_0 - 2\alpha_1 - \alpha_2 + 2\alpha_3 = -2
\end{aligned}$$

It is found that the unique solution to this system of equations is given by $\alpha_0 = 0$, $\alpha_1 = \frac{3}{2}$, $\alpha_2 = -\frac{2}{3}$, $\alpha_3 = \frac{1}{6}$, so that the required partial-fraction expansion is

$$\frac{s^2 + 2s - 2}{s^3 + 2s^2 - s - 2} = \frac{3}{2}\left(\frac{1}{s+1}\right) - \frac{2}{3}\left(\frac{1}{s+2}\right) + \frac{1}{6}\left(\frac{1}{s-1}\right)$$

To check for possible numerical errors, it is always good practice to evaluate this expansion at a convenient value of s to see if the left and right sides agree. If they do not, an error has been made. For example, if we set $s = 0$ in this expansion, it is found that the left and right sides each equal one, which give us some degree of confidence that the α_k coefficients have been properly evaluated.

Although one can always use the above method of equating coefficients for determining the appropriate values of the α_k parameters, it becomes exceedingly unwieldly even for moderate values of n. Fortunately, there exists a very elementary alternate procedure that achieves the same objective in a much simpler fashion. We now demonstrate this procedure by evaluating the parameter α_1. This is begun by first multiplying each side of the partial-fraction expansion (5.51) by the first-order factor $(s - p_1)$ to yield

$$(s - p_1)X(s) = \alpha_0(s - p_1) + \alpha_1 + \alpha_2\frac{(s - p_1)}{(s - p_2)} + \cdots + \alpha_n\frac{(s - p_1)}{(s - p_n)}$$

It is now observed that if s is set equal to p_1, each term on the right side goes to zero except for the α_1 term. Thus, the desired value for α_1 is given by

$$\alpha_1 = (s - p_1)X(s)\Big|_{s = p_1}$$

In performing the evaluation of the right side, it is noted that the term $(s - p_1)$ is cancelled by the $(s - p_1)$ factor of the denominator polynomial of $X(s)$. Using this approach, the remaining parameters may be similarly obtained and are given by

$$\alpha_k = (s - p_k)X(s)\Big|_{s=p_k} \quad \text{for } k = 1, 2, \ldots, n \tag{5.53}$$

and

$$\alpha_0 = b_n$$

The expression for parameter α_0 is obtained by letting s become unbounded (i.e., $s = +\infty$) in expansion (5.51). This results in each right-side term being zero except α_0, whereas from relationship (5.48) it is apparent that $X(s)$ approaches b_n as s becomes bounded.

Example 5.14

Perform a partial-fraction expansion of the transform $X_1(s)$ considered in Example 5.13 using relationship (5.53). The required partial-fraction expansion is of the form

$$X_1(s) = \frac{s^2 + 2s - 2}{(s + 1)(s + 2)(s - 1)}$$

$$= \alpha_0 + \frac{\alpha_1}{s + 1} + \frac{\alpha_2}{s + 2} + \frac{\alpha_3}{s - 1}$$

From relationship (5.53) for $k = 1$, we have

$$\alpha_1 = (s + 1)X_1(s)\Big|_{s=-1}$$

$$= \frac{s^2 + 2s - 2}{(s + 2)(s - 1)}\Big|_{s=-1} = \frac{3}{2}$$

In a similar fashion, it is found that $\alpha_2 = -\frac{2}{3}$, $\alpha_3 = \frac{1}{6}$, and $\alpha_0 = b_3 = 0$. This is in exact agreement with the results of Example 5.13, but a great deal less effort is required in attaining these coefficient evaluations when using expression (5.53).

Case 2: $A(s)$ Has Multiple Roots. When the denominator polynomial $A(s)$ contains multiple roots, an additional complication arises when making a partial-fraction expansion. Fortunately, this case may be easily handled and results in only a moderate amount of additional numerical effort relative to its distinct root case counterpart. To illustrate the procedure to be followed in the multiple root case, let us consider the situation where the root p_1 has a multiplicity of q, that is,

$$X(s) = \frac{B(s)}{A(s)} = \frac{B(s)}{(s - p_1)^q A_1(s)}$$

The appropriate partial-fraction expansion of this rational function is then given by

$$X(s) = \alpha_0 + \frac{\alpha_1}{s - p_1} + \frac{\alpha_2}{(s - p_1)^2} + \cdots + \frac{\alpha_q}{(s - p_1)^q} + n - q \text{ other elementary terms due to the roots of } A_1(s) \tag{5.54}$$

Thus, the root p_1 is seen to give rise to q elementary terms in the expansion (a most noteworthy observation).

The root p_1 of multiplicity q has associated q coefficients $\alpha_1, \alpha_2, \ldots, \alpha_q$, which must be properly evaluated. With the exception of the coefficient α_q, this involves a moderate degree of computational effort, as is shortly made clear. The coefficient α_q is evaluated by first multiplying each side of expansion (5.54) by $(s - p_1)^q$. After properly cancelling the $s - p_1$ factors on each side, we set s equal to p_1 and observe that all terms on the right side go to zero except for α_q. Thus, the α_q coefficient is given by the convenient expression

$$\begin{aligned} \alpha_q &= (s - p_1)^q X(s)\Big|_{s=p_1} \\ &= \frac{B(p_1)}{A_1(p_1)} \end{aligned} \tag{5.55}$$

Unfortunately, the remaining coefficients $\alpha_1, \alpha_2, \ldots, \alpha_{q-1}$ associated with the multiple root p_1 may not be so expendiently evaluated. A relatively efficient procedure for their evaluation is given in the example below.

If the denominator polynomial $A(s)$ contains additional multiple roots, they would be treated in an identical fashion as p_1. Namely, each root of a given multiplicity gives rise to an equivalent number of elementary terms in the expansion as per expression (5.54). The coefficient that corresponds to the elementary multiple term of highest degree can be simply evaluated using relationship (5.55). After appropriately evaluating the coefficients corresponding to all distinct roots and the highest multiple-root coefficients using relationships (5.53) and (5.55), respectively, the remaining additional coefficients need to be determined (e.g., $\alpha_1, \alpha_2, \ldots, \alpha_{q-1}$). Let there be r such unassigned coefficients. The appropriate values for these unassigned coefficients are most efficiently obtained by letting s take on any r distinct values—other than being equal to a root of $A(s)$. Since this expansion must hold for all values of s, this procedure yields a system of r linear equations in the r unassigned coefficients. One simply then solves this system of equations to determine the proper values for the unassigned coefficients. As a final comment, it is often possible to judiciously select the r values of s in order to expedite this solution procedure.

Example 5.15

Make a partial-fraction expansion of the proper-form rational transform

$$X(s) = \frac{s^3 + 8s^2 + 12s + 3}{(s+1)^3(s+2)}$$

This transform is seen to have a simple pole at $s = -2$ and a triple pole at $s = -1$, so that its partial-fraction expansion is of the form

$$\frac{s^3 + 8s^2 + 12s + 3}{(s+1)^3(s+2)} = \alpha_0 + \frac{\alpha_1}{s+1} + \frac{\alpha_2}{(s+1)^2} + \frac{\alpha_3}{(s+1)^3} + \frac{\alpha_4}{s+4} \tag{5.56}$$

To determine the proper values for the coefficient α_4, one simply multiplies each side of this expression by $(s + 2)$ and then lets $s = -2$, that is,

$$\alpha_4 = (s+2)X(s)\Big|_{s=-2} = \frac{(-2)^3 + 8(-2)^2 + 12(-2) + 3}{(-2+1)^3} = -3$$

Similarly, coefficient α_3 is obtained by multiplying each side of the partial-fraction expansion by $(s + 1)^3$ and then letting $s = -1$:

$$\alpha_3 = (s+1)^3 X(s)\Big|_{s=-1} = \frac{(-1)^3 + 8(-1)^2 + 12(-1) + 3}{(-1+2)} = -2$$

The coefficient α_0 may be expediently evaluated by letting s approach infinity, whereby each term on the right side goes to zero except α_0. Thus,

$$\alpha_0 = \lim_{s \to +\infty} X(s) = 0$$

The first stage of the partial-fraction expansion method is thus completed and results in

$$\frac{s^3 + 8s^2 + 12s + 3}{(s+1)^3(s+2)} = \frac{\alpha_1}{s+1} + \frac{\alpha_2}{(s+1)^2} - \frac{2}{(s+1)^3} - \frac{3}{s+2} \tag{5.57}$$

As indicated above, there is actually no simple procedure for properly evaluating the remaining α_1 and α_2 coefficients. Perhaps the best of the known methods is to evaluate relationship (5.57) at two different values of s (other than $s = -1$ and $s = -2$) and solve the resultant set of two linear equations in the unknowns α_1 and α_2. For example, letting $s = 0$ and $s = 1$ is seen to give

$$s = 0: \quad \frac{3}{2} = \alpha_1 + \alpha_2 - 2 - \frac{3}{2}$$

$$s = 1: \quad \frac{24}{24} = \alpha_1\left(\frac{1}{2}\right) + \alpha_2\left(\frac{1}{4}\right) - \frac{2}{8} - 3\,\frac{1}{3}$$

which has the solution $\alpha_1 = 4$ and $\alpha_2 = 1$. The required partial-fraction expansion is completed and results in

$$\frac{s^3 + 8s^2 + 12s + 3}{(s+1)^3(s+2)} = \frac{4}{s+1} + \frac{1}{(s+1)^2} - \frac{2}{(s+1)^3} - \frac{3}{s+2}$$

It should be apparent that the possibility of making a numerical error in this multistage evaluation is not negligible. To guard against this, it is always advisable to perform a simple check by evaluating this expansion equality at any value of s other than those already used (i.e., $s = 0, 1, -1, -2$). If the expansion is correct, then this evaluation should result in equality between the right and left sides. For example, at $s = 2$, it is found that each side equals $\frac{67}{108}$, and no numerical error has been detected.

IV. Causal and Anticausal Components. Up to this point, we have expressed a rational transform function as a linear combination of elementary transform functions. Each of these elementary functions can be the Laplace transform of either a purely causal or anticausal time signal as was pointed out in Section 5.4.[1]

[1]As an example, the elementary transform function $1/(s + a)$ can be generated by either the causal signal $e^{-at}u(t)$ or the anticausal signal $-e^{-at}u(-t)$.

To properly categorize each elementary transform according to its causality characteristic, one must take into account the region of absolute convergence that governs the overall transform function $X(s)$ whose inverse Laplace transform we are seeking. This region is of the general form

$$\sigma_+ < \text{Re}(s) < \sigma_- \tag{5.58}$$

From this information, one is immediately able to categorize each of the elementary transform functions by investigating the real components of each of its poles. If any elementary transform function has a pole whose real component is less than or equal to σ_+ as given by expression (5.58), then that elementary transform function must correspond to a causal time signal. This must be so since, if that elementary transform function is generated by an anticausal time signal, then its region of absolute convergence would have no points in common with interval (5.58), and a contradiction would arise. Similarly, all elementary transform functions whose poles have real parts greater than or equal to σ_- must correspond to anticausal time signals. With this in mind, the following lemma is offered.

Lemma 5.2. In a partial-fraction expansion of a rational Laplace transform $X(s)$ whose region of absolute convergence is given by

$$\sigma_+ < \text{Re}(s) < \sigma_-$$

it is possible to decompose the expansion's elementary transform functions into causal and anticausal functions (and possibly impulse-generated terms). Any elementary function is interpreted as being (1) *causal* if the real component of its pole is less or equal to σ_+, and as being (2) *anticausal* if the real component of its pole is greater than or equal to σ_-.

In performing this decomposition, the rational function being inverse transformed may always be expressed as

$$X(s) = Q(s) + X_+(s) + X_-(s) \tag{5.59}$$

where the entities $X_+(s)$ and $X_-(s)$ correspond to the causal and anticausal components, respectively, whereas $Q(s)$ is a polynomial that arises from putting $X(s)$ into proper form. It is important to note that the poles of $X_+(s)$ fall within the left half plane defined by $\text{Re}(s) \leq \sigma_+$, while those of $X_-(s)$ fall within the right half plane defined by $\text{Re}(s) \geq \sigma_-$. Furthermore, there always exists at least one pole that falls on the boundaries of these half planes as long as σ_+ and σ_- are finite. This allocation of causal and anticausal poles is depicted in Fig. 5.8. With the information given above in mind, we offer the following lemma:

Lemma 5.3. The poles of the rational transform that lie to the left (right) of the associated region of absolute convergence correspond to the causal (anticausal) component of that transform.

The above procedure for properly assigning the elementary functions to either

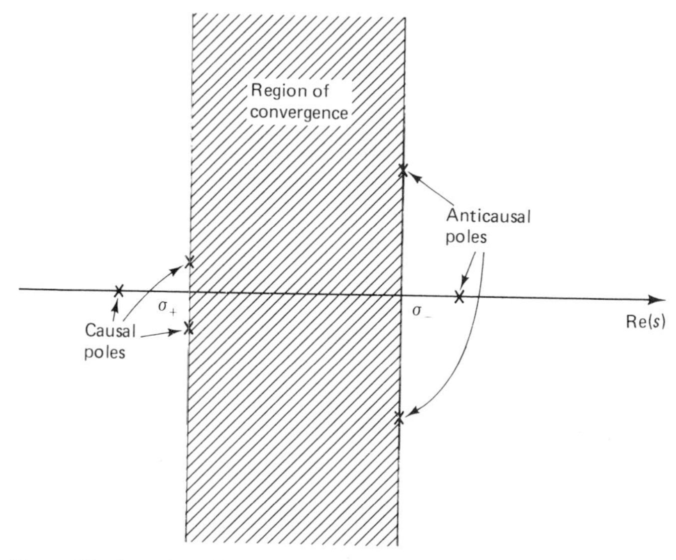

Figure 5.8 Location of causal and anticausal poles of a rational transform.

the causal or anticausal component of the expansion is indeed straightforward. In many practical applications (but certainly not all), the overall transform function being inverted is completely causal in nature and therefore has a region of absolute convergence given by $\sigma_+ < \text{Re}\,(s) < +\infty$. For such cases, one may automatically assign all elementary functions to the causal component. To illustrate the procedure to be followed when the transform is not purely causal in nature, let us consider the following example.

Example 5.16

Make a partial-fraction expansion of the following Laplace transform and decompose this expansion into its causal, anticausal, and impulse-generated functions.

$$X(s) = \frac{s^4 + 2s^3 + 2}{s^3 + 2s^2 - s - 2} \qquad -1 < \text{Re}\,(s) < 1$$

This transform function is observed to contain causal and anticausal components due to the nature of its region of convergence. To begin the expansion process, one first puts this transform into proper form by using long-hand division, to give

$$X(s) = s + \frac{s^2 + 2s + 2}{s^3 + 2s^2 - s - 2} \qquad -1 < \text{Re}\,(s) < 1$$

The partial-fraction expansion of the second term on the right side has already been done in Example 5.13 and resulted in

$$X(s) = s + \underbrace{\left\{\frac{3}{2}\left(\frac{1}{s+1}\right) - \frac{2}{3}\left(\frac{1}{s+2}\right)\right\}}_{\substack{\text{causal component} \\ \text{Re}\,(s) > -1}} + \underbrace{\left\{\frac{1}{6}\left(\frac{1}{s-1}\right)\right\}}_{\substack{\text{anticausal component} \\ \text{Re}\,(s) < 1}}$$

The individual elementary functions of this expansion have been decomposed using Lemma 5.2. As we will shortly see, the term s that resulted from the long-division process corresponds to an impulse signal.

V. Table Look-Up of Inverse Laplace Transform. The transform function to be inversed is now expressed as a linear combination of elementary functions. Furthermore, these elementary functions have been categorized into either causal or anticausal functions. To complete the inverse Laplace transform procedure, one need simply refer to a standard Laplace transform function table to determine the time signals that generate each of the elementary transform functions. The required time signal is then equal to the same linear combination of the inverse Laplace transforms of these elementary transform functions. We are, of course, appealing to the linearity property possessed by the Laplace transform when carrying out this inversion routine.

Example 5.17

Determine the inverse Laplace transform of the function

$$X(s) = \frac{s^4 + 2s^3 + 2}{s^3 + 2s^2 - s - 2} \qquad -1 < \text{Re}(s) < 1$$

From the results of Example 5.16, the required Laplace transform is given by

$$X(s) = s + \left\{\frac{3}{2}\left(\frac{1}{s+1}\right) - \frac{2}{3}\left(\frac{1}{s+2}\right)\right\}_+ + \left\{\frac{1}{6}\left(\frac{1}{s-1}\right)\right\}_-$$

where we have chosen to express the causal and anticausal components of $X(s)$ by the notations $+$ and $-$, respectively, at the lower right-hand corner of the brackets. According to entries 1, 3, and 7 of Table 5.1, the required time signal is

$$x(t) = \frac{d\,\delta(t)}{dt} + \frac{3}{2}e^{-t}u(t) - \frac{2}{3}e^{-2t}u(t) - \frac{1}{6}e^{t}u(-t)$$

5.10 RATIONAL AND STABLE SIGNALS

In much of what is to follow, we are concerned with studying the analytical structure possessed by rational signals. A signal is said to be rational if its corresponding Laplace transform is a rational function of s. From the previous section, it was shown that such signals always consist of a linear combination of elementary exponential signals and perhaps impulse-type signals. It is possible to further characterize these signals by investigating their amplitude behavior as time approaches either plus or minus infinity. In particular, we are concerned with characterizing the salient features of the Laplace transform of rational and stable time signals.

Definition 5.2. A signal is said to be stable if it decays to zero as time approaches both plus and minus infinity.

If a signal has a rational Laplace transform $X(s)$, it is possible to employ the partial-fraction expansion inversion method to recover the generating signal $x(t)$. In this method, it is necessary to decompose the transform $X(s)$ into its natural causal and anticausal components by using the associated region of absolute convergence as given by

$$\sigma_+ < \text{Re}\,(s) < \sigma_- \tag{5.60}$$

The causal component is obtained by selecting all of those elementary expansion terms whose poles have real parts that are less than or equal to σ_+. Similarly, the anticausal component consists of those elementary expansion terms whose poles have real parts that are greater than or equal to σ_-. This decomposition may always be expressed as

$$X(s) = X_+(s) + X_-(s) + Q(s) \tag{5.61}$$

where $X_+(s)$ and $X_-(s)$ correspond to the causal and anticausal components, respectively, and $Q(s)$ is a polynomial in s.[2]

We now investigate the causal component $X_+(s)$ and, more importantly, its associated inverse Laplace transform $x_+(t) = \mathscr{L}[X_+(s)]$ for the requirement that $x_+(t)$ approach zero as t approaches plus infinity.[3] In the partial-fraction expansion of $X(s)$ which led to decomposition (5.61), a typical term of the causal component $X_+(s)$ is given by

$$\frac{1}{(s-p)^{k+1}} \qquad \text{Re}\,(p) \leq \sigma_+$$

where k is a positive integer. Since this is the transform of a causal time signal, it follows that its corresponding inverse Laplace transform is specified by (see entry 2 of Table 5.1).

$$\frac{t^k}{k!}\, e^{pt}u(t)$$

If this time signal is to approach zero as t approaches plus infinity, as is required of a stable signal, it follows that the pole p must have a negative real component. Otherwise, the entity $t^k e^{pt}$ becomes unbounded as t approaches plus infinity. Since this observation must hold for each of the elementary terms constituting $X_+(s)$ if $x_+(t)$ is to be stable, we then conclude that every pole of $X_+(s)$ must have a negative real part. This will be true if and only if the parameter σ_+ that characterizes the causal region of absolute convergence is itself negative. We have thereby proven the following lemma:

Lemma 5.4. The Laplace transform corresponding to a rational and causal time signal is stable if and only if its region of absolute convergence as characterized by $\sigma_+ < \text{Re}\,(s) < +\infty$ is such that $\sigma_+ < 0$.

[2]If $X(s)$ is not in proper form, the decomposition must be supplemented by an additive polynomial $Q(s)$.

[3]It is observed that $x_+(t)$ is identically zero for negative t so the stability condition $x_+(t) \to 0$ as $t \to -\infty$ is automatically satisfied.

Thus, a rational, stable, and causal signal is always characterized by a region of absolute convergence that contains the right half s-plane (i.e., Re $(s) > 0$).

In a parallel manner, one can also establish a similar characterization of the region of absolute convergence for a rational, stable, and anticausal signal. Such a signal always has convergence regions as given by $-\infty < \text{Re}(s) < \sigma_-$, in which σ_- is a positive number. This region is seen to contain the left half of the s-plane, that is, Re $(s) < 0$. From this causal and anticausal characterization, we have thereby established the following general theorem relative to rational and stable signals.

Theorem 5.1. A rational signal is stable if and only if its region of absolute convergence is of the form

$$\sigma_+ < \text{Re}(s) < \sigma_-$$

in which σ_+ is negative and σ_- is positive. This region is seen to contain the imaginary axis of the s-plane.

5.11 PROBLEMS

5.1. Evaluate the Laplace transform of the following signals using the defining relation of expression (5.1); in each case describe the region of absolute convergence.
(a) $x_a(t) = e^{-t}u(t)$
(b) $x_b(t) = e^{t}u(-t)$
(c) $x_c(t) = e^{-|t|}$

5.2. Determine exponential bounds for the causal (positive time) and noncausal (negative time) parts of the following signals (*Hint*: See Appendix 5):
(a) $x_a(t) = e^{-2|t|}\cos 10t$
(b) $x_b(t) = \begin{cases} (5 + e^{-t})e^{-2t} & \text{for } t \geq 0 \\ (2 + \sin t)e^{t} & \text{for } t < 0 \end{cases}$
(c) $x_c(t) = u(t)$

5.3. The Fourier transform is a special case of the two-sided Laplace transform for which $\sigma = 0$ (i.e., $s = j\omega$). The Fourier transform exists if the region of convergence in the s-plane includes $\sigma = 0$ (the $j\omega$-axis). If they exist, find the Fourier transforms of the signals of Problem 5.1.

5.4. Check entry number 5 of Table 5.1 by direct evaluation.

5.5. Calculate the Laplace transform of $x(t) = x_a(t) + x_b(t) + x_c(t)$ where the signals are taken from Problem 5.1.

5.6. Determine the Laplace transform and region of convergence for the following signals:
(a) $x_a(t) = u(t) - u(t-1)$
(b) $x_b(t) = tu(t) - 2(t-1)u(t-1)$
(c) $x_c(t) = 2\,\delta(t) + 3e^{-t}u(t) + 4e^{t}u(-t)$

5.7. Use the differentiation theorem to find the Laplace transform of the unit-doublet, $d\delta(t)/dt$.

5.8. If $\boldsymbol{S}$ represents a unit-time delay (time-shift operator), determine the Laplace transform of $\boldsymbol{S}^N x$ in terms of $X(s)$.

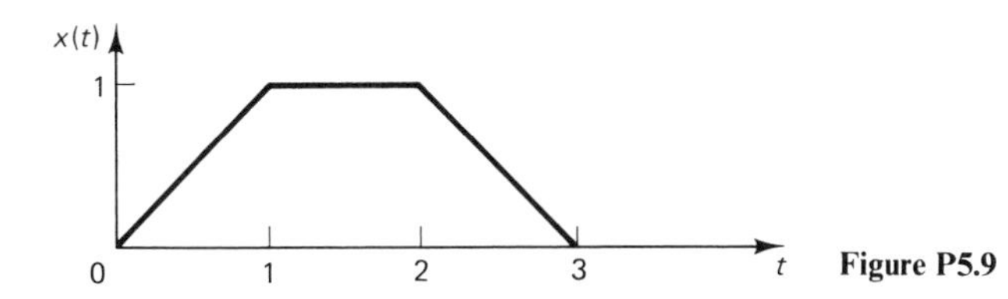

Figure P5.9

5.9. Develop the Laplace transform of the signal x in Fig. P5.9 by first reducing x to a sum of elementary signals.

5.10. Consider the single pulse $p(t) = u(t) - u(t - \tau)$ and its causal periodic replication as given by

$$p_T(t) = \sum_{n=0}^{\infty} p(t - nT) = p(t) * \sum_{n=0}^{\infty} \delta(t - nT)$$

where T is the period of $p_T(t)$.

(a) Sketch $p_T(t)$ for $T = \frac{1}{2}$ and $T = 2$.

(b) Develop the Laplace transform for $p_T(t)$. Put your result in closed form by using the fact that

$$\sum_{n=0}^{\infty} a^n = \frac{1}{1 - a} \qquad \text{for } |a| < 1$$

5.11. Calculate the transfer function of the system shown in Fig. P5.11 where D^{-1} is the integrator operator. Hint: First find the unit-impulse response.

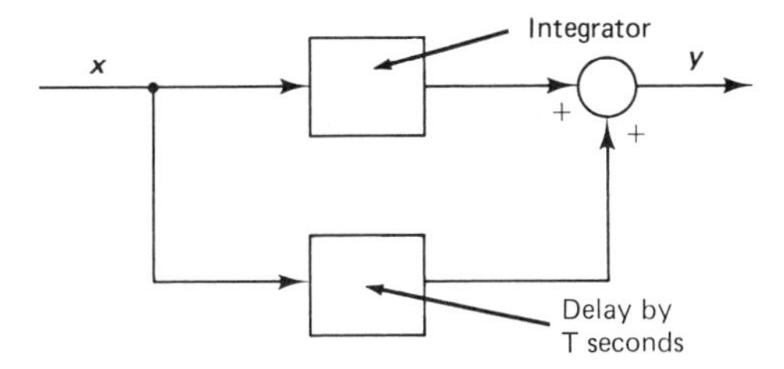

Figure P5.11

5.12. In digital control a clamp (zero-order hold) is often used to recover (approximately) a signal x from its samples. Figure P5.12 illustrates a diagram for such a device.

(a) Calculate the transfer function.

(b) Determine y when the signal x is

$$x(t) = \sum_{n=0}^{\infty} x(nT)\,\delta(t - nT)$$

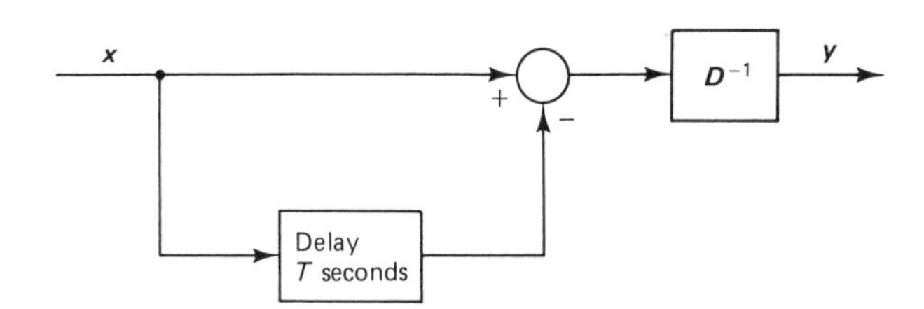

Figure P5.12

5.13. Find the Laplace transform of the autocorrelation functions of the signals x_a, x_b, and x_c of Problem 5.6.

5.14. Determine the Laplace transform of the cross-correlation function ϕ_{xy} where the signals x and y are taken from Fig. P5.11.

5.15. A Laplace transform is given by

$$X(s) = \frac{s^2 + 4s + 8}{(s + 1)(s + 2)(s - 1)}$$

(a) Draw the pole-zero plot for $X(s)$.
(b) Describe all possible regions of convergence for $X(s)$.

5.16. From the inverse transform integral (5.46), derive the inverse Fourier transform by making a change of variables from $s = \sigma + j\omega$ to $s = j\omega$ (with $c = 0$).

5.17. Find $x(t)$ from $X(s)$ in Problem 5.15 for each possible convergence region.

5.18. Given the causal transfer function

$$H(s) = \frac{s + 1}{s^2 + 5s + 6}$$

calculate the output $y(t)$ when the input signal is
(a) $x_a(t) = \cos 2t$
(b) $x_b(t) = e^{-t}u(t)$
(c) $x_c(t) = \cos 2\pi t\ \Delta(t/2)$
where $\Delta(t)$ is defined in Fig. P3.22.

5.19. Often a filter is described by its Fourier domain–transfer function $H(j\omega)$. Since $H(j\omega)$ is generally complex, it can be specified by its magnitude and phase (or angle) characteristics, that is,

$$H(j\omega) = |H(j\omega)| \underline{/H(j\omega)}$$

Figure P5.19 shows a low-pass filter (LPF) described in this manner. Analytically, this function is specified by

$$H(j\omega) = [u(\omega + \omega_c) - u(\omega - \omega_c)]e^{-j\omega\tau}$$

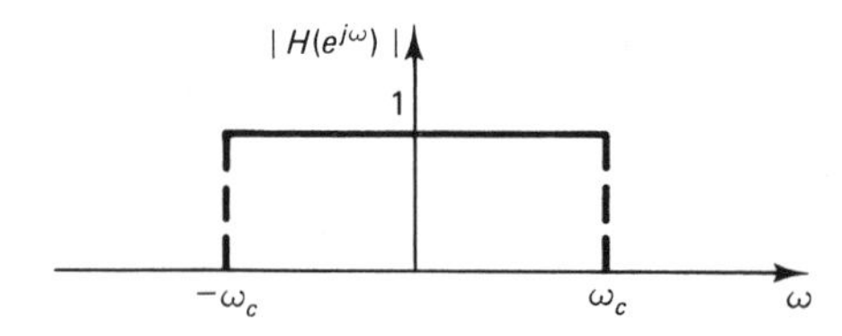

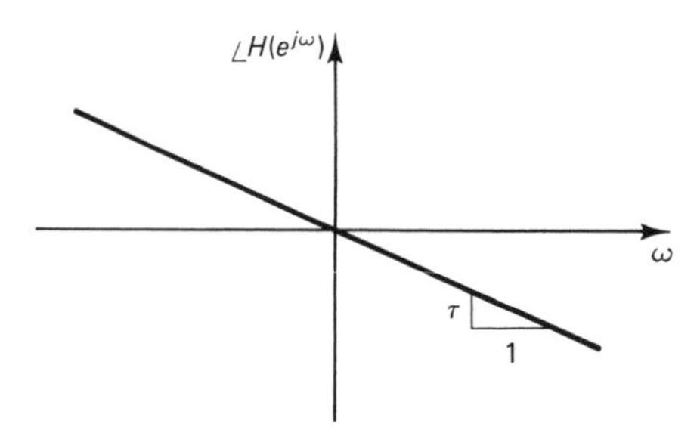

Figure P5.19

Calculate the response $y(t)$ to the input signal $x_a(t) = (10 + \cos 100t)$ when

(a) $\tau = 0, \omega_c = 50$

(b) $\tau = 0.01, \omega_c = 150$

5.20. Invert the following causal transforms:

(a) $X_a(s) = \dfrac{1}{s(s+1)}$

(b) $X_b(s) = \dfrac{s+7}{(s+2)(s+4)}$

(c) $X_c(s) = \dfrac{s}{(s+1)^2(s+2)}$

(d) $X_d(s) = \dfrac{e^{-s}}{s(s+1)}$

5.21. Determine the capacitor voltage $v(t)$ shown in the network of Fig. P5.21 for an initial capacitor voltage of 20 volts (V) and an initial inductor current of 2 amperes (A) with the given polarities.

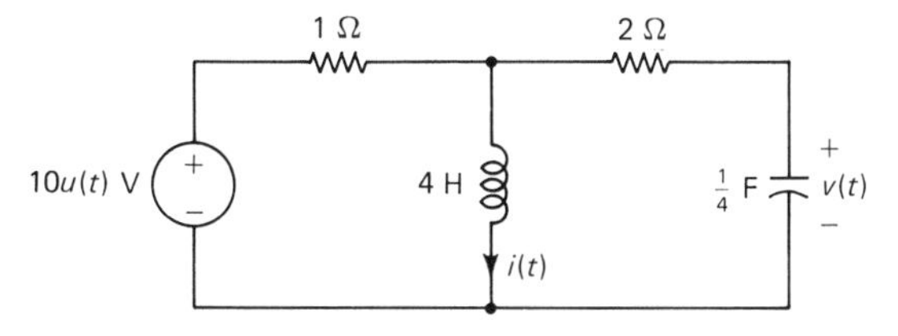

Figure P5.21

5.22. Find $v_0(t)$ in response to $v_{in}(t)$ in the network of Fig. P5.22 for the following conditions:

(a) $v_{in}(t) = \delta(t)$ for $v_0(0^-) = 0$

(b) $v_{in}(t) = 0$ for $v_0(0^-) = 10$

(c) $v_{in}(t) = 10u(t)$ for $v_0(0^-) = 10$

(d) $v_{in}(t) = e^{-10t}u(t)$ for $v_0(0^-) = 10$

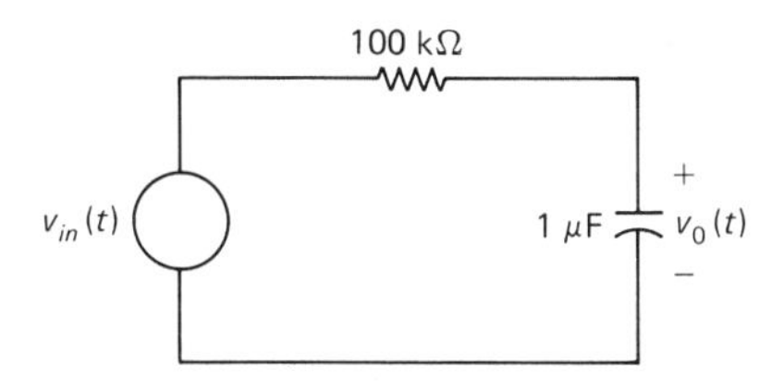

Figure P5.22

5.23. In the network of Fig. P5.23 the input current $i(t) = e^{-t} \cos 2t$. At $t = 0$ the switch is opened. Determine $v(t)$ for $t > 0$.

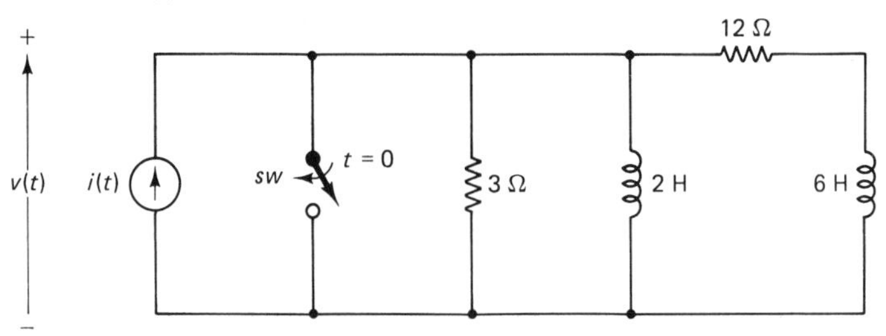

Figure P5.23

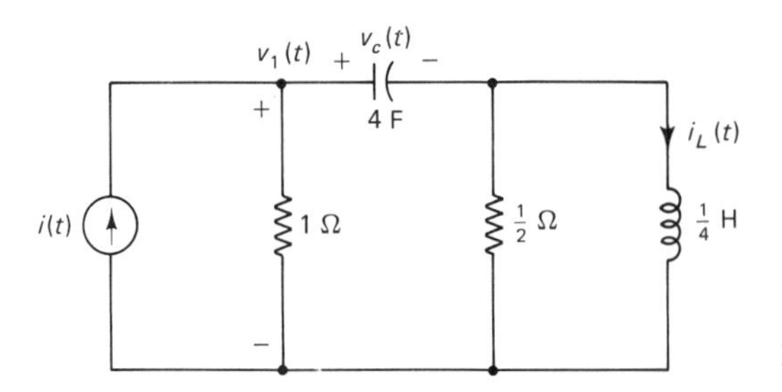

Figure P5.24

5.24. For the network of Fig. P5.24, the input current $i(t)$ is $10u(t)$ and the initial conditions are $v_c(0^-) = 2$ V and $i_L(0^-) = 20$ A.

(a) Write a set of time-domain model equations for the dependent variables $v_1(t)$ and $v_2(t)$.

(b) Transform the equation of part (a) into the s-domain including the effects of the initial conditions.

(c) Determine the transfer function between $i(t)$ and $v_2(t)$.

5.25. In the network of Fig. P5.25 there is the dependent source $e_2(t) = 3v(t)$. Solve for $v_0(t)$, where $t > 0$ if both capacitor voltages are initially zero and

(a) $e_1(t) = u(t)$

(b) $e_1(t) = \sin t$

(c) $e_1(t) = \sin \sqrt{5}\, t$

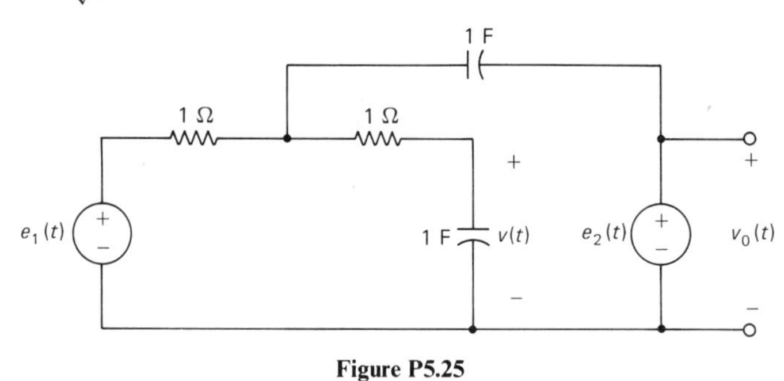

Figure P5.25

5.12 APPENDIX 5: REGION OF CONVERGENCE

To characterize the region of absolute convergence for a specific Laplace transform, let us first take the absolute value of each side of the defining integral transform (5.1). This is found to result in

$$|X(s)| \leq \int_0^{\infty} |x(t)|\, e^{-\sigma t}\, dt + \int_{-\infty}^{0} |x(t)|\, e^{-\sigma t}\, dt \tag{5A.1}$$

where use of the fact that $|e^{-st}| = |e^{-\sigma t}e^{-j\omega t}| = e^{-\sigma t}$ has been made. Furthermore, for reasons that will be now made apparent, we have elected to decompose the expression for $|X(s)|$ into integrals involving the causal and anticausal components of the signal $x(t)$ being transformed. The determination of the region of absolute convergence is seen to be equivalent to that of finding the set of σ (i.e., the real component of s) that causes each of the integrals in expression (5A.1) to be finite. It is important to note that the imaginary component of the complex variable s (i.e., ω) plays no role whatsoever in determining the Laplace transform's region of absolute convergence.

For any signal of practical interest, the integral of the integrand term $|x(t)|$ $e^{-\sigma t}$ in relationship (5A.1) over any finite-time interval is of bounded value. Thus, the boundedness of each of the integrals in expression (5A.1) is generally dependent on the time behavior of $|x(t)|e^{-\sigma t}$ as t approaches either plus or minus infinity. Specifically, it is readily established that integral expression (5A.1) will have a finite value if and only if the parameter σ is selected so as to simultaneously satisfy the following two convergence conditions:

$$\lim_{t \to +\infty} \{|x(t)|e^{-\sigma t}\} = 0 \tag{5A.2a}$$

and

$$\lim_{t \to -\infty} \{|x(t)|e^{-\sigma t}\} = 0 \tag{5A.2b}$$

The first of these conditions is seen to characterize the causal component of the signal being transformed, whereas the second is due to the anticausal component. Not surprisingly, it will be found that a different set of σ satisfies each of these convergence conditions. However, since each of these conditions must be simultaneously satisfied for $|X(s)|$ to be finite, we are really interested in the set of σ common to each of these two sets (i.e., their intersection). With this in mind, let us first characterize the sets of σ that satisfy conditions (5A.2a) and (5A.2b) taken separately and, then take the intersection of these two sets.

Region of Absolute Convergence—The Causal Component

The set of σ that satisfies causal-convergence condition (5A.2a) is readily found for the important class of *exponential bounded signals.* This class of signals is of importance from both a theoretical as well as a practical viewpoint. As a matter of fact, virtually all signals one will ever encounter will be so characterized. The signal $x(t)$ is said to be exponentially bounded for large positive values of time if there exists a real number σ_+, such that

$$|x(t)| \le c_1 e^{\sigma_+ t} \qquad \text{for all } t \text{ sufficiently large and positive} \tag{5A.3}$$

where c_1 is a constant. In this bounding, it is essential that we take σ_+ to be the smallest real number for which this inequality holds.

To determine the set of σ that satisfies causal-convergence condition (5A.2a), we next multiply each side of the exponential bounding inequality (5A.3) by the positive entity $e^{-\sigma t}$ to obtain

$$|x(t)|e^{-\sigma t} \le c_1 e^{(\sigma_+ - \sigma)t} \qquad \text{for all } t \text{ sufficiently large and positive}$$

According to the causal-convergence condition, it is desired to select σ to ensure that $|x(t)|e^{-\sigma t}$ decays to zero as t approaches plus infinity. From the inequality above, this is guaranteed as long as σ is taken to lie in the open interval

$$\sigma > \sigma_+ \tag{5A.4}$$

since in that case $c_1 e^{(\sigma_+ - \sigma)t} \to 0$ as t approaches infinity (i.e., the exponential argument term $\sigma_+ - \sigma$ is negative).

The region of convergence for the causal component of an exponentially bounded signal is then specified by interval (5A.4). In generating the lower bound parameter σ_+, it is important to remember that σ_+ is the *smallest* real number for which $|x(t)|$ is exponentially bounded according to relationship (5A.3). The subscript $+$ on σ_+ has been suggestively used to emphasize the fact that σ_+ specifies the

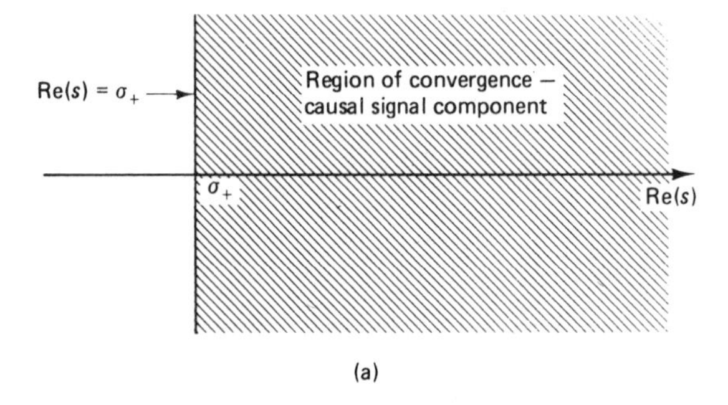

(a)

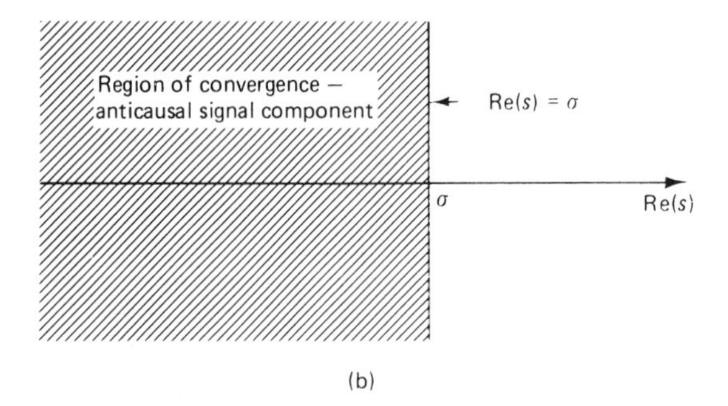

(b)

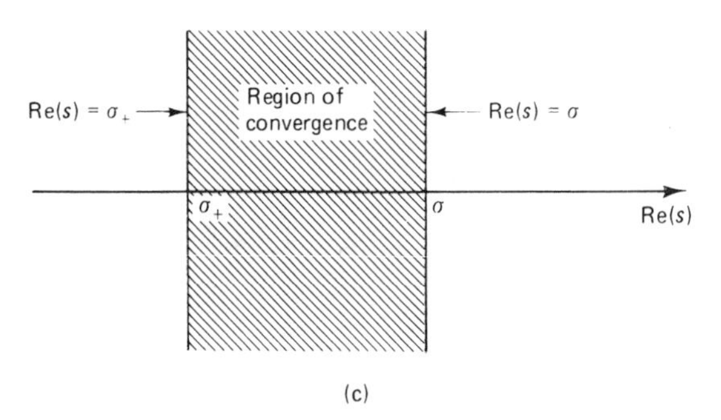

(c)

Figure 5A.1 Regions of absolute convergence for the (a) causal signal component of $x(t)$, (b) anticausal signal component of $x(t)$, and (c) signal $x(t)$.

convergence region corresponding to the causal component (i.e., positive time) of $x(t)$. The set of complex numbers s whose real components satisfy inequality (5A.4) is depicted in Fig. 5A.1a and is observed to be a right half plane with boundary at the vertical line defined by Re $(s) = \sigma_+$.

Region of Absolute Convergence— Anticausal Component

Proceeding in a manner similar to that taken above, the set of σ that satisfies the anticausal convergent condition (5A.2b) is found by first exponentially bounding the signal $|x(t)|$ for t sufficiently small negative (i.e., $t \to -\infty$). Thus, we seek the largest real number σ_- for which

$$|x(t)| \leq c_2 e^{\sigma_- t} \qquad \text{for all } t \text{ sufficiently small and negative} \tag{5A.5}$$

where c_2 is a constant. If it is possible to select such a bounding exponential, it follows that

$$|x(t)| e^{-\sigma t} \leq c_2 e^{(\sigma_- - \sigma)t} \qquad \text{for all } t \text{ sufficiently small and negative}$$

From this inequality it follows that the anticausal-convergence condition is satisfied when σ is selected to lie within the open interval

$$\sigma < \sigma_- \tag{5A.6}$$

since the bound $c_2 e^{(\sigma_- - \sigma)t}$ is seen to approach zero as t approaches negative infinity. The set of complex numbers s whose real components satisfy this inequality is displayed in Fig. 5A.1b and is observed to be a left half plane whose boundary is defined by the vertical line given by Re $(s) = \sigma_-$.

As previously indicated, the region of absolute convergence for the signal $x(t)$ is specified by that set of s that simultaneously satisfies both convergence conditions (5A.2a) and (5A.2b). For the class of exponentially bounded signals (in both the negative and positive time), this set is obviously equal to those real σ that are contained in each of the intervals (5A.4) and (5A.6). If the parameters σ_+ and σ_- that characterize the causal and anticausal components of $x(t)$ are such that $\sigma_+ < \sigma_-$, the region of absolute convergence for the Laplace transform is given by

$$\sigma_+ < \sigma < \sigma_- \tag{5A.8}$$

On the other hand, if $\sigma_+ > \sigma_-$, there exists no σ that simultaneously satisfies the two convergence conditions (5A.2a) and (5A.2b). In such cases, the Laplace transform fails to exist for any choice of s. The region of absolute convergence (5A.8) is depicted in Fig. 5A.1c and is seen to be a vertical strip with right- and left-hand boundaries being given by Re $(s) = \sigma_-$ and Re $(s) = \sigma_+$, respectively.

In view of the results given above, the following lemma is given relative to signals that are strictly causal or strictly anticausal in time behavior. The utility of this lemma becomes apparent when discussing the topic of the inverse Laplace transform.

Lemma 5A.1. The region of convergence for a signal that is strictly causal (i.e., $x(t) \equiv 0$ for $t < 0$) is of the form

$$\text{Re}(s) > \sigma_+$$

Similarly, the region of absolute convergence for an anticausal signal (i.e., $x(t) \equiv 0$ for $t > 0$) is of the form

$$\text{Re}(s) < \sigma_-$$

This lemma is readily proved by noting that the parameter σ_- (σ_+) that characterizes a strictly causal (anticausal) signal is plus (negative) infinity.

Example 5A.1

To demonstrate how a signal may be exponentially bounded, let us consider the specific signal

$$x(t) = \left[\left(\frac{1}{4}\right)e^{-t}\sin 2t - 101e^{-4t}\right]u(t) + \left(\frac{1}{5}\right)e^{2t}u(-t)$$

To determine the parameter σ_+, it is noted that for positive time

$$x(t) = \left[\frac{1}{4}\sin 2t - 101e^{-3t}\right]e^{-t} \qquad \text{for } t > 0$$

If we now take the absolute value of each side of this expression and then multiply that result by $e^{-\sigma t}$, there results

$$|x(t)|e^{-\sigma t} = \left|\frac{1}{4}\sin 2t - 101e^{-3t}\right|e^{-(\sigma+1)t} \qquad \text{for } t > 0$$

Clearly, as the time variable approaches plus infinity, the right side goes to zero only if the exponential argument $(\sigma + 1) > 0$. The region of absolute convergence for the causal component of $x(t)$ is thus given by

$$\sigma > -1$$

where the parameter σ_+ is then identified as being minus one.

The parameter σ_- is obtained similarly, that is,

$$|x(t)|e^{-\sigma t} = \left(\frac{1}{5}\right)e^{(2-\sigma)t} \qquad \text{for } t < 0$$

The only manner in which the right side can go to zero as t approaches $-\infty$ is for the exponential to be positive (i.e., $2 - \sigma > 0$). Thus, the region of absolute convergence for the anticausal component of signal $x(t)$ is

$$\sigma < 2$$

whereby the parameter σ_- is identified as plus two. The region of absolute convergence for the entire signal is thus

$$-1 < \sigma < 2$$

6

The z-Transform

There are two fundamental approaches to the representation of discrete-time signals and systems: (1) The *time-domain* approach involves manipulating terms of the input sequence and those of the difference equations describing the operation directly, and (2) an indirect method, called the *transform* or *frequency-domain* approach, involves the initial step of converting each known signal and signal operation into an equivalent form in the transform domain. The transform approach is limited to the analysis of stationary, linear discrete-time systems. This method is not unlike the use of logarithms to multiply numbers. Once the problem has been redefined in the transform domain, the indicated signal operations may be carried out with relative ease. These resultant transform-domain manipulations may then be translated back to obtain the desired time-domain characterization. The reader familiar with the analysis of continuous-time systems by Laplace transforms will recognize the benefits of the transform approach. In this chapter we develop the important relations that are useful in transforming discrete-time signals.

6.1 INTRODUCTION

Discrete-time signals, as we have seen, can be represented as sequences of numbers. Thus, if x is a discrete-time signal, its values can, in general, be indexed by n as follows:

$$x = \{\ldots, x(-2), x(-1), \underset{\uparrow}{x(0)}, x(1), x(2), \ldots, x(n), \ldots\} \tag{6.1}$$

The arrow denotes the element of the sequence x that corresponds to $n = 0$, which, in turn, may be thought of as the *time-axis origin*. Such signals, as we have seen, can originate from the sampling of a continuous-time signal or as simply a sequence of numbers that can be used to model signals internal to a digital computer. In the first instance we refer to the system as a sampled-data system, and the sequence (6.1) can perhaps be related back to a continuous $x(t)$ by indicating the time scaling of the samples, that is,

$$x = \{\ldots, x(-2T), x(-T), \underset{\uparrow}{x(0)}, x(T), x(2T), \ldots\} \tag{6.2}$$

Here again x is a sequence of numbers, but the effect of the sampling interval is explicitly indicated.

In order to work within a transform domain for discrete-time signals, we define the z-transform as follows.

Definition 6.1. The z-transform of the sequence x in (6.1) is

$$\mathscr{Z}\{x(n)\} = X(z) = \sum_{n=-\infty}^{\infty} x(n)z^{-n} \tag{6.3}$$

in which the variable z can be interpreted as being either a time-position marker or a complex-valued variable. If the former interpretation is employed, the number multiplying the marker z^{-n} is identified as being the nth element of the x sequence, (i.e., $x(n)$. In what is to follow, it will be generally beneficial to take z to be a complex-valued variable. The parameter T can also be included by applying the following definition to (6.2): namely

$$\mathscr{Z}\{x(nT)\} = \sum_{n=-\infty}^{\infty} x(nT)z^{-n} \tag{6.4}$$

This form is useful in studying the effect of varying T, as we will see later. In order for (6.3) or (6.4) to have meaning, the improper sum must converge to a (finite) limit in some region of the complex z-plane. This convergence may be difficult to determine in some cases, but to illustrate the idea, consider taking the z-transform of the unit-step sequence u.

Example 6.1

The z-transform of the unit-step sequence is, using (6.3),

$$\begin{aligned} U(z) &= \sum_{n=-\infty}^{\infty} u(n)z^{-n} \\ &= \sum_{n=0}^{\infty} 1 \cdot z^{-n} \\ &= \frac{1}{1-z^{-1}} \qquad \text{for } |z| > 1 \end{aligned} \tag{6.5}$$

The last equation in (6.5) is a result of utilizing the infinite geometric-sequence identity,

$$\sum_{n=0}^{\infty} a^n = \frac{1}{1-a} \qquad \text{for } |a| < 1 \tag{6.6}$$

To prove (6.6), consider

$$s = 1 + a + a^2 + \cdots + a^n + \cdots$$

then

$$as = a + a^2 + a^3 + \cdots + a^{n+1} + \cdots$$

Combining,

$$s - as = \lim_{n\to\infty} 1 - a^{n+1}$$

Therefore, for $|a| < 1$ so that $\lim_{n\to\infty} a^{n+1} = 0$,

$$(1 - a)s = 1$$

showing that

$$s = \frac{1}{1 - a} \qquad \text{for } |a| < 1$$

This relationship plays a key role in determining regions of convergence for z-transforms. For instance, the closed-form representation of $1/(1 - z^{-1})$ for the transform $U(z)$ in (6.5) is valid in the region $|z^{-1}| < 1$ or $|z| > 1$. For completeness, the region of convergence must be carried along as part of any description of a z-transform function as was done in (6.5).

Since the transform $U(z)$ is simply a function of the complex variable z, algebraic manipulations can be made. For instance,

Example 6.2

The z-transform of the exponential sequence

$$x(n) = \begin{cases} \alpha^n & \text{for } n = 0, 1, 2, \ldots \\ \beta^n & \text{for } n = -1, -2, \ldots \end{cases} \qquad |\beta| > |\alpha|$$

is

$$\begin{aligned} X(z) &= \mathscr{Z}\{x(n)\} \\ &= \sum_{n=-\infty}^{\infty} x(n)z^{-n} \\ &= \sum_{n=0}^{\infty} \alpha^n z^{-n} + \sum_{n=-\infty}^{-1} \beta^n z^{-n} \end{aligned}$$

In order to use (6.6), we rewrite the expression above as

$$X(z) = \sum_{n=0}^{\infty} \left(\frac{\alpha}{z}\right)^n + \sum_{m=0}^{\infty} \left(\frac{z}{\beta}\right)^m - 1 \tag{6.7}$$

In (6.7) the change of summation index $m = -n$ was made, and $\beta^0 = 1$ was included in the second summation and thus was subtracted. It is now shown that the general region of convergence is an annulus, or doughnut-shaped region, in the z-plane. The first term of (6.7) converges for $|\alpha/z| < 1$ and within this region has the value $1/(1 - \alpha/z)$ from

(6.6). Similarly, the second term has the value $1/(1 - z/\beta)$ for $|z/\beta| < 1$. The complete result is therefore

$$\begin{aligned}\mathscr{Z}\{x\} = X(z) &= \frac{1}{1 - \dfrac{\alpha}{z}} + \frac{1}{1 - \dfrac{z}{\beta}} - 1 \\ &= \frac{(\alpha - \beta)z}{(z - \alpha)(z - \beta)}\end{aligned} \tag{6.8}$$

whose convergence region is $|\alpha| < |z| < |\beta|$. Note that if $|\beta| < |\alpha|$, the transform does not exist. The expression (6.8) is a closed-form representation for the two-sided infinite series

$$X(z) = \cdots + \frac{1}{\beta^3} z^3 + \frac{1}{\beta^2} z^2 + \frac{1}{\beta} z + 1 + \alpha z^{-1} + \alpha^2 z^{-2} + \alpha^3 z^{-3} + \cdots \tag{6.9}$$

Notice that the powers of z in the transform X can be used to mark the position of the corresponding sequence value. For instance, the z^{-1} coefficient denotes the $(n = 1)$ position value of the corresponding signal sequence. Compare (6.9) with the original signal,

$$x = \left\{\ldots, \frac{1}{\beta^3}, \frac{1}{\beta^2}, \frac{1}{\beta}, \underset{\uparrow}{1}, \alpha, \alpha^2, \alpha^3, \ldots\right\} \tag{6.10}$$

In the next section we investigate certain basic properties of the z-transform. These properties, together with a moderate table of transform pairs and inversion procedures, which will be derived later, comprise the basic set of transform tools. Much of the sequel can be related to similar developments previously presented for the Laplace transform.

6.2 BASIC PROPERTIES OF THE z-TRANSFORM

The double-sided z-transform of Definition 6.1 should be formally viewed in two parts, namely, those parts described for negative and nonnegative index integers n. Thus,

$$\begin{aligned}X(z) &= \sum_{n=-\infty}^{\infty} x(n)z^{-n} \\ &= \sum_{n=-\infty}^{-1} x(n)z^{-n} + \sum_{n=0}^{\infty} x(n)z^{-n}\end{aligned} \tag{6.11}$$

The second term of (6.11) is a power series in z^{-1}. From the study of complex variables, it is known that the power series $(w = z^{-1})$

$$p(w) = \sum_{n=0}^{\infty} c_n w^n \tag{6.12}$$

converges absolutely inside a circle in the w-plane with radius R, the *radius of convergence*, given by

$$R = \lim_{n \to \infty} \left| \frac{c_n}{c_{n+1}} \right| \tag{6.13}$$

Thus, there is some $R_1 \geq 0$ such that the last sum in (6.11) exists (converges) for $|1/z| < R_1$ or $|z| > 1/R_1$. The first sum, on the other hand, is a power series in z and would converge for $|z| < R_2$ for some $R_2 \geq 0$. Combining these results, $X(z)$ exists (or is a valid transform signal) in a doughnut-shaped region called an *annulus* given by

$$\frac{1}{R_1} < |z| < R_2 \tag{6.14}$$

Clearly, if $R_2 < 1/R_1$, $X(z)$ does not exist anywhere in the plane. Note also that $x(n)$ must be of exponential order for $X(z)$ to exist, for example, a^{n^2} has no transform).

Definition 6.1 reduces to a single-sided z-transform when $x(n) = 0$ for $n < 0$, and (6.14) simplifies to the region exterior to the circle $R_1^{-1} = |z|$. In many applications this reduction is possible since n can represent time in discrete instants and the system is typically initialized at $n = 0$.

Example 6.3

To illustrate the importance of describing the region of convergence, consider $F(z) = 1/(1 - z)$ for the regions of convergence given by (a) $|z| > 1$ and (b) $|z| < 1$. For part (a) we expand in negative powers of z (by long division of $-z + 1\overline{)1}$) to obtain

$$F_a(z) = -(z^{-1} + z^{-2} + \cdots + z^{-k} + \cdots)$$

which has meaning for $|z| > 1$. For part (b) divide $1 - z\overline{)1}$ to get

$$F_b(z) = 1 + z + z^2 + \cdots + z^k + \cdots \qquad |z| < 1$$

Thus, directly from the coefficients of the two series expansions,

$$f_a(n) = \begin{cases} -1 & \text{for } n > 0 \\ 0 & \text{for } n \leq 0 \end{cases} \quad \text{and} \quad f_b(n) = \begin{cases} 0 & \text{for } n > 0 \\ 1 & \text{for } n \leq 0 \end{cases}$$

This example shows that the region of convergence of the form (6.14) is an integral part of the specification of a signal in the z-domain.

Uniqueness. $F(z)$ together with its annulus of convergence represents a unique description of a discrete-time signal in the sense that $f(n)$ can be uniquely determined from $F(z)$ by the inversion procedures discussed later.

Linearity. Both the direct and the inverse z-transform obey the property of linearity. Thus, if $\mathscr{Z}\{f(n)\}$ and $\mathscr{Z}\{g(n)\}$ are denoted by $F(z)$ and $G(z)$, respectively, then

$$\mathscr{Z}\{af(n) + bg(n)\} = aF(z) + bG(z) \tag{6.15}$$

where a and b are constant multipliers. The validity of (6.15) follows directly from the definition of the z-transform of a sequence, i.e., from the linearity of the summation operation. The region of convergence of (6.15) is at least as large as the region given by the intersection of the regions of convergence of F and G.

Example 6.4

Consider

$$f(n) = \begin{cases} 1 & \text{for } n \geq 0 \\ 3^n & \text{for } n < 0 \end{cases} \quad \text{and} \quad g(n) = \begin{cases} 2^n & \text{for } n \geq 0 \\ 4^n & \text{for } n < 0 \end{cases}$$

Thus

$$F(z) = \frac{1}{1 - z^{-1}} + \frac{1}{1 - \frac{1}{3}z} - 1$$

$$= \frac{-2z}{(z-1)(z-3)} \qquad \text{for } 1 < |z| < 3$$

The term -1 is present because the $n = 0$ term is included twice. Similarly,

$$G(z) = \frac{1}{1 - 2z^{-1}} + \frac{1}{1 - \frac{1}{4}z} - 1$$

$$= \frac{-2z}{(z-2)(z-4)} \qquad 2 < |z| < 4$$

The combination (6.19) with $a = 1$ and $b = -1$ is

$$h(n) = f(n) - g(n) = \begin{cases} 1 - 2^n & \text{for } n \geq 0 \\ 3^n - 4^n & \text{for } n < 0 \end{cases}$$

and

$$H(z) = \frac{-2z}{(z-1)(z-3)} + \frac{2z}{(z-2)(z-4)}$$

$$= \frac{2z(2z-5)}{(z-1)(z-2)(z-3)(z-4)} \qquad \text{for } 2 < |z| < 3$$

Translation. An important property when transforming terms of a difference equation is the z-transform of a sequence shifted in time. For a constant shift, we have

$$\mathscr{Z}\{f(n+k)\} = z^k F(z)$$

for positive or negative integer k. The region of convergence of $z^k F(z)$ is the same as that for $F(z)$ for positive k; only the point $z = 0$ need be eliminated from the convergence region of $F(z)$ for negative k.

Proof

$$\mathscr{Z}\{f(n+k)\} = \sum_{n=-\infty}^{\infty} f(n+k) z^{-n}$$

from (6.3). Making a change of the summation variable, that is, with $m = n + k$, we have

$$\begin{aligned}\mathscr{Z}\{f(n+k)\} &= \sum_{m=-\infty}^{\infty} f(m)z^{-(m-k)} \\ &= z^k \sum_{m=-\infty}^{\infty} f(m)z^{-m} \\ &= z^k F(z)\end{aligned}$$

Convolution. As we have seen, the operation of discrete convolution is important in finding the response of a discrete-time system to an arbitrary input sequence. As is now seen, it is possible to avoid carrying out the sometimes cumbersome convolution process by using z-transform theory. In the z-domain, the time-domain convolution operation becomes a simple product of the corresponding transforms, that is,

$$\mathscr{Z}\{f(n) * g(n)\} = F(z)G(z) \tag{6.16}$$

Proof

$$\begin{aligned}\mathscr{Z}\{f(n) * g(n)\} &= \mathscr{Z}\left\{\sum_{k=-\infty}^{\infty} f(k)g(n-k)\right\} \\ &= \sum_{n=-\infty}^{\infty}\left[\sum_{k=-\infty}^{\infty} f(k)g(n-k)\right]z^{-n}\end{aligned}$$

Interchanging summations (this is allowed for all z for which the series is uniformly convergent),

$$\begin{aligned}&= \sum_{k=-\infty}^{\infty} f(k) \sum_{n=-\infty}^{\infty} g(m)z^{-m-k} \qquad m = n - k \\ &= \left[\sum_{k=-\infty}^{\infty} f(k)z^{-k}\right]\left[\sum_{n=-\infty}^{\infty} g(m)z^{-m}\right] \\ &= F(z)G(z)\end{aligned}$$

Multiplication by a^n. This operation corresponds to a rescaling of the z-plane. For $a > 0$,

$$\mathscr{Z}\{a^n f(n)\} = F\left(\frac{z}{a}\right) \qquad \text{for } aR_1 < |z| < aR_2 \tag{6.17}$$

where $F(z)$ is defined for $R_1 < |z| < R_2$.

Proof

$$\begin{aligned}\mathscr{Z}\{a^n f(n)\} &= \sum_{n=-\infty}^{\infty} f(n)a^n z^{-n} \\ &= \sum_{n=-\infty}^{\infty} f(n)\left(\frac{z}{a}\right)^{-n} = F\left(\frac{z}{a}\right) \qquad \text{for } R_1 < \left|\frac{z}{a}\right| < R_2\end{aligned}$$

Time Reversal

$$\mathscr{Z}\{f(-n)\} = F(z^{-1}) \qquad \text{for } R_2^{-1} < |z| < R_1^{-1} \tag{6.18}$$

where $F(z)$ is defined for $R_1 < |z| < R_2$.

Proof

$$\begin{aligned}\mathscr{Z}\{f(-n)\} &= \sum_{n=-\infty}^{\infty} f(-n)z^{-n} \\ &= \sum_{m=-\infty}^{\infty} f(m)z^{m} = F(z^{-1}) \qquad \text{for } R_1 < \left|\frac{1}{z}\right| < R_2\end{aligned}$$

As mentioned earlier, the z-transform can be used for single-sided signals. Thus, if $x(n) = 0$ for $n < 0$,

$$\mathscr{Z}\{x(n)\} = X(z) = \sum_{n=0}^{\infty} x(n)z^{-n} \qquad \text{for } |z| > R \tag{6.19}$$

Definition 6.2. The unilateral z-transform is defined as

$$\mathscr{Z}_+\{x(n)\} = X(z) = \sum_{n=0}^{\infty} x(n)z^{-n} \qquad \text{for } |z| > R$$

just as if the sequence $x(n)$ was in fact single-sided as in (6.19). If there is no ambiguity in the sequel the subscript + is omitted and we use the expression *z-transform* to mean either the double- or the single-sided transform. It is usually clear from context which is meant. By restricting signals to be single-sided, the following useful properties can be proved:

Time Advance. For a single-sided signal $f(n)$,

$$\mathscr{Z}_+\{f(n+1)\} = zF(z) - zf(0). \tag{6.20}$$

Proof

$$\begin{aligned}\mathscr{Z}_+\{f(n+1)\} &= \sum_{n=0}^{\infty} f(n+1)z^{-n} \\ &= \sum_{m=1}^{\infty} f(m)z^{-(m-1)} \\ &= \sum_{m=1}^{\infty} f(m)z^{-m} \\ &= z\left[\sum_{m=0}^{\infty} f(m)z^{-m} - f(0)\right]\end{aligned}$$

The time-advance property can be applied sequentially. For instance, if

$g(n) = f(n+1)$, then

$$\mathscr{Z}_+\{f(n+2)\} = \mathscr{Z}_+\{g(n+1)\} = zG(z) - zg(0)$$

but

$$G(z) = zF(z) - zf(0)$$

so that

$$\mathscr{Z}_+\{f(n+2)\} = z^2F(z) - z^2f(0) - zf(1)$$

More generally,

$$\mathscr{Z}_+\{f(n+k)\} = z^kF(z) - z^kf(0) - z^{k-1}f(1) - \cdots - zf(k-1) \tag{6.21}$$

This result can be used in solving linear constant-coefficient difference equations. Occasionally, it is desirable to calculate the initial or final value of a single-sided sequence without a complete inversion. The following two properties present these results.

Initial Signal Value. If $f(n) = 0$ for $n < 0$,

$$f(0) = \lim_{z\to\infty} F(z) \tag{6.22}$$

where $F(z) = \mathscr{Z}\{f(n)\}$ for $|z| > R$.

Proof

$$\lim_{z\to\infty} F(z) = \lim_{z\to\infty} \sum_{n=0}^{\infty} f(n)z^{-n} = \sum_{n=0}^{\infty} f(n) \lim_{z\to\infty} z^{-n} = f(0)$$

Interchanging the limit and summation is permitted since the series converges uniformly in its region of convergence.

Final Signal Value. If $f(n) = 0$ for $n < 0$ and $\mathscr{Z}\{f(n)\} = F(z)$ is a rational function with all its denominator roots (poles) strictly inside the unit circle except possibly for a first-order pole at $z = 1$,

$$f(\infty) = \lim_{n\to\infty} f(n) = \lim_{z\to 1} (1 - z^{-1})F(z) \tag{6.23}$$

Proof. Express the transform of $[f(n+1) - f(n)]$ as follows:

$$\begin{aligned}\mathscr{Z}\{f(n+1) - f(n)\} &= \lim_{k\to\infty} \sum_{n=0}^{k} [f(n+1) - f(n)]z^{-n} \\ &= [zF(z) - zf(0)] - F(z)\end{aligned}$$

where use of the linearity and the time-advance property has been made. Rewriting, we obtain

$$(1 - z^{-1})F(z) - f(0) = \lim_{k\to\infty} \sum_{n=0}^{k} [f(n+1) - f(n)]z^{-(n+1)}$$

Taking the limit as $z \to 1$ on both sides, we have

$$\lim_{z \to 1} (1 - z^{-1})F(z) = f(0) + [f(1) - f(0)] + [f(2) - f(1)] + \cdots$$
$$= \lim_{k \to \infty} f(k)$$

Unfortunately, the burden of assuring that the hypothesis is satisfied is on the user because (6.23) can be applied to a transform whose corresponding sequence does not settle to a final value.

Example 6.5

Consider the sequence

$$x(n) = \begin{cases} 2^n & \text{for } n \geq 0 \\ 0 & \text{for } n < 0 \end{cases}$$

The transform

$$X(z) = \sum_{n=0}^{\infty} 2^n z^{-n} = \sum_{n=0}^{\infty} \left(\frac{2}{z}\right)^n = \frac{z}{z-2} \qquad \text{for } |z| > 2$$

Clearly, $x(n)$ does not have a finite final value, but from (6.23)

$$x(\infty) = \lim_{z \to 1} \frac{z-1}{z} \cdot \frac{z}{z-2} = 0$$

This illustrates the difficulty (incorrect result) that can occur if the user does not first check that the poles of $(1 - z^{-1})X(z)$ are strictly inside the unit circle. In this example, there is a pole of $X(z)$ at $z = 2$.

6.3 *z*-TRANSFORM INVERSION

Having become acquainted with the z-transform and its basic properties, we now consider ways of obtaining the corresponding sequence from a transformed signal. This is the process of transform inversion, and with it is completed our ability to describe a discrete-time signal either in the sequence (time) domain or in the (frequency) z-domain. Following the notation

$$F(z) = \mathscr{Z}\{f(n)\} \tag{6.24}$$

we operationally denote the inverse transform of $F(z)$ in the form

$$f(n) = \mathscr{Z}^{-1}\{F(z)\} \tag{6.25}$$

There are three useful methods for inverting a transformed signal. They are

1. Expansion into a series of terms in the variables z and z^{-1}
2. Complex integration by the method of residues
3. Partial-fraction expansion and table look-up

We discuss each of these methods in turn.

Method 1. For the expansion of $F(z)$ into a Series, the theory of functions of a complex variable provides a practical basis for developing our inverse transform techniques. As we have seen, the general region of convergence for a transform function $F(z)$ is of the form $a < |z| < b$, i.e., an annulus centered at the origin of the z-plane. See (6.14.) This first method is to obtain a series expression of the form

$$F(z) = \sum_{n=-\infty}^{\infty} c_n z^{-n} \tag{6.26}$$

which is valid in the annulus of convergence. This double-sided series is called a *Laurent series.* When $F(z)$ has been expanded as in (6.26), that is, when the coefficients c_n, $n = 0, \pm 1, \pm 2, \ldots$ have been found, the corresponding sequence is specified by $f(n) = c_n$ by uniqueness of the transform and (6.3).

To illustrate the techniques for expansion, let us consider a single $F(z)$ with three possible regions of convergence. It is shown that $F(z) = z/[(z-1)(z-2)]$ has three possible nonempty regions of convergence, thereby defining three different sequences.

Example 6.6

Invert

$$F(z) = \frac{z}{(z-1)(z-2)} \qquad \text{for } |z| > 2$$

In this case the convergence region is the exterior of a circle, and we expect a single-sided sequence $f(n) = 0$ for $n < 0$. Consequently, we seek an expansion in negative powers of z. Dividing the numerator of $F(z)$ by its denominator yields

$$z^2 - 3z + 2 \overline{\smash{)}\, z \qquad} \quad z^{-1} + 3z^{-2} + 7z^{-3} + 15z^{-4} + 31z^{-5} + \cdots$$

Therefore,

$$f(n) = \{\ldots 0, 0, \underset{\uparrow}{1}, 3, 7, 15, 31, \ldots\}$$

As a second example, we consider the convergence region inside a circle.

Example 6.7

Invert

$$F(z) = \frac{z}{(z-1)(z-2)} \qquad \text{for } |z| < 1$$

In this case, we again expect a single-sided sequence—one that is zero for $n > 0$. Hence, the expansion is a power series in positive powers of z. We now divide the numerator of $F(z)$ by its denominator to obtain positive powers of z, that is,

$$2 - 3z + z^2 \overline{\smash{)}\, z \qquad} \quad \tfrac{1}{2}z + \tfrac{3}{4}z^2 + \tfrac{7}{8}z^3 + \tfrac{15}{16}z^4 + \tfrac{31}{32}z^5 + \cdots$$

Therefore,

$$f(n) = \left\{\ldots, \frac{31}{32}, \frac{15}{16}, \frac{7}{8}, \frac{3}{4}, \frac{1}{2}, 0, 0, \ldots\right\}$$
$$\uparrow$$

Thus, with this $F(z)$ there are three possible nonempty regions of convergence that correspond to all possible circular or annular regions whose boundaries fall on the poles of $F(z)$: $|z| > 2$, $|z| < 1$, and $1 < |z| < 2$. In the next example we consider the annular region case.

Example 6.8

Invert

$$F(z) = \frac{z}{(z-1)(z-2)} \qquad \text{for } 1 < |z| < 2$$

Since $F(z)$ converges in the annulus $1 < |z| < 2$, we look for a way to expand $F(z)$ into a Laurent series valid in this region. To show this explicitly, we expand the partial-fraction expansion representation of $F(z)$:

$$F(z) = \frac{-1}{z-1} + \frac{2}{z-2}$$

Expanding the first term into powers of z^{-1} and the second term into powers of z gives us the correct expansion since a positive power geometric series converges inside some circle and a negative power geometric series converges outside some (other) circle.

$$z-1\overline{)\;-1\;}\quad -z^{-1} - z^{-2} - z^{-3} - \cdots$$

$$-2+z\overline{)\;2\;}\quad -1 - \frac{z}{2} - \frac{z^2}{4} - \frac{z^3}{8} - \cdots$$

Therefore,

$$F(z) = \cdots - \frac{z^3}{8} - \frac{z^2}{4} - \frac{z}{2} - 1 - z^{-1} - z^{-2} - z^{-3} - \cdots$$

From this expansion, we can see that

$$f(n) = \left\{\ldots, -\frac{1}{8}, -\frac{1}{4}, -\frac{1}{2}, -1, -1, -1, -1, \ldots\right\}$$
$$\uparrow$$

In each of the three examples above we found a way to expand the function $F(z)$ into a Laurent series valid in the region of convergence given for $F(z)$. This method provides a direct evaluation of the inverse sequence. But, as can be seen from the examples, the inverse sequence is not in closed form, unless the sequence is simple enough to infer the general term, as in Example 6.8:

$$f(n) = \begin{cases} -1 & \text{for } n \geq 0 \\ -2^n & \text{for } n < 0 \end{cases} \tag{6.27}$$

However, in some cases an open form is sufficient. The next method uses the powerful method of residues from complex variable theory.

Method 2. We now evaluate the inverse transform of $F(z)$ by the method of residues. The method involves the calculation of residues of a function both inside and outside of a simple closed path that lies inside the region of convergence. A number of key concepts are necessary in order to describe the required procedure.

Definition 6.2. A complex-valued function $G(z)$ has a pole of order k at $z = z_0$ if it can be expressed as

$$G(z) = \frac{G_1(z_0)}{(z - z_0)^k}$$

where $G_1(z_0)$ is finite.

Example 6.9

The function

$$G(z) = \frac{4(z-3)}{z(z-1)(z-2)^2}$$

has a pole of order 1 at $z = 0$ and $z = 1$ and a pole of order 2 at $z = 2$.

Definition 6.3. The residue of a complex function $G(z)$ at a pole of order k at $z = z_0$ is defined by

$$\text{Res}\,[G(z), z_0] = \frac{1}{(k-1)!}\frac{d^{k-1}}{dz^{k-1}}[(z-z_0)^k G(z)]\bigg|_{z=z_0} \tag{6.29}$$

Example 6.9

Using the function $G(z)$ of Example 6.9,

$$\text{Res}\,[G(z), 0] = zG(z)\bigg|_{z=0} = \frac{4(-3)}{(-1)(-2)^2} = 3$$

$$\text{Res}\,[G(z), 1] = \frac{4(1-3)}{1(1-2)^2} = -8$$

$$\text{Res}\,[G(z), 2] = \frac{d}{dz}\left[\frac{4(z-3)}{z(z-1)}\right]\bigg|_{z=2} = 5$$

With this brief review of calculating residues, the formula for calculating the inverse transform can now be given.

Inverse Transform Formula (Method 2)

If $F(z)$ is convergent in the annulus $0 < a < |z| < b$ as shown in Fig. 6.1 and C is the closed path shown (the path C must lie entirely within the annulus of convergence), then

$$f(n) = \begin{cases} \text{sum of residues of } F(z)z^{n-1} \text{ at poles of } F(z) \text{ inside } C, & m \geq 0 \\ -(\text{sum of residues of } F(z)z^{n-1} \text{ at poles of } F(z) \text{ outside } C), & m < 0 \end{cases} \tag{6.30}$$

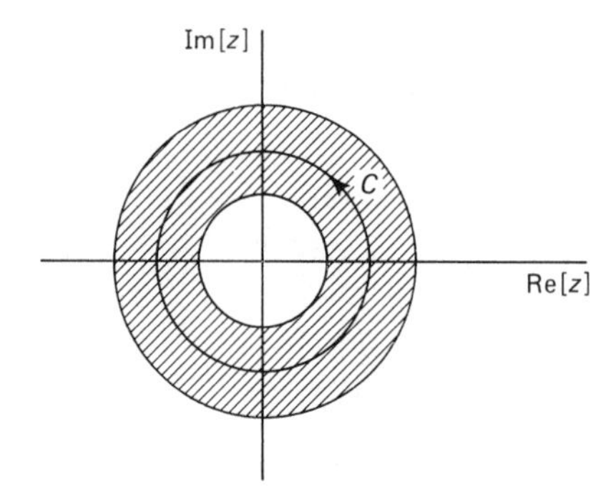

Figure 6.1 Typical convergence region for a transformed discrete-time signal.

where m is the least power of z in the numerator of $F(z)z^{n-1}$, e.g., m might equal $n - 1$. To illustrate (6.30), let us redo the inversion of $F(z)$ in Example 6.8.

Example 6.11

Given that

$$F(z) = \frac{z}{(z-1)(z-2)} \qquad \text{for } 1 < |z| < 2$$

$$f(n) = \begin{cases} \text{Re}[z]\left[\dfrac{z^n}{(z-1)(z-2)}, 1\right] = -1 & \text{for } m = n \geq 0 \\ -\text{Re}[z]\left[\dfrac{z^n}{(z-1)(z-2)}, 2\right] = -(2^n) & \text{for } m = n < 0 \end{cases}$$

which checks the result of Example 6.8 given in (6.27).

To illustrate the case where $m \neq n$, consider the next example.

Example 6.12

To invert

$$F(z) = \frac{3z^2 - 13z - 38}{(z-1)(z+3)(z-4)} \qquad \text{for } 3 < |z| < 4$$

According to the inversion formula (6.30),

$$f(n) = \begin{cases} \text{Res}[F(z)z^{n-1}, 1, -3] & \text{for } m \geq 0 \\ -\text{Res}[F(z)z^{n-1}, 4] & \text{for } m < 0 \end{cases}$$

$$\text{Res}[F(z)z^{n-1}, 1] = \left.\frac{3z^{n+1} - 13z^n - 38z^{n-1}}{(z+3)(z-4)}\right|_{z=1} = 4$$

$$\text{Res}[F(z)z^{n-1}, -3] = \left.\frac{3z^{n+1} - 13z^n - 38z^{n-1}}{(z-1)(z-4)}\right|_{z=-3} = (-3)^{n-1}$$

and

$$\text{Res}[F(z)z^{n-1}, 4 = \left.\frac{3z^{n+1} - 13z^n - 38z^{n-1}}{(z-1)(z+3)}\right|_{z=4} = -2(4)^{n-1}$$

Therefore,

$$f(n) = \begin{cases} 4 + (-3)^{n-1}, & m = n - 1 \geq 0 \quad (n \geq 1) \\ 2(4)^{n-1}, & m = n - 1 < 0 \quad (n < 1) \end{cases}$$

which formally says that $f(0) = \frac{1}{2}$, which is compatible with the expression for the negative terms.

The inverse transform formula is relatively simple to apply to rational $F(z)$ and has the additional advantage of providing a closed-form solution. In the next development, the method of partial-fraction expansion and table look-up is reviewed. The first development closely follows the corresponding inversion technique of Laplace transforms. Subsequently, a minor modification is introduced.

Method 3. The Partial-Fraction expansion method is generally the most preferable procedure for obtaining the sequence that corresponds to a given rational z-transform function because of its ease of application. It is particularly simple when all signals are assumed to be single-sided (positive-time) signals. Readers now familiar with Laplace transform analysis of linear systems will find the general approach here very similar to its continuous-time counterpart.

The following development is not restricted to single-sided signals, but most of the examples fall into this category. For positive-time signals, we know that the z-transform converges outside of some circular region in the z-plane. Thus, it is not necessary to specifically carry along the convergence region in this context.

The partial-fraction expansion technique is used to decompose a more complicated function into elementary terms, each of which can be easily translated into a corresponding sequence, usually with the aid of a table of transform pairs. This procedure requires that the denominator of the function be in factored form. Fortunately, in dealing with the simpler analytic signal and system models this does not present a problem.

Partial-Fraction Expansions

The process of expansion of $F(z)$ into partial fractions assumes that $F(z)$ has a denominator polynomial that is in factored form. Using the standard notation for a rational $F(z)$, we have

$$F(z) = \frac{q(z)}{p(z)} = \frac{k(z - z_1)(z - z_2) \cdots (z - z_M)}{(z - p_1)(z - p_2) \cdots (z - p_N)} \tag{6.31}$$

where q and p are polynomials shown in factored form. The numbers $z_1, z_2, \ldots, z_M$ are called the *zeros* and $p_1, p_2, \ldots, p_N$ the *poles* of the rational function F. Generally, for $M < N$, F is termed a *proper rational function.* An improper $(M \geq N)$ function can always be converted into a proper form by dividing its denominator into its numerator until its remainder order is lower than the denominator order.

Distinct Poles. For the initial development, it is assumed that the poles $p_1, \ldots, p_N$ are all different (distinct), and we seek an expansion of the form

$$F(z) = \sum_{k=1}^{N} \frac{A_k}{z - p_k} \tag{6.32}$$

For this simplest case, the value of A_k is the residue of the function $F(z)$ at the pole $z = p_k$ for each $k = 1, 2, \ldots, N$. Expression (6.32) is called the *partial-fraction expansion* of $F(z)$.

For evaluation of the general coefficient A_k, consider the expansion of (6.32):

$$F(z) = \frac{A_1}{z - p_1} + \frac{A_2}{z - p_2} + \cdots + \frac{A_k}{z - p_k} + \cdots + \frac{A_N}{z - p_N} \tag{6.33}$$

Multiplying (6.33) by $(z - p_k)$ on both sides and evaluating the result at $z = p_k$, all terms on the right go to zero except the kth partial-fraction term. The end result is the formula

$$A_k = (z - p_k)F(z)\bigg|_{z = p_k} \tag{6.34}$$

which gives the value of the single coefficient for any simple pole in the partial-fraction expansion of $F(z)$.

Multiple-Order Poles. For a more general rational function $F(z)$ there may be poles of multiple order. Each multiple-order pole generates several terms in the partial-fraction expansion. Suppose $z = p$ is an mth-order pole of $F(z)$. Thus,

$$F(z) = \frac{g(z)}{(z - p)^m} \tag{6.35}$$

where $g(z)$ will in general have poles $p_k \neq p$ and zeros $z_k \neq p$. This part of the total expansion due only to the singularity at p can be represented as

$$\frac{g(z)}{(z - p)^m} = \frac{B_m}{(z - p)^m} + \frac{B_{m-1}}{(z - p)^{m-1}} + \cdots + \frac{B_1}{(z - p)} + \cdots \tag{6.36}$$

where the coefficients $B_1, \ldots, B_m$ are constants that can be determined as follows

$$B_{m-k} = \frac{1}{k!} \frac{d^k}{dz^k} [g(z)]\bigg|_{z = p} \qquad \text{for } k = 0, 1, \ldots, m - 1 \tag{6.37}$$

Note that B_1 is the "residue of $F(z)$ at p."

To simplify the development of expression (6.37), we consider the specific function $F(z)$ with two poles p_1, order 3 and p_2, simple. If the order of polynomial $g(z)$ is less than four, the required expansion form is

$$F(z) = \frac{g(z)}{(z - p_1)^3(z - p_2)} = \frac{B_3}{(z - p_1)^3} + \frac{B_2}{(z - p_1)^2} + \frac{B_1}{z - p_1} + \frac{A}{z - p_2} \tag{6.38}$$

Since p_2 is a simple pole, A can be found using (6.34):

$$A = (z - p_2)F(z)\Big|_{z=p_2} = \frac{g(p_2)}{(p_2 - p_1)^3}$$

To evaluate the B-coefficients, one first multiplies (6.38) by $(z - p_1)^3$:

$$(z - p_1)^3 F(z) = B_3 + (z - p_1)B_2 + (z - p_1)^2 B_1 + (z - p_1)^3 \frac{A}{z - p_2} \tag{6.39}$$

Thus,

$$B_3 = (z - p_1)^3 F(z)\Big|_{z=p_1}$$

since all other terms drop out. This is the $k = 0$ case in (6.37). To solve for B_2, consider taking the first derivative of (6.39) with respect to z:

$$\frac{d}{dz}(z - p_1)^3 F(z) = 0 + B_2 + 2(z - p_1)B_1 + (z - p_1)^2 G_1(z) \tag{6.40}$$

where $G_1(z)$ has no pole at p_1. Therefore, corresponding to $k = 1$ in (6.37),

$$B_2 = \frac{d}{dz}(z - p_1)^3 F(z)\Big|_{z=p_1} \tag{6.41}$$

Finally, to isolate B_3, we take a second derivative of (6.39) with respect to z.

$$\frac{d^2}{dz^2}(z - p_1)^3 F(z) = 0 + 0 + 2B_1 + (z - p_1)G_2(z) \tag{6.42}$$

where $G_2(z)$ has no pole at p_1. Note the factor of 2.

$$B_1 = \frac{1}{2!}\frac{d^2}{dz^2}(z - p_1)^3 F(z)\Big|_{z=p_1} \qquad 6.43)$$

For higher-order poles, the general expression (6.37) provides all the partial-fraction coefficients.

Complex Poles. Whenever complex poles occur, they are always in complex conjugate pairs so that the transform function $F(z)$ is a ratio of polynomials in z with *real* coefficients.[1] If, for instance, p_1 is complex and $p_2 = p_1^*$, it can be shown that $A_2 = A_1^*$ in (6.33). In this case the two complex-valued partial-fraction terms can be combined into a second-order term that inverts into a real, oscillatory sequence.

Example 6.13

To combine

$$F(z) = \frac{A_1}{z - p_1} + \frac{A_2}{z - p_2}$$

[1]This, of course, assumes that the sequence $\{f(n)\}$ is itself real.

into a second-order expression with real coefficients when $A_2 = A_1^*$ and $p_2 = p_1^*$. Let $A_1 = A + jB$ and $p_1 = a + jb$. Then

$$F(z) = \frac{(A + jB)(z - a + jb) + (A - jB)(z - a - jb)}{(z - a - jb)(z - a + jb)}$$

$$= \frac{2A(z - a) - 2Bb}{(z - a)^2 + b^2} = \frac{2Az - 2(Aa + Bb)}{z^2 - 2az + (a^2 + b^2)}$$

A Table of Elementary Transform Pairs

To illustrate the entire inversion by partial-fraction expansion, we require a table of related functions. For instance, for the single-sided sequence $\{a^n, n = 0, 1, 2, \ldots\}$ the corresponding z-domain function is $z/(z - a)$ for $|z| > |a|$. Since our partial-fraction expansions have constant numerators, the sequence corresponding to $1/(z - a)$, for $|z| > |a|$ is

$$\{\underset{\uparrow}{0}, 1, a, a^2, \ldots\}$$

which can be expressed as $a^{n-1}u(n - 1)$, where u is the unit-step sequence defined earlier.

Similarly, for the negatively indexed sequence

$$\{\ldots, a^{-2}, a^{-1}, \underset{\uparrow}{1}, \ldots\} = a^n u(-n)$$

the z-transform is $-a/(z - a)$ for $|z| < |a|$. Table 6.1 lists these and other useful elementary pairs.

Example 6.14

To invert $F(z) = z/(z - a) = 1 + a/(z - a), |z| > |a|$. From Table 6.1, entries 5a and 1a:

$$f(n) = \delta(n) + a(a^{n-1})u(n - 1) = \{1, a, a^2, \ldots\}$$

Example 6.15

To invert the rational transform

$$Y(z) = \frac{z^3 - 2z^2 + 2z}{(z - 1)^3} \qquad \text{for } |z| > 1$$

we first divide to obtain a proper function

$$Y(z) = 1 + \frac{z^2 - z + 1}{(z - 1)^3} = 1 + \frac{B_3}{(z - 1)^3} + \frac{B_2}{(z - 1)^2} + \frac{B_1}{z - 1}$$

The coefficients B_1, B_2, and B_3 from (6.37) are

$$B_3 = (z^2 - z + 1)\Big|_{z=1} = 1$$

TABLE 6.1 PARTIAL-FRACTION EQUIVALENTS LISTING CAUSAL AND ANTICAUSAL z-TRANSFORM PAIRS

	z-Domain: $F(z)$	Sequence Domain: $f(n)$
1a.	$\frac{1}{z-a}$, for $\lvert z\rvert > \lvert a\rvert$	$a^{n-1}u(n-1) = \{0, \underset{\uparrow}{1}, a, a^2, \ldots\}$
1b.	$\frac{1}{z-a}$, for $\lvert z\rvert < \lvert a\rvert$	$-a^{n-1}u(-n) = \left\{\ldots, \frac{-1}{a^3}, \frac{-1}{a^2}, \underset{\uparrow}{\frac{-1}{a}}\right\}$
2a.	$\frac{1}{(z-a)^2}$, for $\lvert z\rvert > \lvert a\rvert$	$(n-1)a^{n-2}u(n-1) = \{0, \underset{\uparrow}{0}, 1, 2a, 3a^2, \ldots\}$
2b.	$\frac{1}{(z-a)^2}$, for $\lvert z\rvert < \lvert a\rvert$	$-(n-1)a^{n-2}u(-n) = \left\{\ldots, \frac{3}{a^4}, \frac{2}{a^3}, \underset{\uparrow}{\frac{1}{a^2}}\right\}$
3a.	$\frac{1}{(z-a)^3}$, for $\lvert z\rvert > \lvert a\rvert$	$\frac{1}{2}(n-1)(n-2)a^{n-3}u(n-1) = \{0, \underset{\uparrow}{0}, 0, 1, 3a, 6a^2, \ldots\}$
3b.	$\frac{1}{(z-a)^3}$, for $\lvert z\rvert < \lvert a\rvert$	$\frac{-1}{2}(n-1)(n-2)a^{n-3}u(-n) = \left\{\ldots, \frac{-6}{a^5}, \frac{-3}{a^4}, \underset{\uparrow}{\frac{-1}{a^3}}\right\}$
4a.	$\frac{1}{(z-a)^m}$, for $\lvert z\rvert > \lvert a\rvert$	$\frac{1}{(m-1)!}\prod_{k=1}^{m-1}(n-k)a^{n-m}u(n-1)$
4b.	$\frac{1}{(z-a)^m}$, for $\lvert z\rvert < \lvert a\rvert$	$\frac{-1}{(m-1)!}\prod_{k=1}^{m-1}(n-k)a^{n-m}u(-n)$
5a.	z^{-m}, for $z \neq 0$, $m \geq 0$	$\delta(n-m) = \{\ldots, 0, \underset{\substack{\uparrow\\0}}{0}, \ldots, \underset{\substack{\uparrow\\m}}{1}, 0, \ldots, 0, \ldots\}$
5b.	z^{+m}, for $\lvert z\rvert < \infty$, $m \geq 0$	$\delta(n+m) = \{\ldots, 0, \ldots, \underset{\substack{\uparrow\\-m}}{1}, \ldots, \underset{\substack{\uparrow\\0}}{0}, \ldots\}$

$$B_2 = \frac{d}{dz}(z^2 - z + 1)\bigg|_{z=1} = (2z-1)\bigg|_{z=1} = 1$$

$$B_1 = \frac{1}{2}\frac{d^2}{dz^2}(z^2 - z + 1)\bigg|_{z=1} = \frac{1}{2}(2) = 1$$

Therefore,

$$Y(z) = 1 + \frac{1}{z-1} + \frac{1}{(z-1)^2} + \frac{1}{(z-1)^3} \qquad \text{for } |z| > 1$$

Since these forms occur in Table 6.1 in the order 5a, 1a, 2a, and 3a, we write

$$y(n) = \delta(n) + \left[1 + (n-1) + \frac{1}{2}(n-1)(n-2)\right]u(n-1)$$

To compare with Method 2, the residue formula (6.30),

$$y(n) = \text{Res}\,[Y(z)z^{n-1}, 1] \qquad \text{for } m = n \geq 0$$

From (6.29)

$$y(n) = \frac{1}{2}\frac{d^2}{dz^2}(z^{n+2} - 2z^{n+1} + 2z^n)\bigg|_{z=1} = \frac{1}{2}(n^2 - n + 2) \qquad \text{for } n \geq 0$$

By either form,

$$y(n) = \{1, 1, 2, 4, 7, \ldots\}$$

The following example illustrates the case of complex conjugate poles.

Example 6.16

To invert

$$F(z) = \frac{3z^2 + 0.5z}{z^2 - z + 0.5} \qquad \text{for } |z| > \frac{\sqrt{2}}{2}$$

The denominator roots are

$$p_1 = \frac{1 + j1}{2} \quad \text{and} \quad p_2 = \frac{1 - j1}{2}$$

Thus, upon putting the expression into standard form, the partial-fraction expansion is

$$F(z) = 3 + \frac{3.5z - 1.5}{z^2 - z + 0.5} = 3 + \frac{\frac{7-j1}{4}}{z - p_1} + \frac{\frac{7+j1}{4}}{z - p_1^*}$$

and

$$f(n) = 3\,\delta(n) + \left[\frac{7-j1}{4}\left(\frac{1+j1}{2}\right)^{n-1} + \frac{7+j1}{4}\left(\frac{1-j1}{2}\right)^{n-1}\right]u(n-1)$$

Using polar notation to help simplify the expression above,

$$f(n) = 3\,\delta(n) + \left[\frac{\sqrt{50}}{4}\angle\phi\left(\frac{\sqrt{2}}{2}\right)^{n-1}\angle\frac{(n-1)\pi}{4} + \frac{\sqrt{50}}{4}\angle-\phi\left(\frac{\sqrt{2}}{2}\right)^{n-1}\angle\frac{(n-1)\pi}{-4}\right]u(n-1)$$

$$f(n) = 3\,\delta(n) + \frac{\sqrt{50}}{4}\left(\frac{\sqrt{2}}{2}\right)^{n-1} 2\cos\left(\phi + \frac{(n-1)\pi}{4}\right)u(n-1)$$

where $\phi = -\tan^{-1}(1/7)$. Numerically,

$$f(n) = 3\,\delta(n) + 10(0.70711)^n \cos(0.7854n - 0.9273)\,u(n-1)$$

Subsequently, we will use a variation on this method to obtain the inverse sequence as a sampled continuous-time signal.

Example 6.17

A phenomenon may be properly analyzed only after a representative model of that phenomenon is made. Frequently, the investigator postulates a number of properties he or she believes the phenomenon to possess and then generates a model that satisfies these properties. The correctness of the model is then judged by how well the model matches the phenomenon's observed behavior.

To illustrate this notion, let us generate a model that represents the phenomenon of rabbit population. The idealized properties that our rabbit population is hypothesized to have are as follows:

1. A pair (one male and one female) of rabbits is born to each pair of adult rabbits at the end of every month.
2. A newborn pair produces their first offspring at two months of age.
3. Once paired, a pair of rabbits remain true to one another and indefinitely produce rabbits according to properties 1 and 2.

Based on these three properties, it is possible to determine a mathematical model that describes the rabbit population.

To generate a model, let us first assume that a pair of newborn rabbits is placed on an unpopulated (with rabbits) island. Next, we define $y(k)$ as the number of pairs of rabbits present at the end of the kth month after placement of the initial pair. This totality of pairs of rabbits, $y(k)$, is composed of either newborn rabbits or those that are one month or older. From property 2, it follows that the number of newborn pairs at the end of month k must be equal to $y(k-2)$, and the number of rabbits that are one or more months old at the end of month k is obviously $y(k-1)$. Therefore, we must have

$$y(k) = \underbrace{y(k-1)}_{\substack{\text{One or more}\\ \text{months old}\\ \text{at end of}\\ \text{month } k}} + \underbrace{y(k-2)}_{\substack{\text{Newborn at}\\ \text{end of } k\text{th}\\ \text{month}}}$$

This is a linear, second-order difference equation that is a function only of the output signal $y(k)$. To determine the history of rabbit population under the hypothesized model, we take the two initial conditions to be $y(0) = 1$ and $y(-1) = 0$, which reflect the fact that a newborn pair of rabbits were placed on the island initially. Now, letting $k = 1$ in the previous difference equation, we have

$$y(1) = y(0) + y(-1) = 1$$

After the second month ($k = 2$), the number of rabbit pairs is

$$y(2) = y(1) + y(0) = 2$$

Similarly, we can readily determine that $y(3) = 3$, $y(4) = 5$, $y(5) = 8$, $y(6) = 13$, and so on.

To develop a general closed-form solution for $y(k)$, we first write our basic system equation as

$$y(n+2) = y(n+1) + y(n)$$

with $y(0) = 0$ and $y(1) = 1$. The difference equation is simply reindexed by the variable n so that all shifts are advances, rather than delays. The modified initial conditions have been shifted forward by one step, so that we must remember to make the final change that $y(k) = y(n-1)$. Taking the (one-sided) z-transform of the difference equation above, we have

$$z^2\, Y(z) - z^2 y(0) - zy(1) = z\, Y(z) - zy(0) + Y(z)$$

Introducing the initial conditions and solving for $Y(z)/z$,

$$\frac{Y(z)}{z} = \frac{1}{z^2 - z + 1}$$

Using one of the closed-form inversion techniques, we find that (in terms of the original index k),

$$y(k) = \frac{1}{\sqrt{5}}\left[\left(\frac{1+\sqrt{5}}{2}\right)^{k+1} - \left(\frac{1-\sqrt{5}}{2}\right)^{k+1}\right] \qquad \text{for } k = -1, 0, 1, 2, \ldots$$

To demonstrate that this expression is indeed the solution, the reader should show that it satisfies the original difference equation and that the initial conditions $y(0) = 1$ and $y(-1) = 0$ are satisfied.

We may further generalize the given model. Suppose that, at the end of each month, we place pairs of newborn rabbits on the island in addition to the pairs already present. If we define

$U(k)$ = the number of pairs of newborn rabbits placed on the island at the end of the kth month,

then we may argue as before that

$$y(k) = \underbrace{U(k)}_{\text{Newborn at end of } k\text{th month}} + y(k-2) + \underbrace{y(k-1)}_{\text{One or more months old at end of } k\text{th month}}$$

It should be apparent that this more general model contains our previous model as a special case. An implementation of this model is shown in Fig. 6.2. Of course, this example simply illustrates the use of idealized properties and is not intended to be truly representative of the known behavior of a rabbit population.

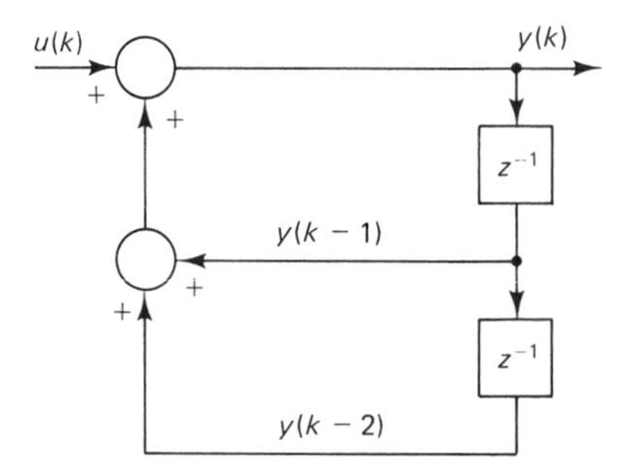

Figure 6.2 Model of a hypothetical rabbit population.

6.4 SAMPLED DATA

The purpose of this section is to emphasize that the sequences can represent sampled versions of actual continuous signals. For instance, if $h(t)$ is some one-sided, continuous-time function, the ideal samples uniformly spaced at T-second intervals can be written as the sequence of numbers $\{h(nT)$ for $n = 0, 1, 2, \ldots\}$. The principal change is the presence of the sampling time T parameter; that is, the sequence values depend on the rate of sampling.

One advantage of working with sampled data is the ability to represent sequences as combinations of sampled time signals. Table 6.2 provides some key z-transform pairs. So that the table can serve a multiple purpose, there are three items per line: The first is an indicated sampled continuous-time signal, the second is the Laplace transform of that continuous-time signal, and third is the z-transform of the uniformly sampled continuous-time signal. To illustrate the interrelation of these entries, consider Fig. 6.3. For simplicity, only single-sided signals have been used in Table 6.2. Consequently, the convergence regions are understood in this context to be Re $[s] > \sigma_0$ and $|z| > \rho_0$ for the Laplace and z-transforms, respectively. The parameters σ_0 and ρ_0 depend on the actual transformed functions, in

TABLE 6.2 *Z*-TRANSFORMS FOR SAMPLED DATA

$f(t)$, $t = nT$, $n = 0, 1, 2, \ldots$	F(s), Re $[s] > \sigma_0$	$F(z)$, $\lvert z \rvert > \rho_0$
1. 1 (unit step)	$\frac{1}{s}$	$\frac{z}{z-1}$
2. t(unit ramp)	$\frac{1}{s^2}$	$\frac{Tz}{(z-1)^2}$
3. t^2	$\frac{2}{s^3}$	$\frac{T^2 z(z+1)}{(z-1)^3}$
4. e^{-at}	$\frac{1}{s+a}$	$\frac{z}{z-e^{-aT}}$
5. te^{-at}	$\frac{1}{(s+a)^2}$	$\frac{Tze^{-aT}}{(z-e^{-aT})^2}$
6. $\sin \omega t$	$\frac{\omega}{s^2+\omega^2}$	$\frac{z \sin \omega T}{z^2 - 2z \cos \omega T + 1}$
7. $\cos \omega t$	$\frac{s}{s^2+\omega^2}$	$\frac{z(z - \cos \omega T)}{z^2 - 2z \cos \omega T + 1}$
8. $e^{-at} \sin \omega t$	$\frac{\omega}{(s+a)^2+\omega^2}$	$\frac{ze^{-aT} \sin \omega T}{z^2 - 2ze^{-aT} \cos \omega T + e^{-2aT}}$
9. $e^{-at} \cos \omega t$	$\frac{s+a}{(s+a)^2+\omega^2}$	$\frac{z(z - e^{-aT} \cos \omega T)}{z^2 - 2ze^{-aT} \cos \omega T + e^{-2aT}}$

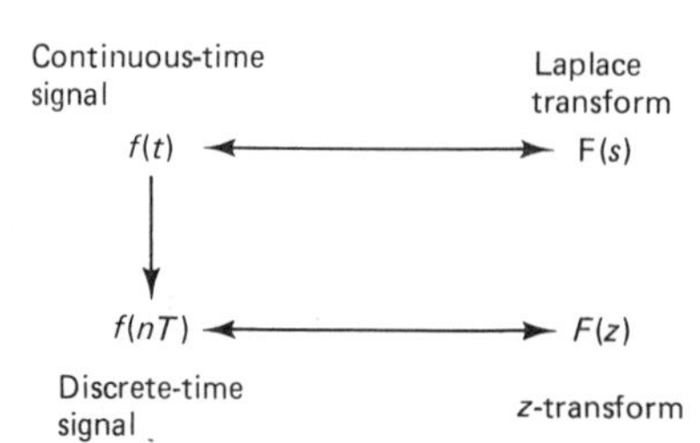

Figure 6.3 Signal and transform relationships for Table 6.2.

that σ_0 is the real part of that pole of F(s) that has the largest real part, and ρ_0 is the magnitude of that pole of $F(z)$ that has the largest magnitude.

It would be of interest to rework Example 6.18 in an effort to formulate the answer as a sampled continuous-time signal.

Example 6.18

To invert

$$F(z) = \frac{3z^2 + \frac{1}{2}z}{z^2 - z + \frac{1}{2}} \qquad \text{for } |z| > \frac{\sqrt{2}}{2}$$

The first step is to translate the denominator coefficients into the form of entry 8 or 9 in Table 6.2; namely, to calculate ωT and aT from

$$e^{-aT} = \frac{1}{2} \quad \text{and} \quad 2e^{-aT} \cos \omega T = 1$$

From these, $\omega T = \pi/4$ and $aT = \ln \sqrt{2}$. Next, using these values, we establish the relative parts of the sine and cosine terms.

$$F(z) = 3\,\frac{z^2 + \dfrac{z}{6}}{z^2 - z + \frac{1}{2}} = 3\left(\frac{z^2 - bz}{z^2 - z + \frac{1}{2}} + \frac{c \cdot \frac{1}{2}z}{z^2 - z + \frac{1}{2}}\right)$$

where $b = e^{-aT} \cos \omega T = \frac{1}{2} = e^{-aT} \sin \omega T$, so that $-b + c/2 = \frac{1}{6}$, which implies that $c = \frac{4}{3}$. Finally, using entries 8 and 9 of Table 6.2,

$$\begin{aligned} f(n) &= e^{-at}(3 \cos t + 4 \sin t)\Big|_{t=n} \\ &= \left(\frac{\sqrt{2}}{2}\right)^n \left(3 \cos \frac{n\pi}{4} + 4 \sin \frac{n\pi}{4}\right) \qquad \text{for } n = 0, 1, 2, \ldots \\ &= \{3, 3.5, 2, 0.25, -0.75, -0.875, -0.5, -0.625, \ldots\} \\ &\quad\ \uparrow \end{aligned}$$

Having developed some proficiency in working with z-transforms, particularly concerning inversion, let us now consider some minor modifications to the methods previously discussed.

Modified Partial-Fraction Expansion. As we have seen, with a normal partial-fraction term such as $1/(z - a)$ for $|z| > |a|$, the inverse is a sequence that is delayed one step (see entry 1a in Table 6.1); whereas if the numerator had a factor z, the inverse sequence would begin at $n = 0$. Thus, we introduce a modified partial-fraction expansion whose terms have this extra z-factor.

Case 1. If $F(z)$ has a zero at $z = 0$, then a standard expansion can be performed on $F(z)/z$ subsequently using the z-factor in the numerator of the expanded terms. Table 6.3 is provided for this method.

Example 6.19

From Example 6.7 we invert

$$F(z) = \frac{z}{(z-1)(z-2)} \qquad \text{for } |z| > 2$$

using the suggested modification.

$$\frac{F(z)}{z} = \frac{1}{(z-1)(z-2)} = \frac{-1}{z-1} + \frac{1}{z-2}$$

$$\therefore \quad F(z) = \frac{-z}{z-1} + \frac{z}{z-2} \qquad \text{for } |z| > 2$$

and from Table 6.3,

$$f(n) = -1 + 2^n \qquad \text{for } n \geq 0$$

Compare this with the result of Example 6.7.

Consider again a general rational function $F(z)$ in factored form, that is, repeating (6.31)

$$F(z) = \frac{q(z)}{p(z)} = \frac{k(z - z_1)(z - z_2) \cdots (z - z_M)}{(z - p_1)(z - p_2) \cdots (z - p_N)} \tag{6.44}$$

where q and p are polynomials in factored form.

Case 2. The poles p_i for $i = 1, 2, \ldots, N$ are all different (possibly complex)

TABLE 6.3 AA SHORT TABLE OF z-TRANSFORM PAIRS

Elementary Terms $F(z)$, $\|z\| > a$	Sequence Values $f(n)$, $n = 0, 1, 2, \ldots$
1. $\dfrac{z}{z-a}$	a^n
2. $\dfrac{z}{(z-a)^2}$	na^{n-1}
3. $\dfrac{z}{(z-a)^3}$	$\dfrac{1}{2} n(n-1)a^{n-2}$
4. $\dfrac{z}{(z-a)^m}$	$\dfrac{1}{(m-1)!} \prod_{k=0}^{m-2} (n-k)a^{n-m+1}$

numbers, but F need not have a zero at $z = 0$ as in Case 1. We now seek the expansion

$$F(z) = a_0 + a_1 \frac{z}{z - p_1} + a_2 \frac{z}{z - p_2} + \cdots + a_N \frac{z}{z - p_N} \tag{6.45}$$

where it has been temporarily assumed that F has no pole at zero. To determine a_0, evaluate $F(z)$ at $z = 0$ (since all other terms drop out).

$$a_0 = F(0) \tag{6.46}$$

Note that a_0 may be nonzero even when $F(z)$ is a proper rational function ($M < N$) using this method. To determine the remaining coefficients a_k for $k = 1, 2, \ldots, N$, use

$$a_k = \frac{z - p_k}{z} F(z) \bigg|_{z = p_k} \tag{6.47}$$

Example 6.20

To invert

$$F(z) = \frac{z + 3}{(z - 1)(z - 2)} \qquad \text{for } |z| > 2$$

Expanding, we obtain

$$F(z) = a_0 + \frac{a_1 z}{z - 1} + \frac{a_2 z}{z - 2}$$

$$a_0 = \frac{3}{2}, \quad a_1 = \frac{z + 3}{z(z - 2)}\bigg|_{z = 1} = -4, \quad a_2 = \frac{5}{2}$$

Therefore, from Table 6.3,

$$f(n) = \frac{3}{2}\,\delta(n) - 4 + \frac{5}{2} \cdot 2^n \qquad \text{for } n \geq 0$$

For the case where $F(z)$ has a pole at the origin, the expansion deviates slightly from (6.45). Consider the following example.

Example 6.21

To invert

$$F(z) = \frac{z + 1}{z^2(z - 1)}$$

we first expand into

$$F(z) = a_0 + \frac{a_1}{z} + \frac{a_2}{z^2} + \frac{b_1 z}{z - 1}$$

where we find that $a_0 = a_1 = -2$, $a_2 = -1$, and $b_1 = 2$. The second two terms can be inverted using entry 5a of Table 6.1.

Multiple-Order Poles. The present method readily generalizes as did the standard expansion.

Assuming as in (6.35) an mth-order pole at $z = p$, that part of the expansion associated with this pole is written as

$$F(z) = \frac{g(z)}{(z-p)^m} = \cdots + \frac{a_1 z}{z-p} + \frac{a_2 z}{(z-p)^2} + \cdots + \frac{a_m z}{(z-p)^m} \tag{6.48}$$

Compare this with expression (6.36). These terms can be found in Table 6.3. In an expansion like that of (6.48), the coefficients can be evaluated by

$$a_{m-k} = \frac{1}{k!}\frac{d^k}{dz^k}\left[\frac{(z-p)^m}{z}F(z)\right]\bigg|_{z=p} \tag{6.49}$$

for $k = 0, 1, 2, \ldots, m-1$. Because of the similarity to the development following (6.37), the proof of (6.49) is omitted.

Example 6.22

To invert

$$F(z) = \frac{z^3 + 2z^2 + z + 1}{(z-2)^2(z+3)} \qquad \text{for } |z| > 3$$

Expanding, we obtain

$$F(z) = a_0 + \frac{a_1 z}{z-2} + \frac{a_2 z}{(z-2)^2} + \frac{b_1 z}{z+3}$$

where

$$a_0 = F(0) = \frac{1}{12}$$

and

$$b_1 = \frac{z^3 + 2z^2 + z + 1}{z(z-2)^2}\bigg|_{z=-3} = \frac{11}{75}$$

From (6.49) ($k = 0$ and 1),

$$a_2 = g(z)\bigg|_{z=2} = \frac{z^3 + 2z^2 + z + 1}{z(z+3)}\bigg|_{z=2} = \frac{19}{10}$$

$$a_1 = \frac{dg(z)}{dz}\bigg|_{z=2} = \frac{77}{100}$$

Thus, from Table 6.3,

$$f(n) = \frac{1}{12}\delta(n) + \frac{11}{75}(-3)^n + \frac{77}{100}(2)^n + \frac{19}{10}n(2)^{n-1} \qquad \text{for } n \geq 0$$

With this example we conclude this chapter on z-transform manipulation. In the next chapter we apply the methods gained here to solutions of difference equations and to analysis of discrete-time systems generally.

6.5 PROBLEMS

6.1. Determine the z-transform of the following discrete-time signals and the associated region of absolute convergence.

(a) $x_a(n) = \begin{cases} 1 & \text{for } n \geq 0 \\ 2^n & \text{for } n < 0 \end{cases}$

(b) $x_b(n) = \begin{cases} 2^n & \text{for } n \geq 0 \\ 3^n & \text{for } n < 0 \end{cases}$

(c) $x_c(n) = nu(n)$

(d) $x_d(n) = n^2u(n)$

6.2. Calculate the z-transform of the sequence

$$x(n) = x_a(n) - x_b(n)$$

where x_a and x_b are taken from Problem 6.1. Include the region of absolute convergence.

6.3. Determine the z-transform for $x(n) = (n + 1)^2u(n + 1)$.

6.4. State all the possible convergence regions for the transform

$$F(z) = \frac{3z(z - 4)}{(z - 1)(z - 2)(z - 3)}$$

For which convergence region is $F(z)$ the transform of a causal (positive-time) sequence?

6.5. Solve the following difference equations by z-transforms:

(a) $y(n + 1) + 10y(n) = 2^nu(n)$ for $y(0) = 1$

(b) $y(n + 2) + 5y(n + 1) + 6y(n) = \delta(n)$ for $y(0) = y(1) = 0$

6.6. Expand $z/(z - 1)$ in a series

(a) that is valid for $|z| > 1$

(b) that is valid for $|z| < 1$

What two sequences have this transform?

6.7. Use the inversion formula, expression (6.30), to find the sequence causal $x(n)$ whose transform is

(a) $X_a(z) = \dfrac{z(z - 3)}{(z - 1)(z - 2)}$

(b) $X_b(z) = \dfrac{4(z - 3)}{z(z - 1)(z - 2)^2}$

(c) $X_c(z) = \dfrac{1}{(z - 1)^3(z - 2)}$

6.8. Repeat Problem 6.7 using partial-fraction expansion and table look-up. Check your results in each case by performing long division to obtain the first three terms of the sequence.

6.9. Determine the z-transform of the causal periodic sequence

$$x = \{\ldots, 0, 0, \underset{\uparrow}{1}, -1, 1, -1, 1, -1, \ldots\}$$

State the region of convergence.

6.10. Given that the unit-pulse response of a discrete-time system is $h(n) = \exp\{-n\}$, determine what input signal is required so that the output $y(n) = \delta(n-2)$.

6.11. An RC network has a transfer function given by

$$H(s) = \frac{1}{\tau s + 1}$$

Find an equivalent discrete-time system whose unit-pulse response sequence values the exact samples of the network's unit-impulse response function.

6.12. Ecologists have found that by measuring both riverflow and rainfall over the watershed region, a model can be established for the "system" that can be used for calculating flood conditions. In particular, with $r(n)$ as a sequence of rainfall measurements and $f(n)$ as a corresponding sequence of riverflow measurements, we can identify the unit-pulse response sequence $h(n)$ by a process of "deconvolution." Since $F(z) = H(z)R(z)$, we recognize deconvolution as finding $h(n)$ such that $H(z) = F(z)/R(z)$. Determine $h(n)$ given that hourly measurements provide the data:

$$r(n) = \{\underset{\uparrow}{0}, 10, 12, 6.1, 2.7, 0.7, 0, 0, \ldots\}$$

and

$$f(n) = \{\underset{\uparrow}{360}, 377, 424, 466, 471, 452, 431, 412, 395, 381, 371, 366, 362, 360, 360, 360, \ldots\}$$

Assume that both sets of data are in compatible units such as cubic meters per second. Hint: First modify $f(n)$ to obtain the increased flow above the nominal flow of 360. How many terms of $h(n)$ are warranted?

6.13. Modeling the national economy is one method that is used to help predict the effects of government control such as changing the prime interest rate, $P(n)$. Let $N(n)$ be the national income (GNP); $C(n)$, the annual consumption; and $I(n)$, the annual investment. A simple model that says that increments in $N(n)$ are proportional to $I(n)$ and $C(n)$ is linearly related to $N(n)$ and $P(n)$ is

$$N(n+1) - N(n) = aI(n)$$

$$C(n) = bN(n) - dP(n)$$

where $N(n) = C(n) + I(n)$. Find the discrete-time transfer function $N(z)/P(z)$.

7

Transfer Functions

In this chapter we emphasize the great utility of the transform concept as applied to physical system models, both continuous- and discrete-time. Although the basic idea of transfer function has already been introduced both in Chapters 5 and 6, we need to develop our facility to use transfer functions effectively in system analysis.

7.1 INTRODUCTION

In subsequent discussions we assume that all signals are single-sided, causal signals whether they be of a continuous- or a discrete-time nature. Specifically, if $x(t)$ is a *single-sided, continuous-time* signal, $x(t)$ is defined for $t \geq 0$. As far as we are concerned, $x(t)$ may also be taken to be identically zero for $t < 0$. Similarly, a *single-sided, discrete-time* signal $x(k)$ is understood to be zero for negative index, $k < 0$.

By restricting our interest to single-sided signals, we can focus attention on realistic analysis problems such as determining response signals from systems that have both internally stored energy at $t = 0$ and excitation signals beginning at $t = 0$. (It is usually convenient to set the origin of the time axis to coincide with the time that excitation begins.)

To accomplish our task, we use the single-sided Laplace transform, which is a special case of that studied in detail in Chapter 5, and the single-sided z-transform, which was introduced in the previous chapter. Following a brief development of these tools, we then look at a wide variety of application problems.

Much of what we present in this chapter will help to provide a bridge between the reader's background in network theory and the more general concepts of system analysis.

7.2 THE SINGLE-SIDED LAPLACE TRANSFORM

Applying the Laplace transform, as studied in Chapter 5, to a single-sided signal $x(t)$ effectively defines the single-sided Laplace transform of $x(t)$ as

$$\mathscr{L}_+[x(t)] = X(s) = \int_{0-}^{\infty} x(t)e^{-st}\,dt \tag{7.1}$$

The lower limit on the integral reflects the fact that $x(t)$ is taken to be zero for negative time. Alternatively, one could view expression (7.1) as the defining relationship of a new transform. In either case, the properties of the $\mathscr{L}_+$ transform (the subscript $+$ is used to emphasize that the signal operated on is defined only for positive time) closely follow those developed in Chapter 5.

Linearity and Convergence

The $\mathscr{L}_+$ transform of expression (7.1) is linear for the same reason that the two-sided transform is linear. There is, fortunately, a dividend to be gained by the restriction to causal signals: The convergence region for $X(s)$ in equation (7.1) is *always* of the form

$$\text{Re}(s) > \sigma_0 \tag{7.2}$$

for exponentially bounded signals $x(t)$. As will be seen, this consistency removes the confusion inherent in specifying the region of convergence to such an extent that many times it can be entirely omitted without ambiguity.

Time Differentiation

Another benefit of the $\mathscr{L}_+$ transform is that initial conditions are handled automatically. For example, in transforming a differential equation term, for instance, $\dot{x}(t)$,

$$\mathscr{L}_+[\dot{x}(t)] = sX(s) - x(0^-) \tag{7.3}$$

where $X(s)$ is the transform of the undifferentiated signal $x(t)$, and $x(0^-)$ is the value of $x(t)$ at time zero minus (0^-). The use of *zero minus* is not universal. Some authors prefer zero plus (0^+) as their initial time. However, 0^- is used here for the reason that certain signals may have impulsive components at $t = 0$, and by using 0^- the transform includes these components.

The proof of expression (7.3) is as follows. From the definition

$$\mathscr{L}_+[\dot{x}(t)] = \int_{0-}^{\infty} \dot{x}(t)e^{-st}\,dt \tag{7.4}$$

Integrating by parts, we obtain

$$\int_{0-}^{\infty} e^{-st}\,dx = e^{-st}x(t)\Big|_{t=0-}^{\infty} + s\int_{0}^{\infty} x(t)e^{-st}\,dt \tag{7.5}$$

And evaluating the first term on the right side,

$$e^{-st}x(t)\Bigg|_{t=\infty} - e^{-st}x(t)\Bigg|_{t=0^-} = -x(0^-) \tag{7.6}$$

In the first instance it is assumed that $|x(t)| < Me^{\sigma_0 t}$ (exponentially bounded) for some constants M and σ_0, so that for Re $(s) > \sigma_0$, the first part vanishes. Note the convergence region as mentioned in (7.2). Finally, introducing the result of expression (7.6) into (7.5), we arrive at the required result as stated in (7.3).

Applying the formula (7.3) to $\dot{g}(t)$, where $g(t) = \dot{x}(t)$, we see that similar expressions may be generated for higher derivatives.

$$\mathscr{L}_+[\dot{g}(t)] = sG(s) - g(0^-)$$

which translates into

$$\mathscr{L}_+[\ddot{x}(t)] = s[sX(s) - x(0^-)] - \dot{x}(0^-) \tag{7.7}$$

Rewriting (7.7), we obtain

$$\mathscr{L}_+[\ddot{x}(t)] = s^2X(s) - sx(0^-) - \dot{x}(0^-) \tag{7.8}$$

More generally, we have

$$\mathscr{L}_+[x^{(n)}(t)] = s^nX(s) - s^{n-1}x(0^-) - s^{n-2}x^{(1)}(0^-) - \cdots - x^{(n-1)}(0^-) \tag{7.9}$$

where $x^{(k)}(t)$ represents the kth derivative of $x(t)$ with respect to time.

Properties of the Laplace Transform

In this section we summarize the remaining properties of $\mathscr{L}_+$ in Table 7.1. Comparing Table 7.1 with Table 5.2, it is clear that the majority of the properties of $\mathscr{L}_+$ carry over directly from the double-sided Laplace transform. However, let us briefly review the important time-convolution property, stressing the single-sided nature of the signals involved.

Recall that the convolution of $h(t)$ and $x(t)$ is represented by

$$y(t) = \int_{-\infty}^{\infty} h(\tau)x(t-\tau)\,d\tau \tag{7.10}$$

or equivalently by making the change of variables $\tau = t - \lambda$,

$$y(t) = \int_{-\infty}^{\infty} x(\lambda)h(t-\lambda)\,d\lambda \tag{7.11}$$

In other words, the roles of x and h may be interchanged. Now, let us assume that both $x(t)$ and $h(t)$ are single-sided, causal time signals, so that

$$x(t) = 0 \quad \text{and} \quad h(t) = 0 \qquad \text{for } t < 0 \tag{7.12}$$

In this context, expression (7.10) may be written as

$$y(t) = \int_{0^-}^{t} h(\tau)x(t-\tau)\,d\tau \tag{7.13}$$

Table 7.1 PROPERTIES OF THE SINGLE-SIDED LAPLACE TRANSFORM

Property	t-Domain	s-Domain
1. Linearity	$ax(t) + by(t)$	$aX(s) + bY(s)$
2. Time differentiation	$\dot{x}(t)$	$sX(s) - x(0^-)$
	$\ddot{x}(t)$	$s^2X(s) - sx(0^-) - \dot{x}(0^-)$
3. Time integration	$\int_0^t x(t)\, dt$	$\frac{1}{s} X(s)$
4. Time shift	$x(t - a)u(t - a)$	$e^{-as}X(s)$
5. s-Domain shift	$e^{at}x(t)$	$X(s - a)$
6. s-Domain differentiation	$-tx(t)$	$\frac{dX(s)}{ds}$
7. Final-value theorem	$\lim_{t \to \infty} x(t) = \lim_{s \to 0} sX(s)$	
provided that $sX(s)$ has only poles with negative real part.		

since $h(\tau) = 0$ for $\tau < 0$ and $x(t - \tau) = 0$ for $(t - \tau) < 0$ or $\tau > t$. Let us develop the $\mathscr{L}_+$ transform of expression (7.13).

$$Y(s) = \int_{0-}^{\infty} \left[\int_{0-}^{\infty} h(\tau)x(t - \tau)\, d\tau \right] e^{-st}\, dt \tag{7.14}$$

The upper limit on the inner integral has been set to ∞ since the integral from t to ∞ has a zero contribution. Based on the convergence of the transform, we can interchange integrations to obtain

$$Y(s) = \int_{0-}^{\infty} h(\tau)e^{-s\tau} \int_{0-}^{\infty} x(t - \tau)e^{-st}e^{s\tau}\, dt\, d\tau \tag{7.15}$$

where the factor $e^{-s\tau}e^{s\tau} = 1$ has been incorporated. Collecting terms appropriately and letting $\lambda = t - \tau$,

$$Y(s) = \int_{0-}^{\infty} h(\tau)e^{-s\tau}\, d\tau \int_{-\tau}^{\infty} x(\lambda)e^{-s\lambda}\, d\lambda \tag{7.16}$$

Finally, the second integral's lower limit can be set to 0^- since $x(\lambda) = 0$ for $-\tau < \lambda < 0$; and we recognize that the result is

$$Y(s) = H(s)X(s) \tag{7.17}$$

just as in the case of the double-sided transform.

Equation (7.17) is a restatement of the transfer-function concept. That is to say, we can consider $x(t)$ as an input to, and $y(t)$ as the resulting output of a system with a transfer function $H(s)$. Thus, the *transfer function* of a system is the Laplace transform of the unit-impulse response of the system, as follows:

$$H(s) = \mathscr{L}\{h(t)\} \tag{7.18}$$

Recall that the impulse response of a causal (or physically realizable) system always satisfies the single-sided criterion

$$h(t) = 0 \qquad \text{for } t < 0 \tag{7.19}$$

so that the transform in (7.18) reduces to $\mathscr{L}_+$.

Looking back to the convolution expression (7.13), we see that, for ordinary signals $h(t)$ and $x(t)$ without impulsive components, the initial output $y(0)$ is zero. In fact, to put the convolution integral into perspective, $y(t)$ in expression (7.13) is the *forced response* (also known as a *particular solution* or the *zero-initial-energy* response) portion of the total response. The other contribution to the total response is derived from any initial energy in the system. This aspect is discussed more fully in the next section.

7.3 INITIAL-VALUE PROBLEMS

Perhaps the strongest motivation for introducing the single-sided $\mathscr{L}_+$ transform is in the use of this transform to analyze problems that involve systems with initial energy. We know from classical solutions of differential equations that the solution is conveniently considered as two separate parts: (1) the *particular solution*, or as preferred in engineering literature, the *forced solution*, which is based on the form of the input, and (2) the *homogeneous solution*, or, more commonly, the *natural solution*, which is the solution to the equation with no input, but with the given initial conditions in force. To be more specific, let us consider a first-order differential equation that might describe the input-output relationship of a simple RC network (see Example 4.5):

$$\tau\dot{y}(t) + y(t) = u(t) \qquad \text{for } y(0) = y_0 \tag{7.20}$$

where τ is a constant (for example $\tau = RC$), $y(t)$ is the output signal (perhaps the voltage across the capacitor), and $u(t)$ represents the input signal (voltage) that is assumed to be a known time variation. In addition, the value y_0 is required to specify the complete solution. We use our new transform tool to solve the equation. By taking the $\mathscr{L}_+$ transform of equation (7.20), we obtain

$$\tau[sY(s) - y_0] + Y(s) = U(s) \tag{7.21}$$

Solving for $Y(s)$,

$$Y(s) = \frac{1}{\tau s + 1}\,U(s) + \frac{\tau y_0}{\tau s + 1} \tag{7.22}$$

This last equation exhibits the two components of the total solution, once we invert transform to get $y(t)$. The first term depends on the input $u(t)$ and not on the initial condition, whereas for the second, the situation is just the reverse. The reader will recognize the coefficient of $U(s)$ in expression (7.22) as the transfer function of the system according to (7.17).

Now, since the inverse Laplace transform techniques of Section 5.9 apply equally well to the special case of the single-sided Laplace transform, we will write down the expression for $y(t)$. Keep in mind that the region of convergence for $Y(s)$ in expression (7.22) is $\text{Re}(s) > \sigma_0$ for some unexpressed value σ_0. Consequently, all of $Y(s)$ is causal (thereby simplifying step IV in the procedure outlined in Table 5.3). As an aid to step V of Table 5.3, the reader is referred to Table 7.2, which lists some of the most useful (single-sided) Laplace transform pairs. Finally, with $\tau = y_0 = 1$ and $U(s) = 1/s$, we have from expression (7.22) that

$$Y(s) = \left[\frac{1}{s(s+1)}\right] + \frac{1}{s+1} \tag{7.23}$$

Expanding the first term,

$$\frac{1}{s(s+1)} = \frac{1}{s} + \frac{-1}{s+1}$$

Table 7.2 A SELECTION OF SINGLE-SIDED LAPLACE TRANSFORM PAIRS[a]

	Time Function	Laplace Transform
1.	$\delta(t)$ unit-impulse	1
2.	$u(t)$ unit-step	$\frac{1}{s}$
3.	t	$\frac{1}{s^2}$
4.	$t^n,\ n = 0, 1, 2, \ldots$	$\frac{n!}{s^{n+1}}$
5.	e^{-at}	$\frac{1}{s+a}$
6.	$\sin \omega_0 t$	$\frac{\omega_0}{s^2+\omega_0^2}$
7.	$\cos \omega_0 t$	$\frac{s}{s^2+\omega_0^2}$
8.	$e^{-at} \sin \omega_0 t$	$\frac{\omega_0}{(s+a)^2+\omega_0^2}$
9.	$e^{-at} \cos \omega_0 t$	$\frac{(s+a)}{(s+a)^2+\omega_0^2}$
10.	$e^{-at}(A \cos \omega t + B \sin \omega t)$	$\frac{A(s+a)+B\omega}{(s+a)^2+\omega^2}$

[a]Used in conjunction with Table 7.1, this set of transform pairs should be found to be adequate.

and from Table 7.2, entries 2 and 5,

$$y(t) = [1 - 3^{-t}] + e^{-t} = 1 \qquad \text{for } t \geq 0$$

The total solution is a constant unity for (all) positive time. This unexpected event is the result of the initial value y_0 being exactly the right value for the natural response to cancel the transient part of the forced solution.

Time Constants

Let us further consider the simple system of expression (7.20). Suppose $y_0 = 0$ and $u(t)$ is again a unit-step input. Working through the solution once more, we find that

$$y(t) = 1 - e^{-t/\tau} \qquad \text{for } t \geq 0 \tag{7.24}$$

The output is obviously approaching a steady-state value of unity since the exponential term vanishes as t becomes large. But it is important to know how fast this happens. Engineers refer to the constant τ as the *time constant* of the system. Figure 7.1 illustrates $y(t)$ in expression (7.24) along with the exponential term itself. From this figure one can see that knowing the time constant provides a great deal of information regarding the response.

Note that the value of τ is the initial slope of $e^{-t/\tau}$ and, more importantly, that at each integer multiple of τ, the amplitude of $e^{-t/\tau}$ decreases by a factor of e^{-1}. For typical engineering calculations, the transient part is said to be significant for about

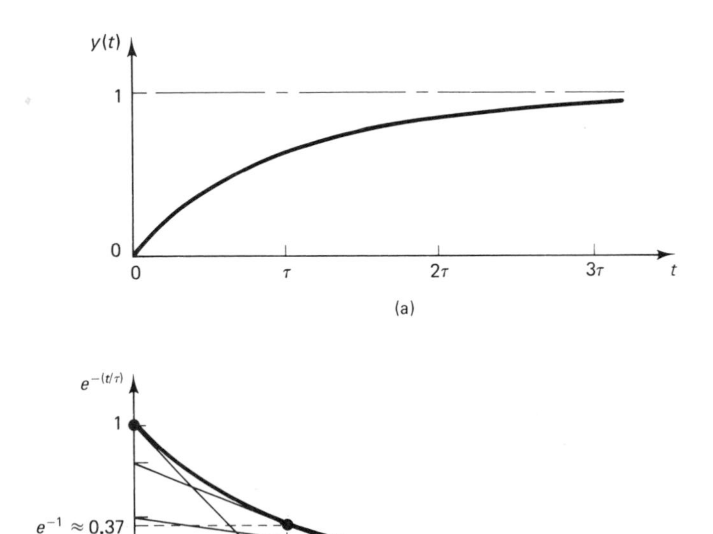

Figure 7.1 (a) The output $y(t)$ of expression (7.24) and (b) the transient term.

four time constants. (At four time constants the amplitude is less than 2 percent of its original value.)

Example 7.1

To illustrate the use of the time constant, consider that a particular device has been given to you, and that from the physical construction of the device, you know that the transfer function has the form

$$T(s) = \frac{K}{(s + a)}$$

(This might apply to a d.c. servo-motor between the amplifier input and the tachometer output.)

One solution is to apply a unit-step input and measure both the steady-state solution and the time constant of the response. With these values, both K and a can be determined: (1) steady-state value $= K/a$ and (2) time constant $= 1/a$.

Check this steady-state value with the final-value theorem given in Table 7.1.

Switched Electrical Networks and Second-Order Response

To build upon the reader's background in network analysis, we first consider the transient analysis of some relatively simple networks whose configuration is altered by a switch (or switches) at $t = 0$. Consider the series R-L-C network of Fig. 7.2.

We assume that the input voltage is a known function of time and that from some unspecified causes, the inductor L and capacitor C have initial current $i(0^-)$ and intial voltage $v(0^-)$, respectively, with polarities shown in the figure. Furthermore, we take the loop current as the output signal. Writing the Kirchhoff voltage law (KVL) equation for $t \geq 0$,

$$u(t) = Ri(t) + L\frac{di}{dt} + \frac{1}{C}\int i(t)\,dt \tag{7.25}$$

Since the indefinite integral may be thought of as an integral from $-\infty$ to t, we write that

$$\frac{1}{C}\int i(t)\,dt = \frac{1}{C}\int_{-\infty}^{0^-} i(t)\,dt + \frac{1}{C}\int_{0^-}^{t} i(t)\,dt \tag{7.26}$$

The first integral on the left side is interpreted as the total charge on the capacitor at

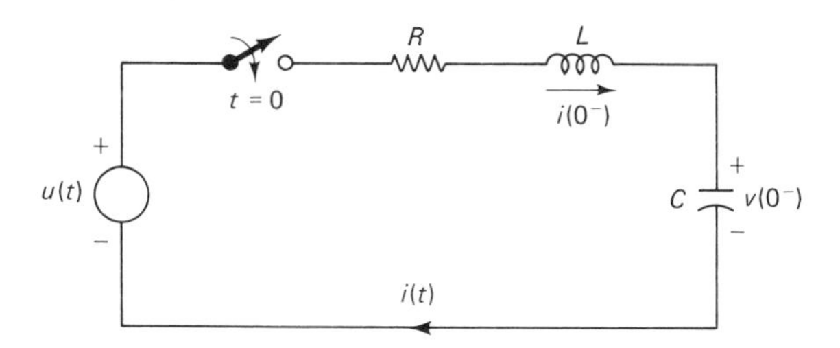

Figure 7.2 A series R-L-C network.

$t = 0^-$ divided by C, but this is just the initial voltage $v(0^-)$. Thus, expression (7.25) becomes

$$u(t) = Ri(t) + L\frac{di(t)}{dt} + v(0^-) + \frac{1}{C}\int_0^t i(t)\,dt \tag{7.27}$$

Taking the Laplace transform of this equation,

$$U(s) = RI(s) + L[sI(s) - i(0^-)] + \frac{v(0^-)}{s} + \frac{1}{sC}I(s) \tag{7.28}$$

Note that $v(0^-)$ is a constant and therefore transforms into a $v(0^-)/s$ term in the transform domain. Solving for $I(s)$,

$$I(s) = \frac{U(s) + \left[Li(0^-) - \dfrac{v(0^-)}{s}\right]}{R + sL + \dfrac{1}{sC}} \tag{7.29}$$

We see that the transfer function is $(R + sL + 1/sC)^{-1}$ and that the numerator is grouped to show the terms that relate to the forced solution and natural solution, respectively. To be more specific, assume that $u(t)$ is a unit-step input and that $i(0^-) = 1$ ampere (A) and $v(0^-) = 1$ volt (V). Introducing these values into (7.29), $I(s)$ reduces to

$$I(s) = \frac{s}{s^2 + as + b} \tag{7.30}$$

where $a = R/L$ and $b = 1/LC$. (In this case the forced response has been cancelled with part of the natural response.) We would expect the current to go to zero as t becomes large because the constant input voltage is eventually counteracted with the charged capacitor voltage, but it is worthwhile to investigate the types of response that we can have from equation (7.30) as a function of the parameters a and b.

Second-order response. By studying the poles of $I(s)$ in expression (7.30), we can see how totally different time responses can be expected. The roots of $s^2 + as + b = 0$ for a and b positive numbers can be categorized as follows:

1. Roots are real and distinct.
2. Roots are real and equal.
3. Roots are complex conjugates.

The key to which type of roots we have is the discriminant of the quadratic formula,

$$D = (a^2 - 4b) \tag{7.31}$$

If D is positive, case 1 holds true; if D is zero, case 2; and if D is negative, we have case 3. Consider the following example.

Example 7.2

Calculate the current waveform $i(t)$ from (7.30) for the three parameter sets

(a) $a = 3, \quad b = 2$

(b) $a = 2, \quad b = 1$

(c) $a = 2, \quad b = 2$

Solution

(a)
$$I(s) = \frac{s}{s^2 + 3s + 2} = \frac{s}{(s + 1)(s + 2)} = \frac{-1}{s + 1} + \frac{2}{s + 2}$$
$$i(t) = (2e^{-2t} - e^{-t}), \text{ for } t \geq 0$$

(b)
$$I(s) = \frac{s}{s^2 + 2s + 1} = \frac{s}{(s + 1)^2} = \frac{1}{s + 1} + \frac{-1}{(s + 1)^2}$$
$$i(t) = e^{-t}(1 - t), \text{ for } t \geq 0.$$

(c)
$$I(s) = \frac{s}{s^2 + 2s + 2} = \frac{s}{(s + 1)^2 + 1} = \frac{(s + 1) - 1}{(s + 1)^2 + 1}$$
$$i(t) = e^{-t}(\cos t - \sin t) \qquad \text{for } t \geq 0$$

These three distinct types of second-order system response are called *overdamped* for case (a), *critically damped* for case (b), and *underdamped* for case (c). We complete this section with one final example of a switched network.

Example 7.3

A network is presented in Fig. 7.3. The battery voltage V_B has been connected for a long time prior to the opening of the switch S at $t = 0$. It is desired to find the capacitor voltage $v_c(t)$ for $t \geq 0$.

Solution. Before the switch is opened, we assume that the network is in a steady state. We can establish the steady-state currents and voltages by considering the capacitor C to be completely charged (and therefore passing no current) and the current through the indicator L to be constant (so that no voltage is developed across L). In this situation, the voltage across R is 9 volts and the current through R is 0.5 ampere.

In this manner we have established the initial inductor current to be $i_L(0^-) = 0.5$ A and the initial capacitor voltage $v_c(0^-) = 9$ V.

Consider what happens at the instant $t = 0$ when the switch S opens. Because $v_L = L \cdot di_L/dt$ cannot be infinite, $i_L(0^+) = i_L(0^-)$, and similarly $i_C = C\ dv_c/dt$ must be finite so that $v_c(0^+) = v_c(0^-)$.

The problem after $t = 0$ is one of the R-L-C circuit shown in Fig. 7.3 with no input. The KVL equation is

$$Ri_L(t) + L\frac{di_L}{dt} + \frac{1}{C}\int_0^t i_L(t)\, dt = v_c(0) \tag{7.32}$$

where we have used our previous method to decompose the capacitor voltage integral into one with limits from 0 to t and a separate term for the initial capacitor voltage.

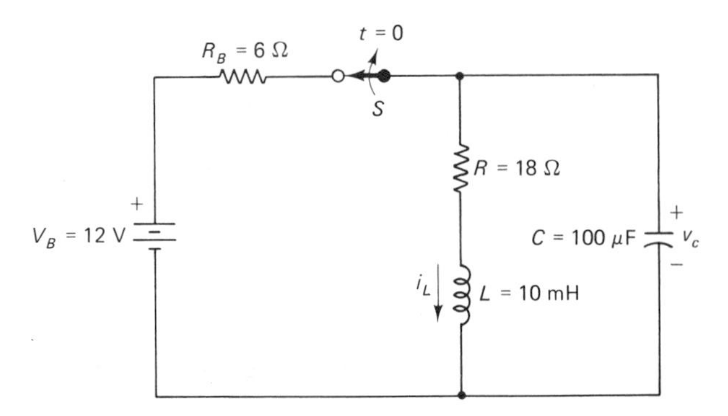

Figure 7.3 Network for Example 7.3.

Note that the initial capacitor voltage occurs on the right-hand side because the inductor current polarity is not consistent with the capacitor voltage polarity. In other words, we had to introduce a negative sign for the capacitor voltage, which equivalently transferred it to the right side of expression (7.32).

Let us first take the Laplace transform of (7.32), then later introduce the numerical values of the elements specified in Fig. 7.3.

$$RI_L(s) + L[sI_L(s) - i_L(0)] + \frac{1}{sC} I_L(s) = \frac{v_c(0)}{s}$$

Solving for $I_L(s)$,

$$I_L(s) = \frac{i_L(0)s + \dfrac{v_c(0)}{L}}{s^2 + \dfrac{R}{L} s + \dfrac{1}{LC}} \tag{7.33}$$

Substituting the numerical values,

$$I_L(s) = \frac{0.5s + 900}{s^2 + 1800s + 10^6} = \frac{0.5(s + 900) + 1.03(435.9)}{(s + 900)^2 + (435.9)^2}$$

From Table 7.2, entry 10,

$$i_L(t) = e^{-900t}[0.5 \cos 435.9t + 1.03 \sin 435.9t] \text{ amperes}$$

But since we want the capacitor voltage $v_c(t)$, we can write

$$V_c(s) = \frac{-1}{sC} I_L(s) + \frac{v_c(0)}{s}$$

Simplifying from expression (7.33),

$$V_c(s) = \frac{9s + 11{,}200}{s^2 + 1800s + 10^6} = \frac{9(s + 900) + 7.11(435.9)}{(s + 900)^2 + (435.9)^2}$$

So that for $t \geq 0$,

$$v_c(t) = e^{-900t}[9 \cos 435.9t + 7.11 \sin 435.9t] \text{ volts}$$

Notice that the small time constant (slightly more than 0.001 second) indicates that this transient condition would vanish in a few milliseconds for some not unreasonable network parameters.

In the next section we investigate transfer functions of continuous-time systems in more detail.

7.4 TRANSFER FUNCTIONS OF CONTINUOUS-TIME SYSTEMS

From Section 7.2 we know that the transfer function of a system $H(s)$ and the system's unit-impulse response $h(t)$ are transform pairs. This could serve as a definition of transfer function, but it is not convenient since $h(t)$ is typically difficult to obtain. Instead, we adopt the following useful definition.

Definition 7.1. The *transfer function*, $H(s)$, of a system with input excitation $x(t)$ and output response $y(t)$ is the ratio of output- to input-signal transforms under the condition of zero initial stored energy in the system:

$$H(s) = \left. \frac{Y(s)}{X(s)} \right|_{\substack{\text{zero initial energy} \\ \text{in the system}}} \tag{7.34}$$

Using this definition, we simply identify the input and output signals, Laplace transform the system equations with *zero initial conditions*, and form the *output- to input*-signal ratio.

Example 7.4

Referring to the first-order system described in expression (7.20), it is reasonable to assume the $u(t)$ and $y(t)$ are the input and output, respectively. In the application of (7.34), we disregard the initial condition to get

$$(\tau s + 1)Y(s) = U(s)$$

Solving for the ratio Y/U, gives us the transfer function for the system

$$\frac{Y(s)}{U(s)} = \frac{1}{\tau s + 1}$$

Network Functions

The concept of network functions such as impedance and admittance functions are special types of transfer functions. Briefly reviewing these basic ideas, an *impedance* is a ratio of a voltage transform to a current transform, which, in turn, implies that we think of the current being the input variable and the voltage, the output variable.

Example 7.5

A pure capacitance C, as we know, relates the current through it to the voltage across it by

$$v_c(t) = \frac{1}{C} \int i_c(t)\, dt \tag{7.35}$$

Transforming (with zero initial value of v_c),

$$V_c(s) = \frac{1}{sC} I_c(s)$$

It follows that the impedance of a capacitor is $1/sC$.

Example 7.6

Find the input impedance of the network in Fig. 7.4 in terms of the parameters R_1, R_2, L, and C.

Solution. The impedance of L is given by sL, which can be developed from $v_L(t) = L\ [di_L(t)]/dt$ in a manner similar to that in the previous example. We employ the technique of converting each element to an equivalent impedance, writing appropriate network equations, and solving for the requried signal ratio.

The input shown in Fig. 7.4 is the current signal $i_1(t)$, and the response signal required is the voltage $v_1(t)$ at the same port. For the analysis, we choose the "window loop" currents i_1 and i_2 and write for loop 2 (in the s-domain)

$$\left(R_2 + \frac{1}{sC}\right) I_2(s) - \frac{1}{sC} I_1(s) = 0 \tag{7.36}$$

Note that there is no need to write a KVL equation for loop 1 since $i_1(t)$ is assumed to be known. The expression for the output is, however, taken from loop 1, as follows:

$$V_1(s) = \left(R_1 + sL + \frac{1}{sC}\right) I_1(s) - \frac{1}{sC} I_2(s) \tag{7.37}$$

I_2 is an intermediate variable and may be eliminated. From equation (7.36),

$$I_2(s) = \left(R_2 + \frac{1}{sC}\right)^{-1} \left(\frac{1}{sC}\right) I_1(s) \tag{7.38}$$

Introducing $I_2(s)$ into expression (7.37), we can finally obtain the required ratio

$$\frac{V_1(s)}{I_1(s)} = R_1 + sL + \frac{R_2}{1 + R_2 Cs} \tag{7.39}$$

This expression is then the desired input impedance.

Remark. If $v_2(t)$ is defined as the voltage across R_2 in Fig. 7.4, the impedance $V_2(s)/I_1(s)$ would be called a *transfer impedance*. The reciprocal ratio of current to

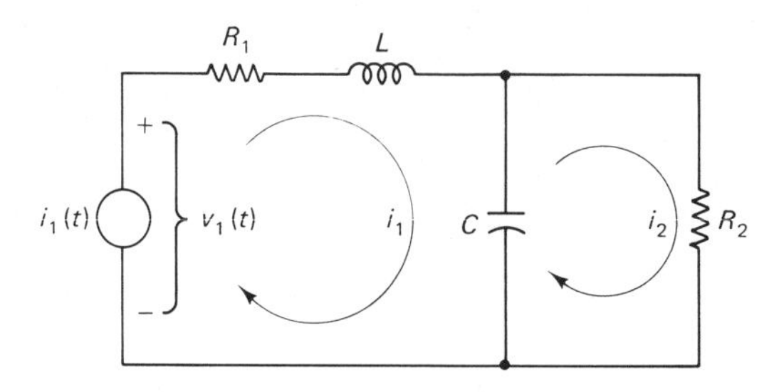

Figure 7.4 Network for Example 7.6.

voltage is called an *admittance function.* In this type of transfer function, the voltage would be considered as the input signal, with the current as the response.

Interconnected Systems and Block Diagrams

One of the most useful aspects of transfer functions is the ability to construct more complex systems out of simpler "building blocks." Linear systems, such as electrical networks and mechanical systems, are typically modeled as simultaneous differential equations (or integro-differential equations). And the Laplace transform can reduce such equations to algebraic functions of the complex variable s. To take advantage of the corresponding simultaneous equation that results in the s-domain, engineers have developed and extensively use a graphical representation of the equations that is referred to as a *block diagram.* This natural tool was introduced in Chapters 1 and 2.

Figure 7.5a illustrates a simple *gain block* with transfer function $G(s)$, input $X(s)$, and output $Y(s)$. A second operation, *summation,* is shown in Fig. 7.5b, with several inputs $X_1(s)$, $X_2(s)$, and $X_3(s)$, which sum algebraically (that is, the signs of the input connections are taken into account) to produce the output $Y(s)$.

Block diagrams are graphical representations of interdependent signals and systems. They provide an efficient means of displaying the connection information and individual cause-and-effect relationships. The signals present in one block may be used to connect to other blocks. Figure 7.6 presents a block diagram of an elementary feedback system; the term *feedback* is used to denote that the output signal $Y(s)$ is being returned to the input side through the feedback gain $G_2(s)$.

The equivalent set of simultaneous equations can be obtained by writing out the individual block and summing junction relationships. (A number of intermediate variables such as $U(s)$ and $V(s)$ may have to be added.)

$$\begin{aligned} U(s) &= X(x) - V(s) \\ V(s) &= G_2(s)Y(s) \\ Y(s) &= G_1(s)U(s) \end{aligned} \tag{7.40}$$

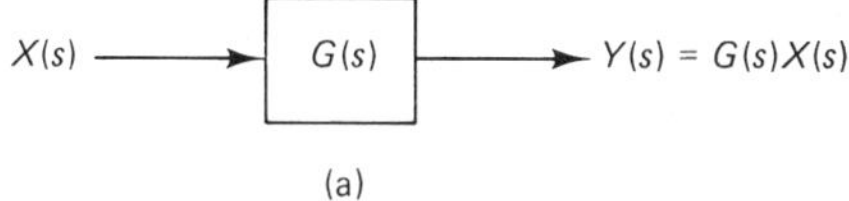

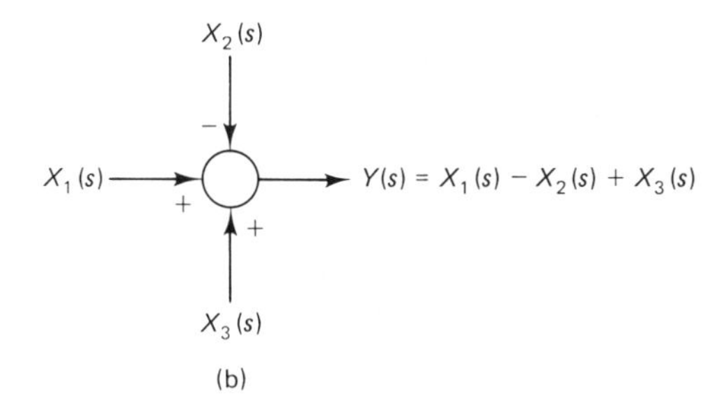

Figure 7.5 (a) Gain block and (b) summation junction.

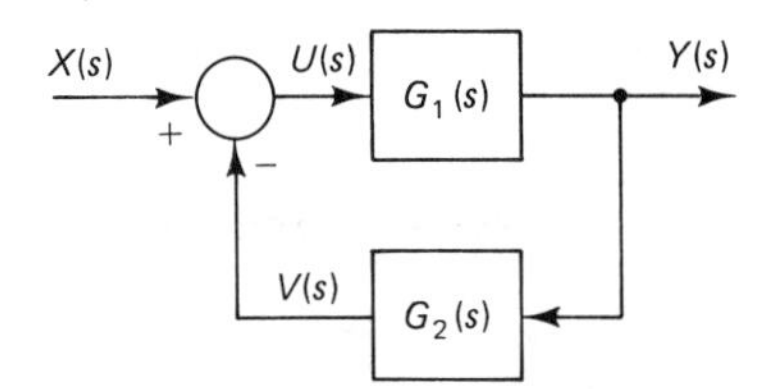

Figure 7.6 Block diagram of a feedback system.

Eliminating $U(s)$ and $V(s)$, we can finally arrive at the overall transfer function

$$\frac{Y(s)}{X(s)} = \frac{G_1(s)}{1 + G_1(s)G_2(s)} \tag{7.41}$$

A complicated system can be diagrammed by considering small subsystems and their interconnections. To illustrate this procedure as well as to show that the functional operation of a system is often clearer from the block diagram than from the basic physical system, consider the following example about a field-controlled d.c. motor.

Example 7.7

Figure 7.7a presents a schematic drawing of a d.c. motor and its mechanical load. It is assumed that external circuitry provides a constant current I_a through the armature winding of the motor. The motor then delivers a torque T, which is proportional to the field current with proportionality constant K. Three related equations may be written:

$$E(s) = (R + sL)I(s) \tag{7.42}$$

where R and L are the field-winding resistance and inductance, respectively;

$$T(s) = KI(s) \tag{7.43}$$

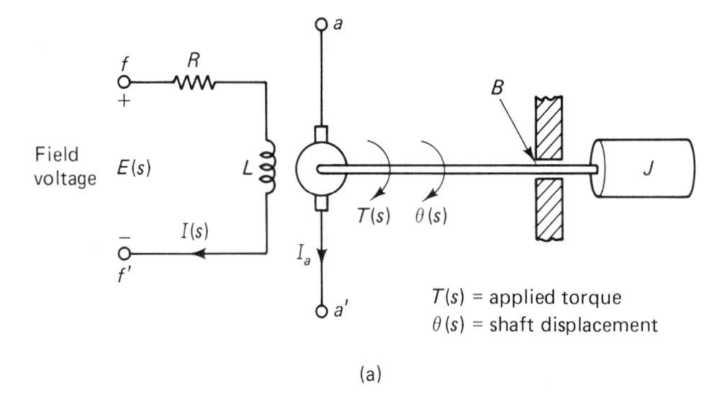

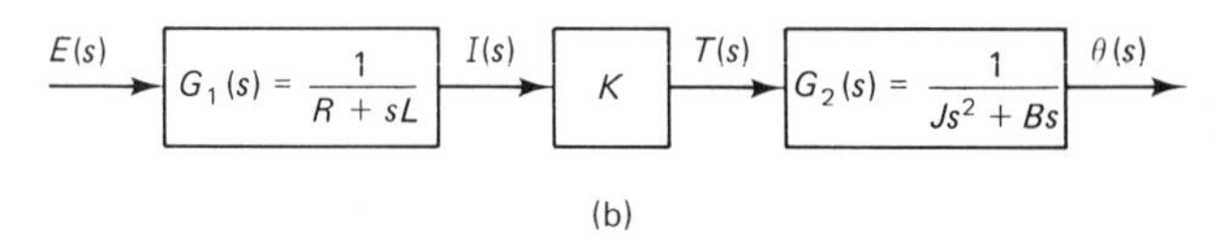

Figure 7.7 (a) A field-controlled d.c. motor and (b) the equivalent block diagram.

as mentioned above; and

$$T(s) = (Js^2 + Bs)\theta(s) \tag{7.44}$$

which represents the torque equation for the rotational inertia J with linear friction coefficient B. Separately, these equations are easily obtained, and they may be diagrammed as shown in Fig. 7.7b. Once diagrammed, the overall transfer function is given by the product of the three cascaded blocks:

$$\frac{\theta(s)}{E(s)} = KG_1(s)G_2(s) = \frac{K}{s(R + sL)(Js + B)} \tag{7.45}$$

This result may be obtained by eliminating $I(s)$ and $T(s)$ from expressions (7.42), (7.43), and (7.44).

In a later section we illustrate the basic concept of feedback as a design method for changing the response of a system.

7.5 DISCRETE-TIME TRANSFER FUNCTIONS

In Chapter 6 we introduced the single-sided z-transform. The development of the z-transform applied to causal signals proceeds in a similar fashion to that of the single-sided Laplace transform of Section 7.2.

For (causal) discrete-time signals $f(k)$, which are taken to be zero for negative time, we define the *single-sided z-transform* as follows:

$$F(z) = \mathscr{Z}\{f(k)\} = \sum_{n=0}^{\infty} f(n)z^{-n} \tag{7.46}$$

As an example of direct calculation, the unit-step signal $u(k)$ has the transform

$$\sum_{n=0}^{\infty} z^{-n} = \frac{1}{1 - z^{-1}} = \frac{z}{z - 1} \tag{7.47}$$

where the sum of an infinite geometric series was used, thereby restricting z to be in the (convergence) region $|z| > 1$. In like manner to the single-sided Laplace transform, all transforms of causal discrete-time signals converge outside some circle centered at the origin in the z-plane. Therefore, the actual convergence region is usually omitted. Table 6.2 provides a number of transform pairs for sequences $f(k)$ derived from continuous-time signals when $T = 1$.

Properties of the Single-Sided z-Transform

In this section we present the more useful properties in Table 7.3 for quick reference. Three of those properties listed are proved; the others follow from similar developments in the previous chapter. The most useful property for application to difference equations is the translation property, which we now investigate.

TABLE 7.3 PROPERTIES OF THE Z-TRANSFORM

Sequence Domain	z-Domain
$f(k)$	$F(z)$
$af(k)$	$aF(z)$
$f_1(k) + f_2(k)$	$F_1(z) + F_2(z)$
$f(k+1)$	$zF(z) - zf(0)$
$f(k+2)$	$z^2F(z) - z^2f(0) - zf(1)$
$f(k+N)$	$z^NF(z) - z^Nf(0) - \cdots - zf(N-1)$
$a^kf(k)$	$F(z/a)$
$f(0)$	$\lim_{z\to\infty} F(z)$[a]
$f(\infty)$	$\lim_{z\to 1} (z-1)F(z)$[b]
$\sum_{n=0}^{k} f(n)g(k-n)$	$F(z)G(z)$
$f(k-N)$	$z^{-N}F(z)$

[a]If the limit exists.

[b]If $(1 - z^{-1})F(z)$ has all its poles in the region $|z| < 1$.

Translation. The terms of a difference equation typically involve a time-shifted signal of the form $f(k+1)$ or, more generally, $f(k+N)$ for some integer N. We will prove that

$$\mathscr{Z}\{f(k+1)\} = zF(z) - zf(0) \tag{7.48}$$

where $F(z)$ is the transform of $f(k)$, and $f(0)$ is the initial value of $f(k)$.

Proof. From the definition in expression (7.46),

$$\mathscr{Z}\{f(k+1)\} = \sum_{n=0}^{\infty} f(n+1)z^{-n}$$

For $m = n + 1$,

$$\mathscr{Z}\{f(n+1)\} = \sum_{m=1}^{\infty} f(m)z^{-m+1}$$

Adding snd subtracting $zf(0)$,

$$\mathscr{Z}\{f(n+1)\} = z\left[\sum_{m=0}^{\infty} f(m)z^{-m} - f(0)\right]$$

And from the definition

$$\mathscr{Z}\{f(n+1)\} = zF(z) - zf(0)$$

thus completing the proof. The result just proved is easily extended to higher-order time shifts. For example, if $g(k) = f(k+1)$, then $g(k+1) = f(k+2)$ and

$$\mathscr{Z}\{f(k+2)\} = \mathscr{Z}\{g(k+1)\} = zG(z) - zg(0)$$

And using the result of expression (7.48),

$$\mathscr{Z}\{f(k+2)\} = z[zF(z) - zf(0)] - zf(1)$$

Finally,

$$\mathscr{Z}\{f(k+2)\} = z^2F(z) - z^2f(0) - zf(1) \tag{7.49}$$

Similarly, the general Nth-order time shift takes the form

$$\mathscr{Z}\{f(k+N)\} = z^NF(z) - [z^Nf(0) + z^{N-1}f(1) + \cdots + zf(N-1)] \tag{7.50}$$

The collection of terms within the brackets emphasizes the contribution of initial or boundary conditions for the shifted signal. The reader should carefully note the similarities and differences for the corresponding Laplace transform result given in expression (7.9).

Example 7.8

To solve a first-order difference equation, let us consider the difference equation

$$y(k+1) - uy(k) = u(k) \qquad \text{for } a \neq 1$$

where $u(k)$ denotes the unit-step sequence. Solve for the response sequence $y(k)$ if $y(0) = 0$.

Solution. Taking the z-transform of the difference equation and recognizing the linearity property of the transform,

$$\mathscr{Z}\{y(k+1)\} - a\mathscr{Z}\{y(k)\} = \mathscr{Z}\{u(k)\}$$

Incorporating expressions (7.48) and (7.47) for the right-hand side,

$$zY(z) - aY(z) = \frac{z}{z-1}$$

Solving for $Y(z)$,

$$Y(z) = \frac{z}{(z-1)(z-a)} \qquad \text{for } |z| > \max\{1, a\}$$

Expanding, we obtain

$$\frac{Y(z)}{z} = \frac{(1-a)^{-1}}{z-1} - \frac{(1-a)^{-1}}{z-a}$$

Finally, from Table 6.3 the solution is

$$y(k) = \frac{1}{1-a}(1-a^k) \qquad k = 0, 1, 2, \ldots$$

We now rederive the key property for the z-transform of a convolution summation.

Convolution. As we have seen in previous chapters, the operation of discrete convolution is important in finding the response of a discrete-time system to an arbitrary input sequence. We will show that

$$\mathscr{Z}\{f(k) * g(k)\} = F(z)G(z) \tag{7.51}$$

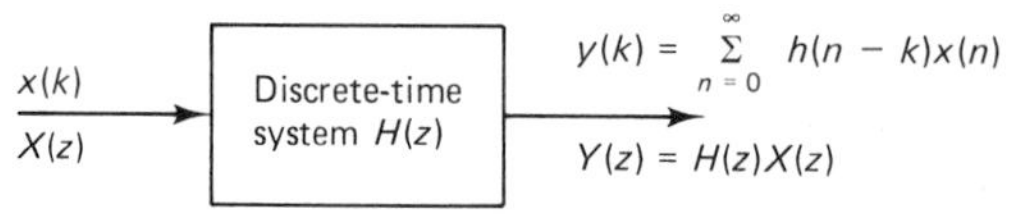

$x(k)$ / $X(z)$ → Discrete-time system $H(z)$ → $y(k) = \sum_{n=0}^{\infty} h(n-k)x(n)$ / $Y(z) = H(z)X(z)$

Figure 7.8 Discrete-time system transfer function.

where the "star product" ($*$) represents the convolution operation and $F(z)$ and $G(z)$ are the transforms of $f(k)$ and $g(k)$, respectively.

Proof. By the definition of the convolution operation

$$\mathscr{Z}\{f(k) * g(k)\} = \mathscr{Z}\left\{\sum_{k=0}^{\infty} f(k)g(n-k)\right\}$$

$$= \sum_{n=0}^{\infty}\left[\sum_{k=0}^{\infty} f(k)g(n-k)\right]z^{-n}$$

Interchanging summations (since the infinite series are uniformly convergent in some region of the z-plane) and letting $m = n - k$,

$$\mathscr{Z}\{f(k) * g(k)\} = \sum_{k=0}^{\infty} f(k) \sum_{m=-k}^{\infty} g(m)z^{-m-k}$$

$$= \left[\sum_{k=0}^{\infty} f(k)z^{-k}\right]\left[\sum_{m=0}^{\infty} g(m)z^{-m}\right] = F(z)G(z)$$

as desired. Note that the lower limit on the second summation is set to zero since $g(m) = 0$ for $m < 0$.

The convolution property provides us with the transfer function concept for discrete-time systems. Figure 7.8 illustrates this idea. To reenforce the concept of a discrete-time transfer function, consider the following example about compound interest.

Example 7.9

Here we view the process of compound interest as a discrete-time system. An initial amount of money $u(1)$ is invested in a savings account at time zero and a 10 percent interest rate accrues annually. Additional deposits $u(k)$ are made at the beginning of year k for $k = 2, 3, 4, \ldots$.

If $p(k)$ is the amount on deposit during year k, it can be shown that

$$p(k) = (1.10)p(k-1) + u(k) \qquad \text{for } p(0) = 0 \tag{7.52}$$

Equation (7.52) represents a discrete-time system with input $u(k)$ and output $p(k)$. To describe the transfer function, we take the z-transform and solve for $P(z)/U(z)$:

$$P(z) = 1.10z^{-1}P(z) + U(z) \tag{7.53}$$

$$\frac{P(z)}{U(z)} = \frac{1}{1 - 1.10z^{-1}} = \frac{z}{z - 1.10} \tag{7.54}$$

Now, for any sequence of annual deposits we could use the transfer-function concept to solve for the corresponding sequence of savings. In the next section we investigate various applications of transfer functions both for continuous-time and discrete-time systems.

Linear Difference Equations

Sequence operations, as we know, can be typically described by difference equations. These operations form an important part of signal processing such as filtering and communication. When we restrict our attention to *linear* difference equations, the z-transform can be used to obtain a solution as we have seen.

Let us consider a general second-order difference equation (for $k = 0, 1, 2, \ldots$), as follows:

$$y(k) + a_1 y(k-1) + a_2 y(k-2) = b_0 u(k) + b_1 u(k-1) + b_2 u(k-2) \qquad (7.55)$$

where the coefficients a_1, a_2, b_0, b_1, and b_2 are constants. For our purposes, the signal $u(k)$, $k = 0, 1, \ldots$ is known and corresponds to the input to the system. The dependent variable $y(k)$ is interpreted, in like manner, as the system output. We are, perhaps, more familiar with the system described in its transfer-function form

$$\frac{Y(z)}{U(z)} = \frac{b_0 + b_1 z^{-1} + b_2 z^{-2}}{1 + a_1 z^{-1} + a_2 z^{-2}} \qquad (7.56)$$

To solve for y in equation (7.55), we can break the solution into two parts: (1) the forced response (also known as a *particular solution*), y_f; and (2) the natural response (also known as the *homogeneous* or *zero-input solution*), y_n. Thus,

$$y(k) = y_f(k) + y_n(k) \qquad (7.57)$$

The forced response is the solution obtained from the transfer-function approach, completely disregarding any initial conditions. On the other hand, the natural response is the solution of expression (7.55) with $u(k)$ identically equal to zero. In this case we can write

$$y_n(m+2) + a_1 y_n(m+1) + a_2 y_n(m) = 0 \qquad (7.58)$$

where $m = k - 2$.

The general solution for $y_n(k)$ contains two constants determined by $y_n(0)$ and $y_n(1)$. Once both y_f and y_n have been found, the arbitrary constants of y_n can be evaluated from knowledge of $y(0)$ and $y(1)$, for instance, which could be specified with the original problem.

Example 7.10 (Difference Equation)

Consider the problem of solving for the signal $y(k)$ if

$$y(k+1) - \frac{1}{2} y(k) = x(k+1) + \frac{1}{3} x(k)$$

$$x(k) = \left(\frac{1}{4}\right)^k \qquad \text{for } k = 0, 1, 2, \ldots \qquad \text{and} \qquad y(0) = 2$$

Solution. We will solve the problem two ways: a. using a single-step z-transform operation, and b. finding the component parts y_f and y_n separately.

a. Taking z-transforms, we obtain

$$zY(z) - 2z - \frac{1}{2}Y(z) = zX(z) - zx(0) + \frac{1}{3}X(z)$$

Recognizing that $X(z) = z/[z - (1/4)]$ and $x(0) = 1$, we can solve for $Y(z)$:

$$Y(z) = \frac{z(2z + \frac{1}{12})}{(z - \frac{1}{2})(z - \frac{1}{4})} = \frac{(\frac{13}{3})z}{z - \frac{1}{2}} + \frac{-(\frac{7}{3})z}{z - \frac{1}{4}}$$

Consequently,

$$y(k) = \frac{13}{3}\left(\frac{1}{2}\right)^k - \frac{7}{3}\left(\frac{1}{4}\right)^k \qquad \text{for } k = 0, 1, 2, \ldots$$

b. Using the transfer-function approach, we solve for y_f:

$$Y_f(z) = \frac{z + \frac{1}{3}}{z - \frac{1}{2}}X(z) = \frac{(z + \frac{1}{3})z}{(z - \frac{1}{2})(z - \frac{1}{4})}$$

$$y_f(k) = \frac{10}{3}\left(\frac{1}{2}\right)^k - \frac{7}{3}\left(\frac{1}{4}\right)^k \qquad \text{for } k = 0, 1, 2, \ldots$$

And with zero input,

$$y_n(k + 1) - \frac{1}{2}y_n(k) = 0$$

So that

$$y_n(k) = c\left(\frac{1}{2}\right)^k \qquad \text{for } k = 0, 1, 2, \ldots$$

where c is an arbitrary constant at this point. To evaluate c, we write

$$y(k) = y_f(k) + y_n(k) = \left(\frac{10}{3} + c\right)\left(\frac{1}{2}\right)^k - \frac{7}{3}\left(\frac{1}{4}\right)^k$$

The initial condition on $y(k)$ is $y(0) = 2$, which specifies that $c = 1$. Note that the total solution is the same either way.

Remark. If the initial-condition terms are kept separate in the one-step method, the two components are evident. In this case,

$$Y(z) = \left[\frac{(z + \frac{1}{3})}{(z - \frac{1}{2})}\right]X(z) + \frac{zy(0) - zx(0)}{(z - \frac{1}{2})}$$

The first term inverts to $y_f(k)$, and the second to $y_n(k)$.

7.6 SYSTEM APPLICATIONS

In this section various examples and procedures are presented to illustrate some system modeling or analysis technique related to transfer functions. We begin with two examples of modeling nonelectrical systems.

Nonelectrical Systems

Most of our illustrations have been related to electrical networks. Here, let us consider some examples of a wider class of systems. Perhaps the most familiar is that of a mechanical translational system. Figure 7.9 shows a simple mass-spring system with an applied force $f(t)$ and a response variable $x(t)$, which is the displacement of the mass. This system is conveniently described by the balance of forces acting on the mass unit set equal to the inertial force $M\ddot{x}$ by Newton's law:

$$M\ddot{x}(t) = f(t) - Kx(t) \tag{7.59}$$

where K is the spring constant and $x(t)$ is the mass displacement measured from the equilibrium position of the mass when $f(t)$ is zero. We have assumed no friction acting on the system.

Once the system parameters M and K are specified, we can write the specific transfer function as $X(s)/F(s)$, output over input transforms with zero initial conditions on the system. Transforming and solving, we obtain

$$\frac{X(s)}{F(s)} = \frac{\dfrac{1}{M}}{s^2 + \left(\dfrac{K}{M}\right)} \tag{7.60}$$

More generally, if linear friction is present due to the mass sliding on the frame, then the equation corresponding to expression (7.59) could be (after rearranging)

$$M\ddot{x}(t) + B\dot{x}(t) + kx(t) = f(t) \tag{7.61}$$

with the corresponding transfer function

$$\frac{X(s)}{F(s)} = \frac{\dfrac{1}{M}}{s^2 + \left(\dfrac{B}{M}\right)s + \left(\dfrac{K}{M}\right)} \tag{7.62}$$

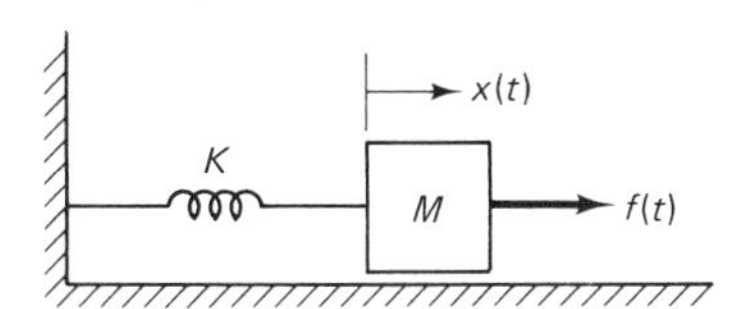

Figure 7.9 A mass-spring system.

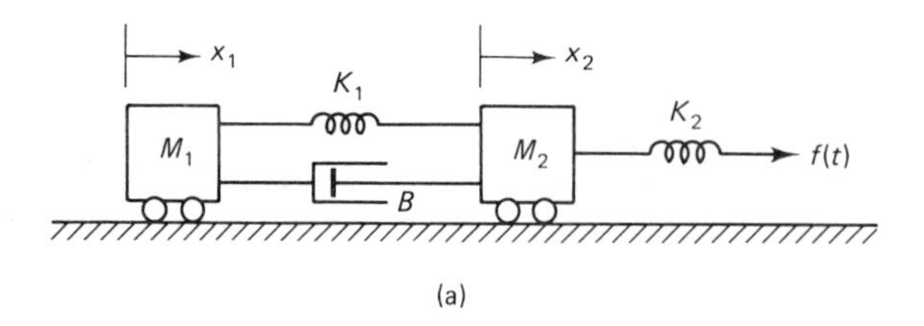

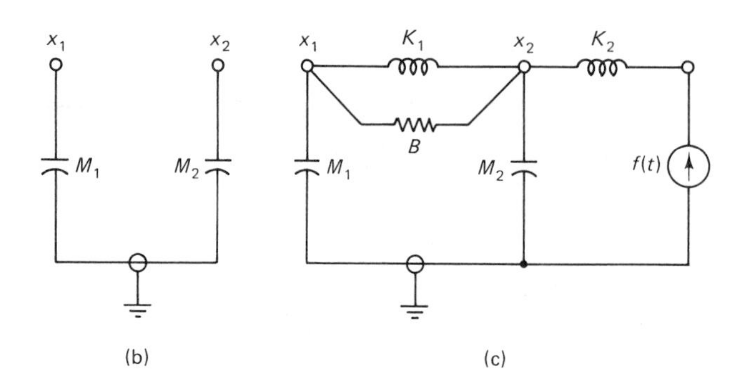

Figure 7.10 (a) A mechanical system, (b) the first step of the analog network construction and (c) the complete analog network.

When the mechanical system is more complicated, such as the two-mass system shown in Fig. 7.10a, then two separate "free body diagrams" are needed for the equations. Alternatively, one may draw an *analog network* shown in Fig. 7.10c using the force-current analog. In an analog system, the mathematical equations describing the system are of identical form as that of the equations describing the original system. For example, a parallel *R-L-C* network with current source $f(t)$ can be described by the equation

$$C\ddot{x}(t) + \left(\frac{1}{R}\right)\dot{x}(t) + \left(\frac{1}{L}\right)x(t) = f(t) \tag{7.63}$$

where $\dot{x}(t) = v(t)$, the common element voltage. Note the similarity between expressions (7.61) and (7.63). The analog variables and parameters are listed in Table 7.4. All units are taken to be in one consistent system such as the *M-K-S* system. As an illustration of using the analog network to write a set of mechanical equations, consider the following example.

Example 7.11

From the system given in Fig. 7.10a, we first draw the capacitor network shown in Fig. 7.10b. The capacitors are referenced to ground just as the inertia terms are referenced to a fixed Newtonian reference frame. The capacitor voltages would correspond to the mass velocities, but mechanical equations are more typically described in terms of displacements (as shown in the figure). The next step is to connect the other analog elements such as a conductance of B units and springs of K_1 and K_2 reciprocal inductance units, as shown in Fig. 7.10c. Finally, the current of $f(t)$ is added with the polarity to increase the voltage $\dot{x}_2$, which corresponds to the velocity of M_2. Notice that each mechanical element such as K_1 is connected between two mass points (M_1 and M_2), so that the analog element K_1 is also connected between the nodes corre-

TABLE 7.4 A SET OF ANALOG VARIABLES

Mechanical	Electrical
Force	Current
Velocity	Voltage
Mass (M)	Capacitance (C)
Spring constant (K)	Reciprocal inductance ($1/L$)
Friction constant (B)	Conductance ($1/R$)

sponding to the mass units. Of course, the force must be applied with respect to the frame. Consequently, the current source is between the reference node and the node corresponding to the "massless" and linkage of the spring K_2. The equations of motion are simply the node equations for the analog network:

$$\begin{aligned} M_1\ddot{x}_1 + B(\dot{x}_1 - \dot{x}_2) + K_1(x_1 - x_2) &= 0 \\ M_2\ddot{x}_2 + B(\dot{x}_2 - \dot{x}_1) + K_1(x_2 - x_1) &= f(t) \end{aligned} \tag{7.64}$$

Remarks

- The polarity of the terms in expression (7.64) arises from the sum of the currents leaving the node set equal to the source current into the node.
- Each mechanical coefficient occurs in its proper, rather than reciprocal, form; i.e., one need not worry about that aspect of the analogy.
- If $x_3(t)$ is the displacement of the massless end of spring K_2 taken as positive to the right, then equating the currents

$$f(t) = K_2(x_3 - x_2) \tag{7.65}$$

By transforming the expressions in (7.64), we obtain

$$\left.\begin{aligned} (M_1s^2 + Bs + K_1)X_1(s) - (Bs + K_1)X_2(s) &= 0 \\ -(Bs + K_1)X_1(s) + (M_2s^2 + Bs + K_1)X_2(s) &= F(s) \end{aligned}\right\} \tag{7.66}$$

Clearly, $F(s)$ is taken as the input signal, and we could solve for transfer functions such as $X_1(s)/F(s)$ or $X_2(s)/F(s)$, or even $X_3(s)/F(s)$ using equation (7.65) by algebraic reduction of the transformed equations.

As a second system type, consider the two-tank reservoir system shown in Fig. 7.11. The total inflow is $Q_1(t)$, the flow between tank 1 and tank 2 is $Q_{12}(t)$, and the outflow is $Q_2(t)$, all in common units, say cubic feet per second. Assuming uniform cross-sectional areas A_1 and A_2 for tanks 1 and 2, respectively, we can write the following "flow-balance" equations:

$$\begin{aligned} A_1\dot{H}_1(t) &= Q_1(t) - Q_{12}(t) \\ A_2\dot{H}_2(t) &= Q_{12}(t) - Q_2(t) \end{aligned} \tag{7.67}$$

where $H_1(t)$ and $H_2(t)$ are the water levels shown in Fig. 7.11.

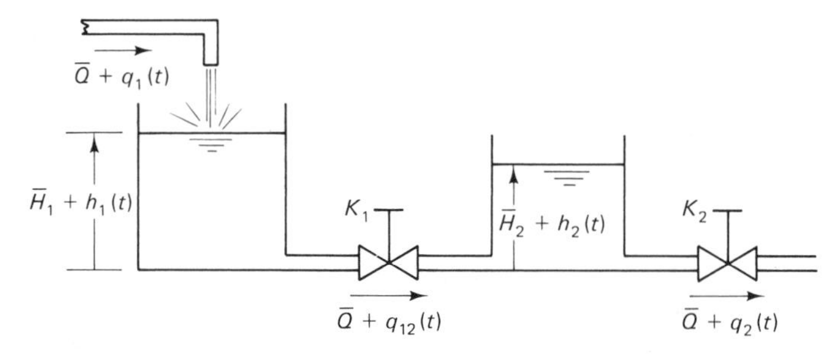

Figure 7.11 A two-tank reservoir system.

If a steady flow through $\bar{Q}$ corresponds to the steady levels $\bar{H}_1$ and $\bar{H}_2$, then we are typically interested in small variations from this steady-state condition. Thus, if

$$H_1(t) = \bar{H}_1 + h_1(t), \qquad H_2(t) = \bar{H}_2 + h_2(t)$$
$$Q_1(t) = \bar{Q} + q_1(t), \qquad Q_{12}(t) = \bar{Q} + q_{12}(t) \tag{7.68}$$

and

$$Q_2(t) = \bar{Q} + q_2(t)$$

we can reasonably assume that, in general, an incremental flow rate $q(t)$ is proportional to the difference in water head (levels); that is, the ohm's law analog

$$q(t) = K\ \Delta h(t) \tag{7.69}$$

where K depends on the constriction between the two levels, such as a valve controlling the flow. If we introduce expressions (7.68) and the relation (7.69) (for the effect of the valves) into expressions (7.67), we obtain (disregarding steady-state values) the incremental flow equations

$$\left.\begin{aligned} A_1\dot{h}_1(t) &= q_1(t) - K_1[h_1(t) - h_2(t)] \\ A_2\dot{h}_2(t) &= K_1[h_1(t) - h_2(t)] - K_2 h_2(t) \end{aligned}\right\} \tag{7.70}$$

where K_1 and K_2 are the valve parameters.

Taking $q_1(t)$ as an input, we can transform relationships (7.70) and solve for $H_1(s)/Q_1(s)$, $H_2(s)/Q_1(s)$, or $Q_2(s)/Q_1(s)$ as individual transfer functions of the system, depending on which variable was considered as the system output.

In addition, expressions (7.70) may be used directly with integrator blocks to draw a *simulation diagram* of the system. These diagrams exhibit the structure used for an analog computer simulation. The method for constructing the diagrams is quite simple:

- Use the s^{-1} blocks (integrators) to derive signals from their derivatives.
- Follow the system equations to obtain the remaining structure consisting of gain blocks and summation junctions.

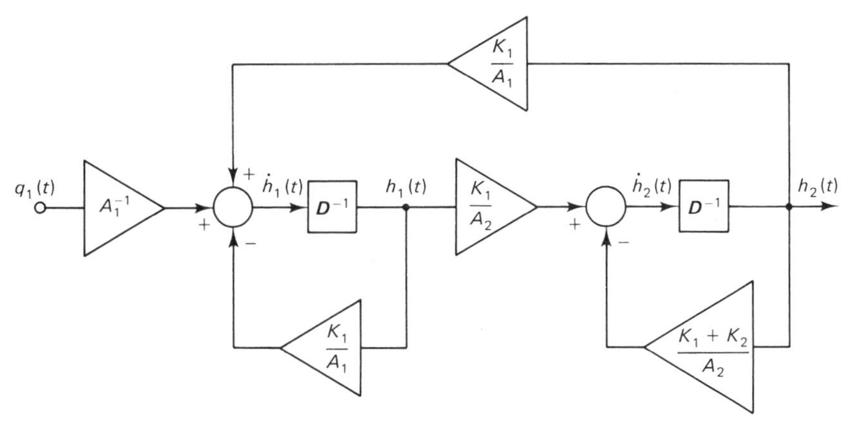

Figure 7.12 Simulation diagram for equations (7.70).

Expressions (7.70) are shown in a simulation diagram in Fig. 7.12. In the next section we introduce the basic concept of feedback systems and how feedback can be used to improve system performance.

Feedback Control

The idea of controlling a system is not new. The reader can, no doubt, think of many systems that are "controlled." Some examples are

- the temperature in a room
- the position of an elevator
- the altitude of an airplane
- the speed of an automobile

Underlying each control system are certain excitation and response (or controlled) variables. We will take a particular system to illustrate the basic concepts of feedback control.

In Example 7.7 we derived the transfer function of a d.c. motor from its field excitation voltage $E(s)$ to its mechanical shaft displacement $\theta(s)$; this was given in result (7.45). For simplicity, let us assume that the time constant $L/R = \frac{1}{2}$ and the time constant $J/B = 1$. We write the transfer function as

$$G(s) = \frac{2}{s(s+1)(s+2)} \tag{7.71}$$

where the amplifier gain K has been adjusted to equal BR.

We refer to the amplifier input as the control variable $e(t)$. Figure 7.13 shows the electro-mechanical network and the resulting transfer function. Controlling the shaft displacement $\theta(t)$ from the amplifier input $e(t)$ is difficult. One solution to this difficulty is to use feedback where a measurement of the controlled variable θ is

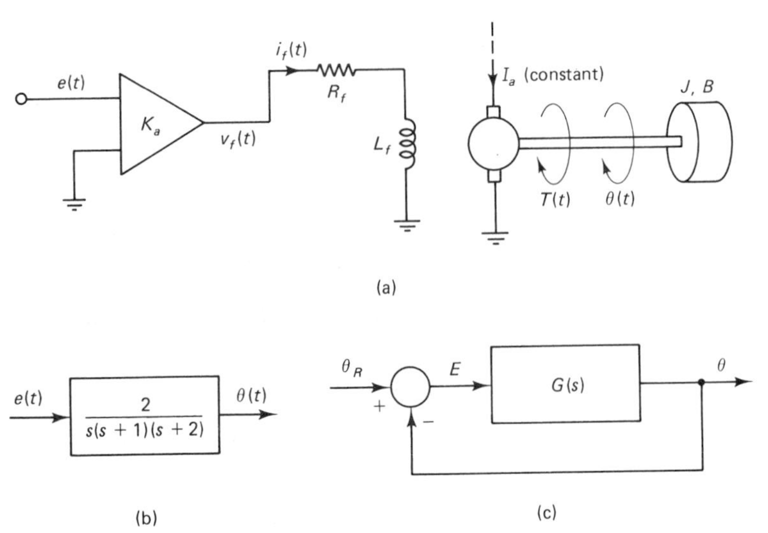

Figure 7.13 A position-control system.

subtracted from the command (or reference) input θ_R to generate an error voltage $e(t)$, which, in turn, feeds the servo-amplifier. This arrangement is diagrammed in Fig. 7.13c. In this way, any difference between the desired value θ_R and the actual value θ drives the motor in the proper direction to bring the error ($e = \theta_R - \theta$) to zero.

Another function that the "unity feedback" configuration of Fig. 7.13c might serve is that of remote control; that is, the reference command θ_R might be a potentiometer knob in a control panel at some distance from the motor location.

The use of feedback permits one to achieve accurate control because of the direct comparison between the desired value and the current value of the controlled variable. However, for this example we want to focus on the change in the pole locations of the system caused by feedback.

From Fig. 7.13b we see that the poles of the open-loop system are

$$\text{Open-loop poles} = \{s = 0, \ -1, \ -2\}$$

And from Fig. 7.13c we can write that

$$\frac{\theta(s)}{\theta_R(s)} = \frac{G(s)}{1 + G(s)} \tag{7.72}$$

Expression (7.13) is the *closed-loop transfer function*, and for our example

$$\frac{G(s)}{1 + G(s)} = \frac{2}{s^3 + 3s^2 + 2s + 2} \tag{7.73}$$

The closed-loop pole locations are the roots of the denominator polynomial in expression (7.73). Solving for these numerically, we find that the

$$\text{closed-loop poles} = \{s = -2.52138, \ -0.23931 \ \pm j0.85787\} \tag{7.74}$$

Without actually solving for a particular response, we can imagine that the closed-loop (that is, step) response is quite different from the corresponding open-loop response.

Remark. Although it is not obvious, the closed-loop response matches the level of the input step in steady state; that is, the error $e(t)$ between θ_R and $\theta(t)$ goes to zero with time.

As an example of calculating the unit-step response for a closed-loop system, consider the following example.

Example 7.12 (Unit-Step Response)

A control system has an open-loop transfer function $G(s)$ given by

$$G(s) = \frac{K}{s(s+2)} \tag{7.75}$$

where K represents an adjustable amplifier gain. When this system is used in the feedback configuration of Fig. 7.13c, the closed-loop transfer functon with K set to a value of unity is

$$\frac{G(s)}{1+G(s)} = \frac{1}{s^2+2s+1} = \frac{1}{(s+1)^2} \tag{7.76}$$

The unit-step response (with zero initial energy in the system) is the inverse transform of the causal signal

$$Y(s) = \frac{1}{s(s+1)^2} = \frac{1}{s} + \frac{-1}{(s+1)^2} + \frac{-1}{s+1} \tag{7.77}$$

From the partial-fraction expansion and Table 5.1,

$$y(t) = 1 - (1+t)e^{-t} \qquad \text{for } t \geq 0 \tag{7.78}$$

This response is graphed in Fig. 7.14.

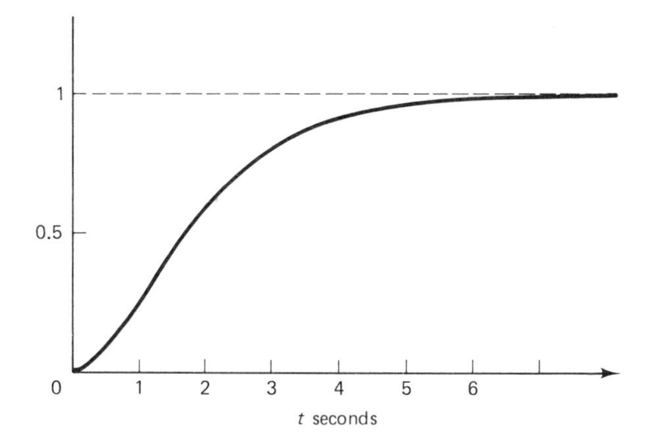

Figure 7.14 Step response for Example 7.12.

Example 7.13 (Temperature Regulation)

In Fig. 7.15 is shown a closed tank containing a liquid. It is known that the thermal energy stored in the liquid is proportional to its temperature in degrees Kelvin. Let us assume that the energy stored in this system is

$$W = 2T \qquad \text{joules}$$

where T is the temperature of the liquid. In addition, the heat loss through the container walls is proportional to the difference in temperature as follows:

$$Q = 4(T - T_o) \qquad \text{joules/second}$$

where T_o is the (constant) ambient temperature. If a resistance heater can supply a power $P(t)$, the dynamics of the system can be described by the power balance equation:

$$\dot{W}(t) = 2\dot{T}(t) = P(t) - 4T(t) + 4T_o$$

or, in terms of $T(t)$,

$$\dot{T}(t) = -2(T - T_o) + \frac{1}{2} P(t)$$

Suppose now that we need to regulate the temperature T to be 20° above the ambient temperature. Thus in steady state, $\dot{T}(t) = 0$ and $T = T_{ss} = T_o + 20$:

$$2(T_{ss} - T_o) = \frac{1}{2} P_{ss}$$

Therefore, the steady-state power required is $P_{ss} = 80$ watts. Defining $\Delta T = T - T_{ss}$ and $u(t) = P(t) - 80$,

$$\Delta\dot{T}(t) = -2\,\Delta T(t) + 2u(t)$$

From the similarity with the network equation

$$\frac{di}{dt}(t) = \left(\frac{R}{L}\right)i(t) + \frac{1}{L}\,w(t)$$

we can write a description for our system in terms of the unit-impulse response function

$$h(t) = 2e^{-2t} \qquad 0 \le t$$

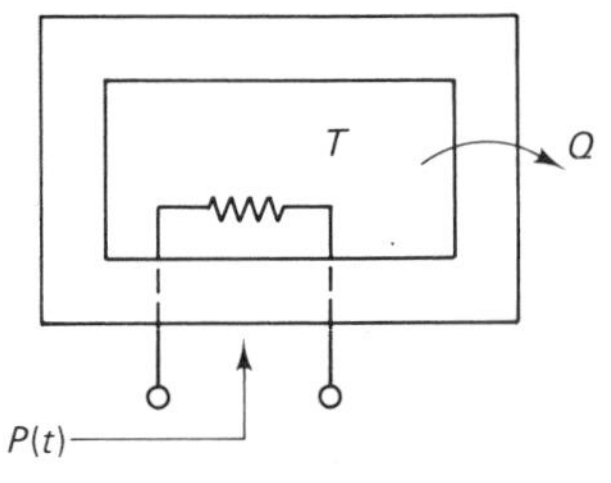

Figure 7.15 A temperature-regulation system.

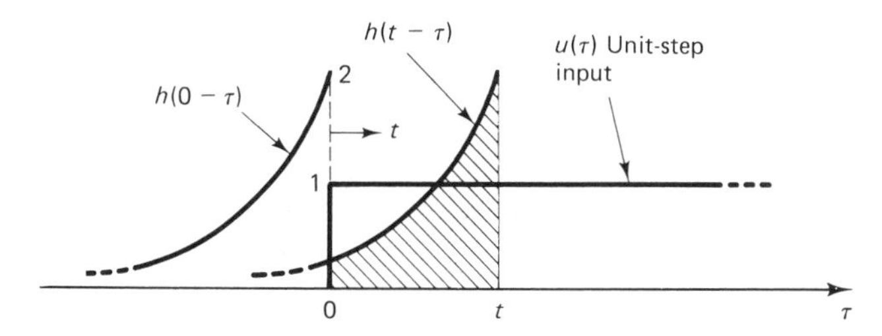

Figure 7.16 Graphical convolution for Example 7.13.

The unit-step response can be calculated as

$$\Delta T(t) = \int_{0-}^{\infty} u(\tau) 2e^{-2(t-\tau)}\, d\tau$$

Figure 7.16 illustrates this operation graphically. The area of the cross-hatched region (and therefore the unit-step response of the system) is

$$T(t) = \int_{0}^{t} 2e^{-2\lambda}\, d\lambda = (1 - e^{-2t}) \qquad \text{for } 0 \le t$$

The time constant 0.5 second is characteristic of the response as it varies in response to a change in steady input power levels.

The use of feedback provides several advantages: (1) greater accuracy of control, (2) greater tolerance for parameter variations (such as not modeling the correct inductance or friction coefficient in the d.c. motor example), and (3) greater tolerance for extraneous (noise) signals coming into the system, to mention a few. However, these advantages are not without some price. And this price is the possibility of inadvertently obtaining an "unstable system." This is the topic of the next section.

System Stability—The Routh and Jury Tests

One of the most important single qualities associated with system models relates to the question of stability. A natural notion of system stability used by engineers is that if a system is stable, its output signal remains bounded whenever the corresponding input is bounded. Recall that $x(t)$ is said to be bounded if there exists some fixed number M (however large), such that

$$|x(t)| < M \qquad \text{for all } t$$

Thus, we define system stability as follows:

Definition. A system or its transfer function T is bounded-input–bounded-output (BIBO) stable if for every bounded input $x(t)$, the corresponding response $y(t)$ is also a bounded signal ($Y = TX$ in either the s-domain or the z-domain).

With the previous methods studied in this chapter, let us investigate the implications of BIBO stability.

Discrete-Time Systems. We use the characteristic unit-pulse response sequence $\{h(n)\}$ to specify the system. Therefore, the response sequence is related to the input sequence by the convolutional sum

$$y(n) = \sum_{k=0}^{\infty} x(k)h(n-k) \tag{7.79}$$

Since the magnitude of a sum is less than or equal to the sum of the magnitudes,

$$|y(n)| \leq \sum_{k=0}^{\infty} |h(k)| \cdot |x(n-k)|$$

By assumption, $|x(n)| < M$ for any n (x-bounded), the output $y(n)$ must satisfy

$$|y(n)| < M \sum_{k=0}^{\infty} |h(k)| \tag{7.80}$$

Thus, the system is BIBO stable only if

$$\sum_{k=0}^{\infty} |h(k)| < \infty \tag{7.81}$$

that is to say, the sum of all the magnitudes of the terms of the sequence $\{h(k)\}$ must be finite. The reader may recall from Chapter 2 that (7.81) states that the ℓ_1 norm of $\{h(k)\}$ is finite.

Continuous-Time Systems. The requirement for a continuous-time system to be BIBO stable can be obtained in a similar manner. Recalling the continuous-time convolution,

$$y(t) = \int_0^{\infty} x(\tau)h(t-\tau)\, d\tau \tag{7.82}$$

we introduce the change of variables $u = t - \tau$, to obtain

$$y(t) = \int_0^{\infty} h(u)x(t-u)\, du \tag{7.83}$$

If $x(t)$ is bounded, there exists a number M such that $|x(t)| < M$ for all t. Thus, we obtain

$$|y(t)| < M \int_0^{\infty} |h(t)|\, dt \tag{7.84}$$

We conclude that the continuous-time operator is BIBO stable only if

$$\int_0^{\infty} |h(t)|\, dt < \infty \tag{7.85}$$

We next introduce some simpler means by which we can test the stability of linear systems.

If we think of the BIBO stability criterion in terms of continuous-time transfer functions, we see that for expression (7.85) to hold for causal systems with rational transfer functions, all poles of the corresponding transfer function $H(s)$ must have negative real parts (so that the terms in the partial-fraction expansion translate to signals whose amplitudes decay with time).

Similarly for discrete-time systems, expression (7.81) is true only if the poles of the corresponding discrete-time transfer $H(z)$ lie completely inside the unit circle of the z-plane. This follows from consideration of the terms in a partial-fraction expansion of $H(z)$; that is, the z-transform terms must translate to signals that decay with time. In particular, $\{\lambda^k, k = 0, 1, 2, \ldots\}$ decays to zero only if $|\lambda| < 1$.

With this introduction, we now present two tests on polynomials: (1) the Routh test and (2) the Jury test. The first test examines a polynomial for roots in the right-half plane and the second gives necessary and sufficient conditions for the roots of a polynomial to lie inside the unit circle.

A Test for Stability—The Routh Test. As we have seen from the previous discussion, a necessary and sufficient condition that a continuous-time linear, time-invariant system be stable is that all its poles have negative real parts. A useful procedure for obtaining this information is the Routh test, which uses the coefficients of the system's characteristic polynomial to generate an array (the Routh array or table) whose first column tells us how many roots of the polynomial have positive real parts, as we shall see in more detail subsequently.

Routh Array. From the characteristic polynomial of an Nth order system

$$A(s) = a_N s^N + a_{N-1} s^{N-1} + \cdots + a_1 s + a_0 \tag{7.86}$$

we write the array

$$\begin{array}{c|cccc}
s^N & a_N & a_{N-2} & a_{N-4} & \cdots \\
s^{N-1} & a_{N-1} & a_{N-3} & a_{N-5} & \cdots \\
s^{N-2} & b_{N-1} & b_{N-3} & b_{N-5} & \cdots \\
s^{N-3} & c_{N-1} & c_{N-3} & c_{N-5} & \cdots \\
 & & \cdots & & \\
\vdots & & & & \\
s^1 & p_{N-1} & p_{N-3} & & \\
s^0 & q_{N-1} & & &
\end{array}$$

where, starting with row 3, the coefficients are calculated from the two preceding rows as follows:

$$b_{N-1} = \frac{-1}{a_{N-1}} \begin{vmatrix} a_N & a_{N-2} \\ a_{N-1} & a_{N-3} \end{vmatrix} \quad \text{and} \quad b_{N-3} = \frac{-1}{a_{N-1}} \begin{vmatrix} a_N & a_{N-4} \\ a_{N-1} & a_{N-5} \end{vmatrix}$$

and so on until the end of the first two rows. Following this, row 4 is calculated using the identical format; that is,

$$c_{N-1} = \frac{-1}{b_{N-1}} \begin{vmatrix} a_{N-1} & a_{N-3} \\ b_{N-1} & b_{N-3} \end{vmatrix} \quad \text{and} \quad c_{N-3} = \frac{-1}{b_{N-1}} \begin{vmatrix} a_{N-1} & a_{N-5} \\ b_{N-1} & b_{N-5} \end{vmatrix}$$

This procedure is continued until q_{N-1} is obtained to complete the array.

The Stability Criterion. The number of sign changes in the coefficients of the first column of the Routh array is equal to the number of roots of $A(s)$ with positive real parts.

Example 7.14

Determine if $A(s)$ below is the characteristic polynomial of a stable system:

$$A(s) = s^3 + 6s^2 + 11s + 6$$

Solution. Calculating the Routh array,

$$\begin{array}{c|cc} s^3 & 1 & 11 \\ s^2 & 6 & 6 \\ s^1 & 10 & 0 \\ s^0 & 6 & \end{array}$$

From column 1—1, 6, 10, 6—there are no sign changes; therefore, there are no roots of $A(s)$ with positive real parts. Thus $A(s)$ represents a stable polynomial, that is, a characteristic polynomial of a stable system.

There are two situations that occasionally arise when using the Routh test which require slight modifications to the construction of the Routh array.

Case I. A zero appears in column 1, but not a zero row. Let us illustrate a simple method to handle this case using an example.

Example 7.15

Determine if $A(s)$ has any right-half plane roots and, if so, how many?

$$A(s) = s^3 + 11s + 6$$

Solution. Constructing the Routh array,

$$\begin{array}{c|cc} s^3 & 1 & 11 \\ s^2 & 0 & 6 \\ s^1 & ? & \end{array}$$

Since a zero occurs in column 1, we are uncertain whether or not there is a sign change from the first to the second element, as well as how to calculate the third row! To overcome this difficulty, we can assume a value $\varepsilon > 0$ for zero, procede with the calculations, and finally take the limit as $\varepsilon \to 0$. Thus, our array becomes

$$\begin{array}{c|cc} s^3 & 1 & 11 \\ s^2 & \varepsilon & 6 \\ s^1 & c & 0 \\ s^0 & 6 & \end{array} \qquad c = \frac{11\varepsilon - 6}{\varepsilon} = 11 - \frac{6}{\varepsilon}$$

In the limit as $\varepsilon \to 0$, $c \to -6/\varepsilon$ (a large, negative value). Therefore, there are two right-half plane roots, since $\varepsilon \to c$ and $c \to 6$ are two sign changes.

The other exceptional case is when $A(s)$ has a factor $Q(s)$ that has "quadrantal symmetry"; this means that the roots of $Q(s)$ have 4-quadrant symmetry, that is, mirror symmetry about both the real axis and the imaginary axis.

Case II. A zero row occurs in the Routh array. Consider an example where

$$A(s) = [(s+1)(s-1)](s+2)(s+3) = Q(s)(s+2)(s+3)$$

Example 7.16

$$A(s) = s^4 + 5s^3 + 5s^2 - 5s - 6$$

The Routh array is

$$\begin{array}{c|ccc} s^4 & 1 & 5 & -6 \\ s^3 & 5 & -5 & 0 \\ s^2 & 6 & -6 & 0 \longrightarrow \hat{Q}(s) \\ s^1 & 0 & 0 \longleftarrow \text{zero row!} & \end{array}$$

Once a zero row is encountered, the factor $\hat{Q}(s)$ can be determined from the preceding row. In this case,

$$\hat{Q}(s) = 6s^2 - 6s^0 = 6(s+1)(s-1)$$

remembering that the coefficients are associated with alternate powers of s beginning with the power that precedes, or "labels," the row. Note that $\hat{Q}(s)$ differs from the original $Q(s)$ by, at most, a constant factor (independent of s). Thus, our factor $Q(s) = (s+1)(s-1)$ and $A(s)$ is an unstable polynomial.

Remark. If it is desired to continue the investigation of the right-half plane roots of $A(s)$, we can replace the zero row with the coefficients corresponding to the derivative of $\hat{Q}(s)$ (with respect to s). Thus, in this instance, $d\hat{Q}/ds = 12s$, and the array is completed with the last lines:

$$\begin{array}{c|cc} s^1 & 12 & 0 \\ s^0 & -6 & \end{array}$$

Finally, we have the indication of one right-half plane root of $A(s)$, which we already knew from $Q(s)$.

The Routh test for continuous-time systems provides a direct test on the coefficients of a polynomial to determine the number of roots of the polynomial that lie in the right-half plane. Clearly, this test is useful in determining the BIBO stability of a continuous-time system since it can be applied to the characteristic polynomial of the system (whose roots are the poles of the system).

The Jury Test. A second test call the *Jury stability test* was designed to indicate if any roots of a polynomial lie outside of the unit circle in the complex plane. The nature of the test is similar to the Routh test in that an array is constructed from the original coefficients of the polynomial. However, the array is constructed differently from the Routh array, and for this reason the reader may initially have some difficulty in keeping the details of the two tests straight.

The Jury test provides a set of necessary and sufficient conditions for the roots of a polynomial to lie inside the unit circle, that is, in the region $|z| < 1$. Let us assume that a discrete-time linear system has a characteristic polynomial (denominator of its pulse transfer function) given by

$$F(z) = a_n z^n + a_{n-1} z^{n-1} + \cdots + a_1 z + a_0 \tag{7.87}$$

where $a_n > 0$. Usually, it is convenient to normalize the polynomial by dividing through by a_n, in which case the coefficient z^n is unity. Before discussing the criterion, we first show how to construct the tabular array.

Jury Array. From the (given) coefficients in expression (7.87), we construct a table of coefficients (see Table 7.5). Beginning with row 3, each new entry is calculated as the determinant of a two-by-two array. For the entries of row 3,

$$b_k = \begin{vmatrix} a_0 & a_{n-k} \\ a_n & a_k \end{vmatrix} \qquad \text{for } k = 0, 1, 2, \ldots, n-1 \tag{7.88}$$

These determinants are formed from the previous two rows of known entries. The

TABLE 7.5 ARRAY CONSTRUCTION FOR THE JURY TEST[a]

Row	z^0	z^1	z^2	—	z^{n-k}	—	z^{n-2}	z^{n-1}	z^n
1	a_0	a_1	a_2	—	a_{n-k}	—	a_{n-2}	a_{n-1}	a_n
2	a_n	a_{n-1}	a_{n-2}	—	a_k	—	a_2	a_1	a_0
3	b_0	b_1	b_2	—	b_{n-k}	—	b_{n-2}	b_{n-1}	
4	b_{n-1}	b_{n-2}	b_{n-3}	—	b_{k-1}	—	b_1	b_0	
5	c_0	c_1	c_2	—	c_{n-k}	—	c_{n-2}		
6	c_{n-2}	c_{n-3}	c_{n-4}	—	c_{k-2}	—	c_0		
	$\cdots$								
$2n-5$	g_0	g_1	g_2	g_3					
$2n-4$	g_3	g_2	g_1	g_0					
$2n-3$	h_0	h_1	h_2						

[a]Note that the required number of rows in the table is $2n - 3$. Thus, for $n = 2$, only one row is needed and if $n = 3$, only three rows. The first two rows require no calculation and are simply the original coefficients of $F(z)$ arranged first backward, then forward.

first column stays fixed whereas the second column begins at the far right end with $k = 0$ and progresses to the left. Once the third row's entries have been found, the fourth row is a reverse replication of these same values, as shown in Table 7.5. The remaining rows follow this same pattern. Thus, for row 5,

$$c_k = \begin{vmatrix} b_0 & b_{n-1-k} \\ b_{n-1} & b_k \end{vmatrix} \qquad \text{for } k = 0, 1, 2, \ldots, n-2 \tag{7.89}$$

By continuing on in like manner until the entries of the row $(2n - 3)$ have been determined, the table is completed. The actual test may now be stated in terms of the calculated values.

Stability Criterion. The discrete-time system with characteristic polynomial $F(z)$ of order n in expression (7.87) is (asymptotically) stable if and only if all of the $(n + 1)$ conditions below are satisfied:

$$\begin{array}{ll} (1) & F(1) > 0 \\ (2) & F(-1) \begin{cases} >0 & \text{for } n \text{ even} \\ <0 & \text{for } n \text{ odd} \end{cases} \\ (3) & |a_0| < a_n > 0 \\ (4) & |b_0| > |b_{n-1}| \\ (5) & |c_0| > |c_{n-2}| \\ (6) & |d_0| > |d_{n-3}| \\ \cdots & \\ (n+1) & |h_0| > |h_2| \end{array} \tag{7.90}$$

It is worth noting that the set of conditions above must *all* be satisfied for a stable system. If even *one* condition fails, the system is unstable. Since the first three conditions do not require the Jury table, it is useful to check these conditions before troubling to construct the table. Conditions 4 and higher do involve the Jury table and in each case call for the magnitude of the first entry in a row to be greater than the magnitude of the last entry in the row, these rows being the "calculated" elements of rows 3, 5, 7, ..., $(2n - 3)$. An example helps to illustrate how the test can be applied.

Example 7.17

$$F(z) = z^2 + z + 0.2 \qquad \text{for } n = 2$$

(1) $F(1) = 2.2 > 0$ ✓

(2) $F(-1) = 0.2 > 0$ ✓ (since $n = 2$ is even)

(3) $|a_0| = 0.2 < a_2 = 1$ ✓

Only $(2n - 3) = 1$ row was required, rendering any table calculations unnecessary. The polynomial $F(z)$ has all its roots inside the unit circle, since every condition was satisfied.

Introduction to Digital Control

In a feedback control system the difference between a desired response R and a controlled variable C is called an *error signal* $E = R - C$. The purpose of the system is to force this error signal to zero. This can be done using a digital computer that accepts E as an input through a sampling device. The program in the computer, which constitutes the "control law," processes the error samples to obtain an "actuation" signal. The actuation signal U then becomes the input signal to the actuator, such as a servo-amplifier driving a d.c. motor whose output shaft is directly related to the variable C. (Indeed, C may be the motor shaft position.) The structure of such a system is shown in Fig. 7.17.

When a computer interfaces with analog devices, the signal converters [analog-to-digital (ADC) and digital-to-analog (DAC)] provide the necessary operations such as converting sample values to a binary word (ADC) and converting binary words back to analog voltages (DAC).

A useful control law, called *proportional-integral* (*PI*) *control*, effectively creates an actuation signal U, which is in part proportional to E and in part proportional to the integral of E. However, a numerical approximation must be used for the "integration." The end result is a program in the digital computer that implements the following sequence operation:

$$u(n) = K_p e(n) + K_I T \sum_{i=0}^{n} e(i) \tag{7.91}$$

where K_p and K_I are control gains to be adjusted for particular types of performance, and $T = t_{k+1} - t_k$ is the (uniform) interval between sample instants. Note that T times the summation approximates the integral of the error signal. Figure 7.18 illustrates the sequence operation of expression (7.91) using the elementary operations previously discussed.

It is of interest to study the combination of shift operation and summation connected in the loop as shown in Fig. 7.18. The reader should persevere until convinced that the summation operation in (7.91) is implemented by this combination.

Example 7.18 (Computer Control)

An increasing amount of attention is being given to computer control, where a properly programmed computer generates an appropriate actuation signal $u(k)$ based on an error signal $e(t)$ between the desired response $r(t)$ and the controlled response $c(t)$ of

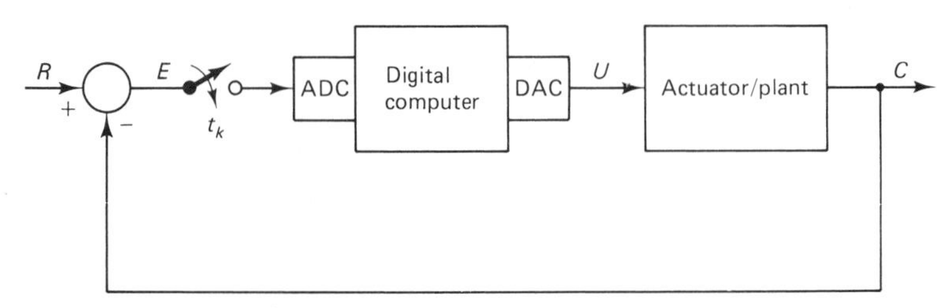

Figure 7.17 A computer control system.

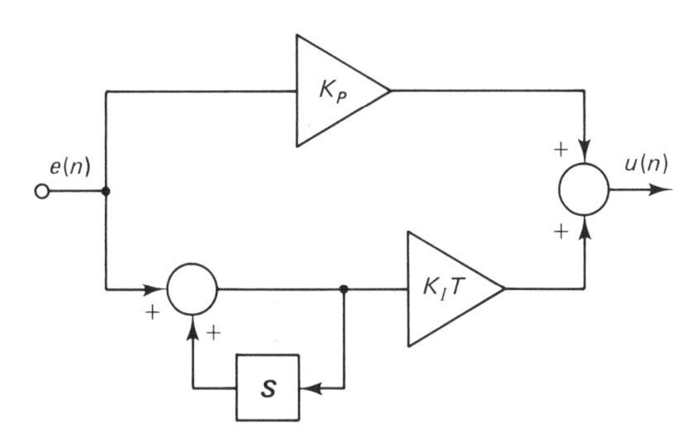

Figure 7.18 Diagram of the control law for equation (7.91).

some analog device. Such a configuration (called *direct-digital control*) is shown in Fig. 7.19.

It is necessary to provide electronic signal converters to interface the analog system with the digital computer. A convenient model for the analog-to-digital converter (ADC) is a simple sampler switch that operates every T seconds (T = the sample period). Similarly, the digital-to-analog converter (DAC) is a simple "holding device" that clamps or latches the incoming sample values to create a staircase signal $u(t)$ to drive the plant. The model for this analog conversion is the transfer function

$$G_H(s) = \frac{1 - e^{-sT}}{s}$$

For this example we have assumed that the controller has been programmed to implement the discrete-time transfer function

$$\frac{U(z)}{E(z)} = \frac{z}{z - 1}$$

to control the analog plant whose dynamics are represented by the transfer function

$$\frac{C(s)}{U(s)} = \frac{1}{s + 1}$$

To analyze this system, we first recognize that the effect of $c(t)$ on the control is only through its samples. We obtain the equivalent discrete-time system from $u(k)$ to

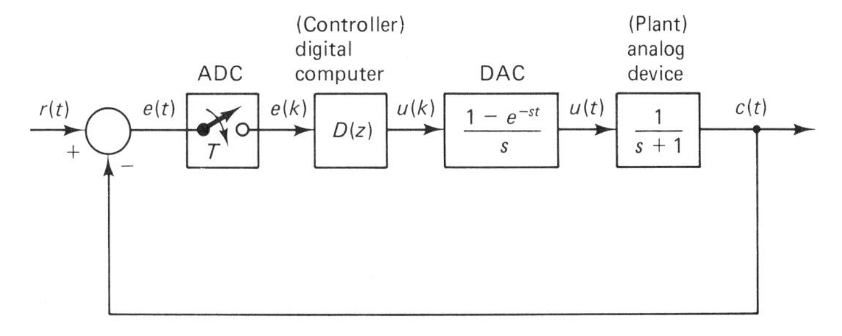

Figure 7.19 Diagram for direct-digital control.

$c(k)$ as follows:

$$G(z) = \frac{C(z)}{U(z)} = \mathscr{Z}\left[\frac{1 - e^{-sT}}{s(s+1)}\right] = \mathscr{Z}\left[\frac{1}{s(s+1)} - \frac{e^{-sT}}{s(s+1)}\right]$$

$$G(z) = \mathscr{Z}[(1 - e^{-kT}) - z^{-1}\mathscr{Z}(1 - e^{-kT})]$$

$$G(z) = (1 - z^{-1})\mathscr{Z}[(1 - e^{-kT})]$$

where the unit time-step delay e^{-sT} was replaced by its z-domain equivalent, z^{-1}. Finally, we have

$$G(z) = \frac{1 - e^{-T}}{z - e^{-T}}$$

as the discrete-time equivalent of the DAC and plant together. Combining with $D(z)$, the forward-gain equivalent is

$$D(z)G(z) = \frac{(1 - e^{-T})z}{(z - 1)(z - e^{-T})}$$

And the closed-loop equivalent is

$$\frac{D(z)G(z)}{1 + D(z)G(z)} = \frac{(1 - e^{-T})z}{z^2 - 2e^{-T}z + e^{-T}}$$

This transfer function describes the response in terms of the samples of $c(t)$. For example, if $r(t)$ is a unit-step input, the output samples are given by

$$c(kT) = \mathscr{Z}^{-1}\left[\frac{(1 - a)z^2}{(z - 1)(z^2 - 2az + a)}\right]$$

where $a = e^{-T}$, and $k = 0, 1, 2, \ldots$. Thus, if $T = 1$ second, and with no initial energy stored in the system, the first few output samples will be

$$[c(kT)] = \{\underset{\uparrow}{0}, 0.632, 1.097, 1.207, 1.116, \ldots\}$$

The one-step delay in the response is caused by the numerator order being one less than the denominator order. This is apparent when carrying out the long-division method of generating the sequence terms $c(kT)$ from the transform $C(z)$.

7.7 PROBLEMS

7.1. Investigate the following operators to determine if the operation is causal. Justify your answer.

(a) $T_a[x(t)] = \sin x(t)$

(b) $T_b[x(t)] = \int_{t-10}^{t+10} x(t)\, dt$

(c) $T_c[x(t)] = x(t - 5)$

(d) $T_d[x(t)] = ax(t) + b$

(e) $T_e[x(t)] = \int_0^t x(t)\, dt$

(f) $T_f[x(t)] = \int_{-\infty}^{t} a(t)x(t)\,dt$

(g) $T_g[x(t)] = \frac{d}{dt}x(t) + 5x^2(t)$

7.2. Determine if the operators in Problem 7.1 are linear. Justify your answer.

7.3. Determine if the operators in Problem 7.1 are time-invariant. Justify your answer.

7.4. Determine if the operators in Problem 7.1 are stable. Justify your answer.

7.5. A linear, causal, time-invariant system has a transfer function given by

$$H(s) = \frac{s+3}{(s+1)(s+2)(s-1)}$$

(a) Write a differential equation between input x and output y. Recall that $H(s) = Y(s)/X(s)$ with zero initial conditions on the system.

(b) Determine if the system is stable.

7.6. For the system given in Fig. P7.6, determine the output signal $y(t)$ if $x(t) = u(t)$ (a unit-step) and $y(0) = 2$.

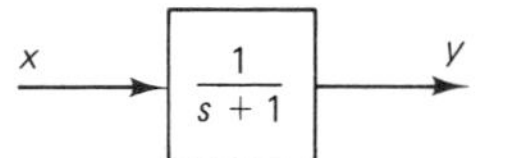

Figure P7.6

7.7. A causal system operation is defined implicitly by the relationship

$$\frac{dy}{dt} + y(t) = x(t)$$

Find an equivalent convolutional representation for this system.

7.8. The voltage across a capacitor of C farads is given by

$$v_C(t) = \frac{1}{C}\int_{-\infty}^{t} i_C(t)\,dt$$

where $i_C(t)$ is the current through the capacitor. Assume that $t > 0$ and write $v_C(t)$ in two parts, one of which describes the initial voltage $v_C(0) = (1/C)q_C(0)$, where $q_C(0)$ is the electric charge on the capacitor.

7.9. **(a)** Write a differential equation relating $y(t)$ to $x(t)$ from Fig. P7.9. (*Hint*: First develop a transfer function using circuit analysis techniques.)

(b) Write $y(0)$ and $Dy(0)$ in terms of $v_C(0)$ and $i_L(0)$.

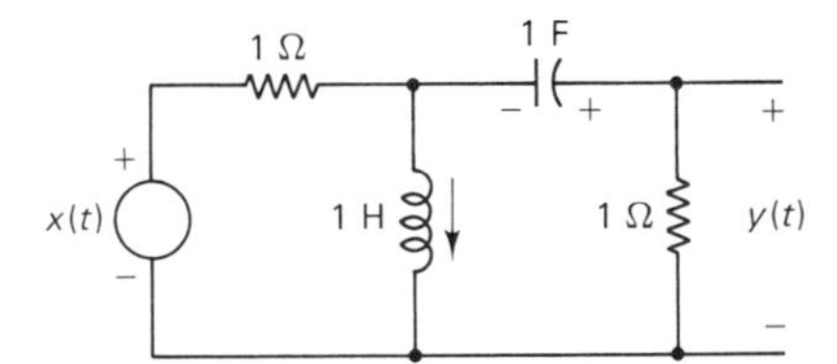

Figure P7.9

7.10. For a system with transfer function

$$H(s) = \frac{s+1}{(s+2)(s+3)} = \frac{Y(s)}{X(s)}$$

(a) Describe the particular solution $y_P(t)$ to the input $x(t) = \cos 3t\, u(t)$.
(b) Describe the homogeneous solution form. You should have two arbitrary constants.
(c) Find the complete solution if $y(0) = 0$ and $\boldsymbol{D}y(0) = 1$.

7.11. Determine the unit-impulse response $h(t)$ for a system with input x and output y related by the differential equation

$$\boldsymbol{D}^2 y + 5\boldsymbol{D}y + 6y = 2\boldsymbol{D}x + 2x$$

using the methods of Section 7.2.

7.12. Find the total response $y(t)$ by calculating $y_P(t)$ and $y_h(t)$ if the input is a unit-step response and

$$H(s) = \frac{s+3}{(s+1)(s+2)}$$

You may leave two constants of integration present in your answer.

7.13. Repeat Problem 7.12 for the input $x(t) = 4e^{-t}u(t)$.

7.14. Determine a particular solution for the system defined in Problem 7.11 when $x(t) = (e^{-t} - e^{-2t})u(t)$.

7.15. Use the result of Problem 7.14 to write a complete solution $y(t)$ in terms of two arbitrary integration constants.

7.16. For the system described in Problem 7.11, write a complete solution $y(t)$ when $x(t) = 3e^{-t} + 4\delta(t)$. Evaluate the constants of integration for the initial conditions $y(0) = \dot{y}(0) = 0$.

7.17. Find the response $y(t)$ of Fig. P7.17 when the input $x(t) = (2 \cos 2t + e^{-t})u(t)$ and $y(0) = 1$.

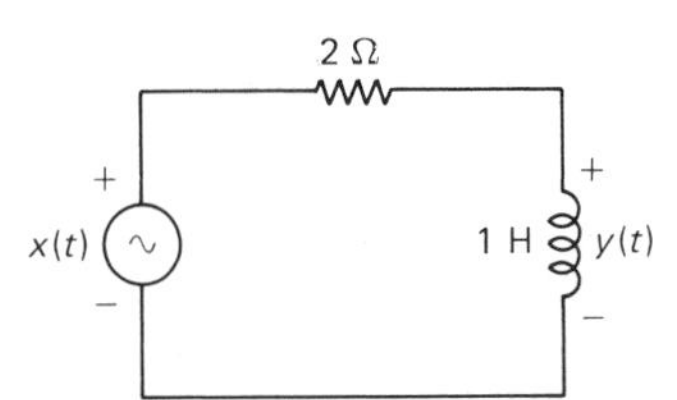

Figure P7.17 A simple RL network.

7.18. Repeat Problem 7.17 for an input $x(t) = (\sin t + \cos 2t)u(t)$ and $y(0) = 1$.

7.19. For the system of Fig. P7.17, suppose $x(t)$ is a constant 10 volts.
(a) Determine the time constant of the system.
(b) Sketch the response if the initial inductor current is zero.

7.20. For the switched network of P7.20, the switch S is opened for one second and closed for one second, repeatedly.
(a) Determine the time constant associated with the charging of the capacitor.
(b) Determine the time constant associated with the capacitor discharge.
(c) Sketch a typical $v(t)$ starting with S open and $v(0) = 0$, over a few switching cycles.

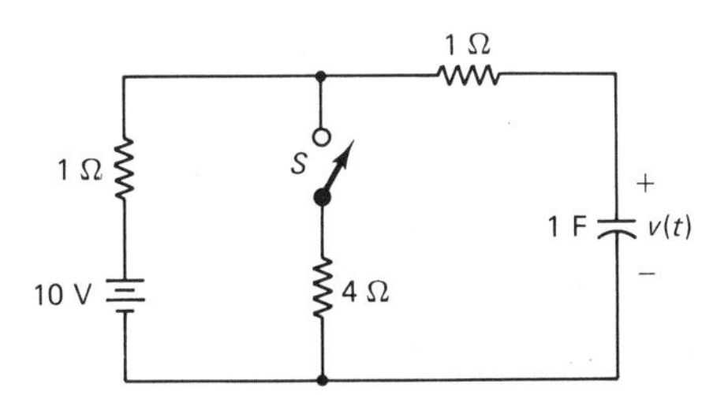

Figure P7.20

7.21. Determine the transfer function $X_2(s)/F(s)$ in Figure P7.21 if the indicated values represent standard units of mass, spring constant, and viscous friction.

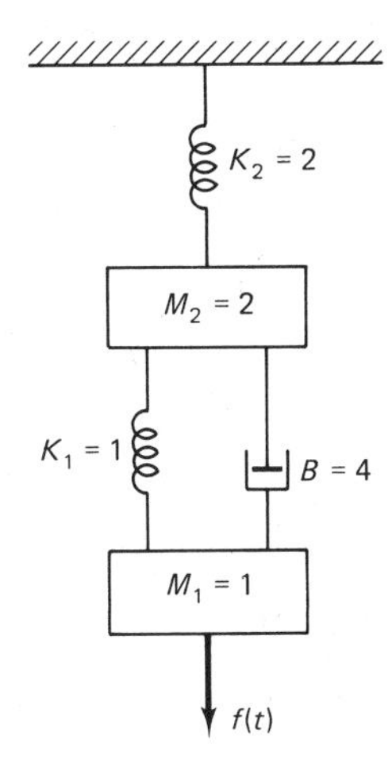

Figure P7.21

7.22. For the network given in Fig. P7.22, calculate the following network functions:

(a) $V_1(s)/I_1(s)$

(b) $I_2(s)/I_1(s)$

(c) $V_2(s)/V_1(s)$

Make appropriate assumptions on the type of source.

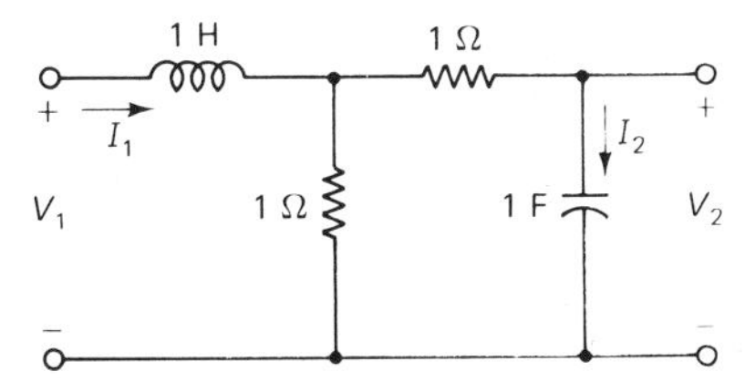

Figure P7.22

7.23. For the system described by the block diagram in Fig. P7.23, determine the transfer functions $T_1(s)$ and $T_2(s)$ in terms of G_1 snd G_2, such that

$$C(s) = T_1(s)R(s) + T_2(s)D(s)$$

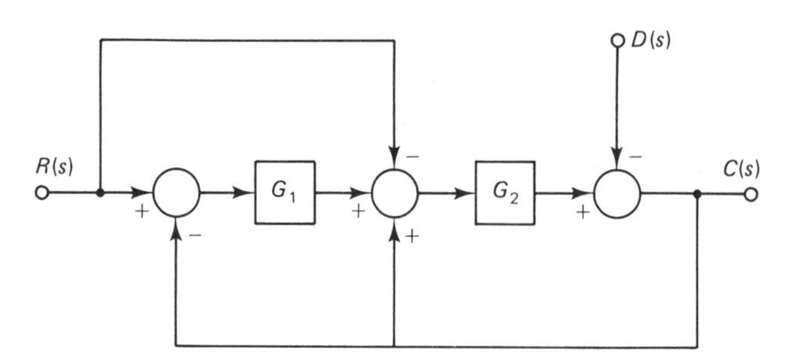

Figure P7.23

7.24. State which values of y are needed to correspond to the boundary (or initial) conditions for the following difference equations:

(a) $y_a(n+2) + 2y_a(n+1) + y_a(n) = \delta(n)$

(b) $y_b(n) = -5y_b(n-1) - 6y_b(n-1) + u(n-1)$

7.25. What is the initial nonzero value of y in Problem 7.24 if the system is initially at rest?

7.26. Solve for y in Problem 7.24 by recursion when the system is initially at rest.

7.27. Evaluate the systems in Problem 7.24 for the sequence y that satisfies the boundary conditions $y(0) = 0$ and $y(5) = 1$.

7.28. For a discrete-time system described by the difference equation

$$y(n+2) - 5y(n+1) + 6y(n) = x(n)$$

determine the total solution sequence $y(n)$ when $x(n) = u(n) - u(n-5)$, such that $y(0) = y(1) = 1$.

(a) Use the z-transform approach.

(b) Determine the unit-pulse sequence and use convolution to obtain the particular solution, then add on the appropriate homogeneous solution terms.

(c) Check your results by obtaining the first 5 terms of $y(n)$ by recursion.

7.29. Find the sequence $w(n)$ such that with $x(n) = u(n)$, a unit-step sequence

$$w(n) = 2x(n) + 3x(n-1) + 4x(n-2)$$

7.30. Repeat Problem 7.29 when $x(n) = (0.5)^n u(n)$.

7.31. Solve the equation

$$y(n) + y(n-1) = w(n)$$

where $w(n)$ is given in Problem 7.29 and $y(0) = 2$.

7.32. A discrete-time transfer function

$$H(z) = \frac{b_0 + b_1 z^{-1} + \cdots + b_m z^{-m}}{1 + a_1 z^{-1} + \cdots + a_n z^{-n}} = \frac{N(z)}{D(z)}$$

is sometimes referred to as an *ARMA model* (autoregressive, moving-average model), where $N(z)$ is the transfer function of the moving-average part (MA) and $D^{-1}(z)$ is the transfer function of the autoregressive part (AR). Determine the AR and MA parts of the system of Problem 7.31.

7.33. Find the discrete-time transfer function of the system described by

$$y(n+2) - 5y(n+1) + 6y(n) = 2x(n+2) + 4x(n+1) + 2x(n)$$

where $x(n)$ is the input signal and $y(n)$ is the response.

7.34. **(a)** Solve the system in Problem 7.33 if $y(0) = 0$, $y(1) = 1$, and $x(n)$ is a unit-step input.
(b) Describe the natural and forced response components.

8

Fourier Series Representation

8.1 INTRODUCTION

The concept of *signal* is central to any contemporary system application. Thus, it is not surprising that one of the main objectives of system theory is that of characterizing the salient features possessed by a given signal. Undoubtedly, one of the most frequently used tools for this purpose is that of a *Fourier series representation*. In making a Fourier series representation of a signal, one is equivalently expressing that signal as an infinite, weighted sum of prespecified "basis signals." The individual basis signals are generally very simple in time behavior (e.g., the exponentials) and are selected with analytical considerations in mind. A Fourier series can then be thought of as a procedure for systematically decomposing an arbitrary signal into a sum of "elementary" signals that are much simpler to use and analyze.

Before developing the general aspects of Fourier analysis, it is worth mentioning that the fundamental ideas in this section are basically an extension of steady-state network analysis.

The reader is familiar with the impedance concept of network theory from Chapter 7. As a brief review, consider the *R-L-C* network in Fig. 8.1. When the input is sinusoidal, the voltages across and currents through each element very quickly become sinusoidal. And since the relationships of the elements are

$$v_R = Ri_R, \qquad v_L = L\frac{di_L}{dt}, \quad \text{and} \quad v_C = \frac{1}{C}\int i_C\, dt \tag{8.1}$$

the derivative and integral operations applied to sinusoid currents reduce the funda-

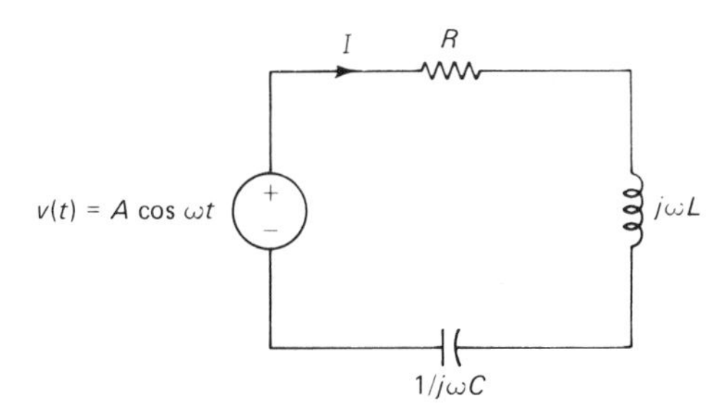

Figure 8.1 An *R-L-C* network.

mental element relations to the "complex resistance" form:

$$V_R = RI_R, \quad V_L = (j\omega L)I_L, \quad \text{and} \quad V_C = \left(\frac{1}{j\omega C}\right)I_C \tag{8.2}$$

where the voltages and currents now represent the magnitudes and phase angles of the original time-varying sinusoids. These complex quantities are called *phasors.*

Example 8.1

For the problem indicated in Fig. 8.1, the input voltage can be represented by the complex-valued (phasor) quantity $V = A\underline{/0^\circ}$, which carries the amplitude A and 0° phase information of the input signal. By treating each element shown as a complex resistance, we can write the Kirchhoff voltage law (KVL) equation, as follows:

$$A = \left(R + j\omega L + \frac{1}{j\omega C}\right)I \tag{8.3}$$

It follows that the loop current (phasor) I is

$$I = \frac{A}{R + j\left(\omega L - \dfrac{1}{\omega C}\right)} \tag{8.4}$$

And, by manipulating the complex expression, we can write that

$$I = V + jW \tag{8.5}$$

where

$$V = \frac{AR}{D}, \quad W = \frac{-A\left(\omega L - \dfrac{1}{\omega C}\right)}{D}$$

and

$$D = R^2 + \left(\omega L - \frac{1}{\omega C}\right)^2$$

Since V and W are both real-valued and the j-multiplier corresponds to a 90° shift in phase,

$$i(t) = V \cos \omega t + W \cos (\omega t + 90^\circ) \tag{8.6}$$

is the "translation" or interpretation of expression (8.5) back to the time domain. The current in equation (8.6) is the steady-state current in the loop and neglects any exponential terms in the solution under the assumption that they decay away very fast.

It is clear from expression (8.4) that the loop current is different for different frequencies ω of the input signal. In this chapter we develop the basic Fourier technique of describing periodic signals (which are very common in signal analysis) in terms of a sine-cosine series of terms at different frequencies, whose effect may then be calculated on a particular system.

Frequency Response

Our simple example above illustrates the basic idea that the input-output response changes with frequency. The transfer function from input voltage to output (loop) current is

$$T(s) = \frac{1}{R + sL + \dfrac{1}{sC}} = \frac{Cs}{LCs^2 + RCs + 1} \tag{8.7}$$

Replacing the Laplace variable s with $j\omega$,

$$T(j\omega) = \frac{1}{R + j\left(\omega L - \dfrac{1}{C\omega}\right)} \tag{8.8}$$

relates the output to the input phasor variables. $T(j\omega)$ is complex-valued, hence to display T versus ω we require two graphs. By convention, the magnitude and phase plots are preferred over the real and imaginary parts. See Fig. 8.2.

When R is relatively small, the frequency response curves of $T(j\omega)$ have the general shape shown in Fig. 8.2. For even smaller R the magnitude curve is more peaked and the phase-change form $+90^\circ$ to -90° occurs over a shorter range of frequencies around the critical frequency ω_0. This phenomenon is called *resonance*, and ω_0 is the *resonant frequency*. A system at resonance can exhibit a very dramatic response compared with its response at other frequencies.

In the past much emphasis was placed on graphical techniques for obtaining frequency response graphs from transfer functions in the Laplace variable s. Today it is a simple matter to write a short program to calculate these curves.

Example 8.2

Let us illustrate the computation of the curves of Fig. 8.2 from expression (8.7) with the specific values of $R = 0.1$ and $L = C = 1$.

$$T(s)\bigg|_{s=j\omega} = \frac{j\omega}{(j\omega)^2 + 1 + 0.1(j\omega)}$$

After substituting $s = j\omega$, the numerator and denominator terms are collected into real and imaginary parts.

$$T(j\omega) = \frac{0 + j\omega}{(1 - \omega^2) + j(0.1\omega)}$$

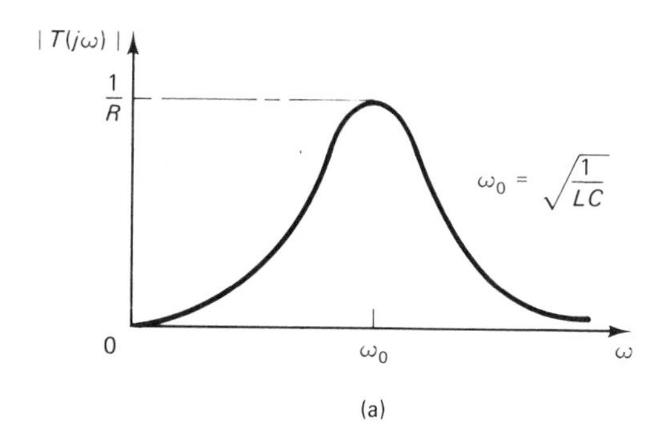

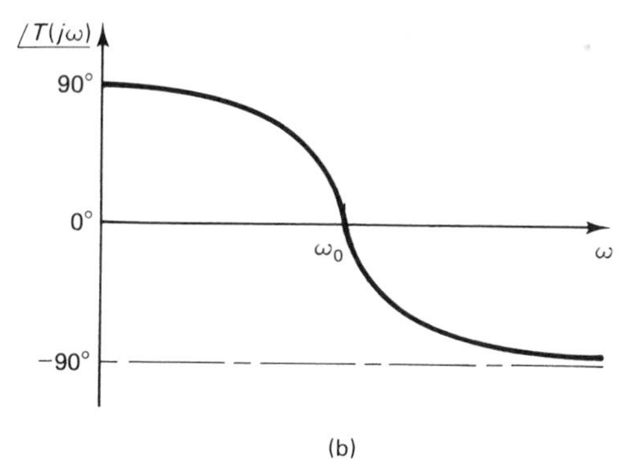

Figure 8.2 (a) Magnitude and (b) phase plots of $T(j\omega)$.

Using the notation that

$$T = \frac{A + jB}{C + jD}$$

one simply picks a value for ω, evaluates A, B, C, and D numerically, and changes the numerator and denominator from rectangular to polar form, so that

$$T = \frac{R\underline{/\theta}}{Q\underline{/\phi}} = \frac{R}{Q}\underline{/\theta - \phi} = |T|\,\underline{/T}$$

and divides magnitudes to obtain the magnitude of T and subtracts the denominator angle from the numerator angle to obtain the phase of T. This provides one point on each graph for the value of ω chosen. Typically, this calculation can be programmed and ω used in a loop so that following one calculation, ω is automatically incremented to the next value to start the calculation cycle over again.

Multiple Sinusoidal Inputs

Using the superposition principle, the steady-state analysis approach discussed above can be used on each sinusoidal input to obtain a total response as the sum of all the individual responses. Figure 8.3 illustrates a system with three sinusoidal

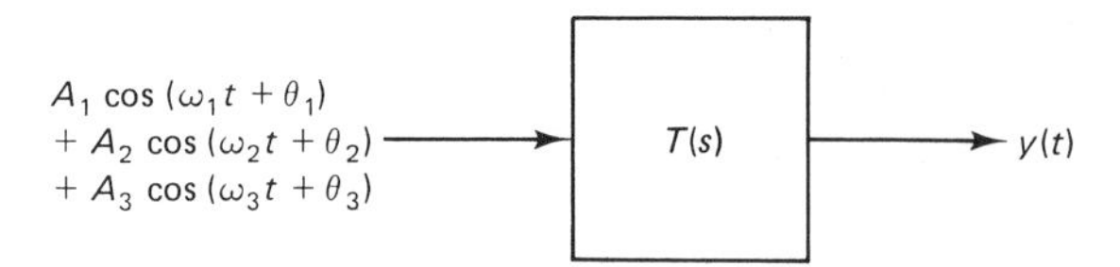

Figure 8.3 System with three sinusoidal sources.

sources applied to the same port of a system with transfer function $T(s)$. If we denote

$$\left.\begin{aligned} M(\omega) &= \text{Magnitude of } T(j\omega) \\ &\text{and} \\ P(\omega) &= \text{Phase of } T(j\omega) \end{aligned}\right\} \tag{8.9}$$

then for each phasor input $A_i \underline{/\theta_i}$, for $i = 1, 2, 3$, there is an output phasor given by

$$Y_i = A_i M(\omega_i) \underline{/\theta_i + P(\omega_i)} \qquad \text{for } i = 1, 2, 3 \tag{8.10}$$

Note that for each sinusoidal input its magnitude and phase are modified by the system transfer function evaluated at the sinusoidal frequency. From (8.10) we can write that

$$y(t) = y_1(t) + y_2(t) + y_3(t) \tag{8.11}$$

where

$$y_i(t) = A_i M(\omega_i) \cos [\omega_i t + \theta_i + P(\omega_i)]$$

With this development as motivation, we now investigate sinusoidal representation of periodic signals.

Sine-Cosine Series

There are several technical details that we are leaving to our subsequent development of more general series expansions. Here we investigate sinusoidal expansions of periodic signals. The signal $x(t)$ is said to be *periodic with period T* if T is the smallest number for which

$$x(t + T) = x(t) \qquad \text{for all } t \tag{8.12}$$

For example, $\sin 2t$ is periodic with period $T = \pi$.

The properties of sines and cosines that are important to us are given below. These results are easily verified using integral tables. For $T = 2\pi/\omega_0$,

$$\int_0^T \sin m\omega_0 t \, dt = 0 \qquad \text{for } m = 1, 2, 3, \ldots \tag{8.13}$$

$$\int_0^T \cos m\omega_0 t \, dt = 0 \qquad \text{for } m = 1, 2, 3, \ldots \tag{8.14}$$

$$\int_0^T \sin m\omega_0 t \cos n\omega_0 t \, dt = 0 \qquad \text{for } m, n = 1, 2, 3, \ldots \tag{8.15}$$

$$\int_0^T \sin m\omega_0 t \sin n\omega_0 t \, dt = \begin{cases} 0 & \text{for } m \neq n \\ T/2 & \text{for } m = n \end{cases} \tag{8.16}$$

$$\int_0^T \cos m\omega_0 t \cos n\omega_0 t \, dt = \begin{cases} 0 & \text{for } m \neq n \\ T/2 & \text{for } m = n \end{cases} \tag{8.17}$$

Although the integration is shown starting at $t = 0$, expressions (8.13) through (8.17) hold for the integration taken over one full period starting at any point in the cycle.

The basic Fourier series representation for a periodic signal $x(t)$ with period T is written as follows:

$$x(t) = a_0 + \sum_{n=1}^{\infty} (a_n \cos n\omega_0 t + b_n \sin n\omega_0 t) \tag{8.18}$$

where the coefficients $a_0, a_1, a_2, \ldots$ and $b_1, b_2, \ldots$ are calculated from $x(t)$ and are called the *Fourier coefficients of* $x(t)$. The fundamental frequency ω_0 is $2\pi/T$, where T is the period of $x(t)$.

In developing expressions for the Fourier coefficients, one may recognize that a_0 is the average value of $x(t)$:

$$a_0 = \frac{1}{T} \int_0^T x(t) \, dt \tag{8.19}$$

Multiplying expression (8.18) by a particular sine or cosine and integrating over one period gives us the means of calculating the remaining coefficients. For instance, to obtain a_3, multiply $x(t)$ in equation (8.18) by $\cos 3\omega_0 t$ and integrate over a full period:

$$\int_0^T x(t) \cos 3\omega_0 t \, dt = a_3 \frac{T}{2} \tag{8.20}$$

The right-hand side contains only one term, since all the other integrals vanish by equations (8.14), (8.15), and (8.17). The one nonvanishing term is the integral of $\cos^2 3\omega_0 t$, which equals $T/2$ from relationship (8.17). In a similar manner, we can derive the general expressions

$$a_k = \frac{2}{T} \int_0^T x(t) \cos k\omega_0 t \, dt \qquad \text{for } k = 1, 2, 3, \ldots \tag{8.21}$$

and

$$b_k = \frac{2}{T} \int_0^T x(t) \sin k\omega_0 t \, dt \qquad \text{for } k = 1, 2, 3, \ldots \tag{8.22}$$

It is worthwhile for the reader to develop these expressions from equation (8.18) by multiplying, in turn, by $\cos k\omega_0 t$ and $\sin k\omega_0 t$ and using the results presented in (8.13) to (8.17) to simplify the right-hand sides.

Example 8.3 (A Symmetric Square Wave)

Let us define a periodic signal $x(t)$ as follows:

$$x(t) = \begin{cases} A & \text{for } 0 \leq t < \pi \\ 0 & \text{for } \pi \leq t < 2\pi \end{cases} \qquad \text{for } T = 2\pi$$

This signal is presented in Fig. 8.4a. To develop the Fourier series for $x(t)$, we calculate the Fourier coefficients using (8.21) and (8.22). Since $T = 2\pi$, $\omega_0 = 1$, and

$$a_0 = \frac{1}{2\pi}\int_0^{\pi} A\,dt = \frac{A}{2}$$

$$a_k = \frac{1}{\pi}\int_0^{\pi} A\cos kt\,dt = \frac{A}{k\pi}\sin kt\Big|_{t=0}^{t=\pi} = 0$$

$$b_k = \frac{1}{\pi}\int_0^{\pi} A\sin kt\,dt = \frac{A}{k\pi}(-\cos kt)\Big|_{t=0}^{t=\pi} = \begin{cases} 2A/k\pi & \text{for } k \text{ odd} \\ 0 & \text{for } k \text{ even} \end{cases}$$

Thus, to write out the first few terms of our expression,

$$x(t) = \frac{A}{2} + \frac{2A}{\pi}\left(\sin t + \frac{1}{3}\sin 3t + \frac{1}{5}\sin 5t + \cdots\right)$$

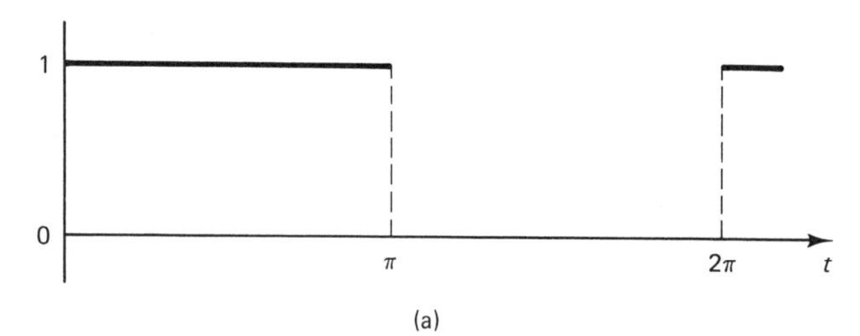

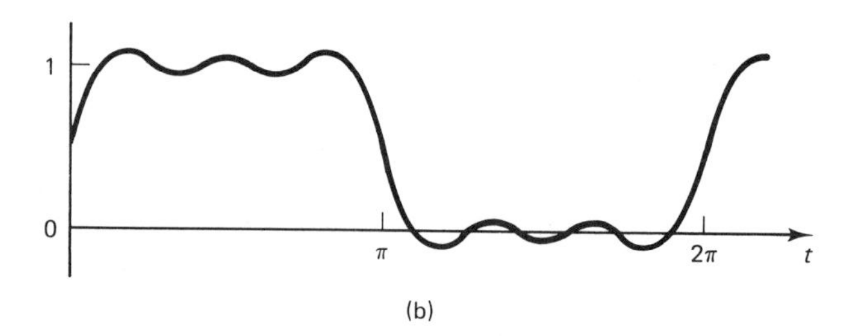

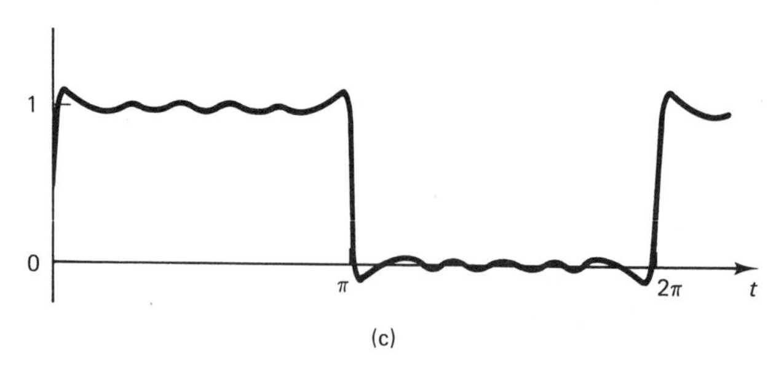

Figure 8.4 (a) Square-wave signal, (b) partial sum through fifth harmonic, and (c) partial sum through eleventh harmonic.

Figure 8.4b illustrates the partial sum of the Fourier series including the sin $5t$ term, and Fig. 8.4c shows the partial sum carried through the sin $11t$ term. For convenience, A is taken to be unity. The various frequency terms can be referred to as the d.c. term for the constant a_0; the *fundamental* frequency for the terms with frequency ω_0, and the *Nth harmonic frequency* for the terms having frequency $N\omega_0$. Example 8.3 indicates several interesting facts about Fourier series which will be discussed further in subsequent sections, for instance,

1. There are no cosine or even harmonic terms.
2. The more terms, the better the approximation generally.
3. The series converges to the midpoint of a discontinuity.
4. Near a discontinuity there is always an overshoot of about 9 percent of the step.

The symmetry of $x(t)$ is responsible for the simplification mentioned in the first observation. We discuss this in more detail later. Regarding the second remark, the series is said to converge pointwise, but not uniformly. That is, at any fixed point t, not at a discontinuity, the partial sums converge to the true function value, but for any finite number of terms, there will always be a 9 percent overshoot close to a discontinuity. At a discontinuity t_0 (like $t = 0$ or $t = \pi$ in Example 8.3), the series converges to

$$\frac{x(t_0^-) + x(t_0^+)}{2} \tag{8.23}$$

The fourth observation is a statement of the effect of a discontinuity known as *Gibb's phenomenon.* The reader may be interested to try a relatively large number of terms of our result in Example 8.3 and plot the partial sum for small t to verify this effect.

Example 8.4

If in a series R-L-C network similar to that shown in Fig. 8.1, $R = 1$ ohm, $C = 10^{-6}$ farad, $L = (2\pi)^{-2}$ henry, and $v(t)$ is periodic with period $T = 10^{-3}$ second, such that

$$v(t) = \frac{200}{T}\, t \qquad \text{for } 0 \le t < T$$

determine the first two nonzero terms of the Fourier series for the loop current $i(t)$.

Solution. The Fourier series for $v(t)$ is

$$v(t) = v_0 + \sum_{n=1}^{\infty} (a_n \cos n\omega_0 t + b_n \sin n\omega_0 t)$$

From expressions (8.21) and (8.22) we find that

$$a_n = 0 \qquad \text{for } n = 1, 2, 3, \ldots$$

and by using the normalized time $x = \omega_0 t$ and subtracting away the d.c. value $v_0 = 100$,

$$b_n = \frac{100}{\pi} \int_0^{2\pi} \left(\frac{1}{\pi} x - 1\right) \sin nx \, dx$$

Using the integral result that

$$\int u \sin u \, du = \sin u - u \cos u$$

we find that

$$b_1 = \frac{-200}{\pi}, \quad b_2 = \frac{-100}{\pi}, \quad \text{and } b_3 = \frac{-200}{3\pi}$$

Thus,

$$v(t) = \frac{100}{\pi}\left(\pi - 2 \sin \omega_0 t - \sin 2\omega_0 t - \frac{2}{3} \sin 3\omega_0 t - \cdots\right)$$

The network impedance at d.c. is infinite. At the fundamental frequency of 2000 hertz (2000π radians/second),

$$Z(\omega_0) = R + j\left(\omega_0 L - \frac{1}{\omega_0 C}\right)$$

$$Z(\omega_0) = 1 \underline{/0^\circ}$$

At the second-harmonic frequency $2\omega_0$,

$$Z(2\omega_0) = 1 + j\,\frac{3000}{4\pi} \simeq j\,\frac{3000}{4\pi}$$

Therefore, for each frequency component,

$$k_{\text{d.c.}} = \frac{v_0}{\infty} = 0$$

$$I_1 = \frac{-200}{\pi} \bigg/ 1 = \frac{-200}{\pi}$$

$$I_2 = \frac{-100}{\pi}\left(j\,\frac{3000}{\pi}\right)^{-1} = j\,\frac{4}{30}$$

so that the first two nonzero current terms are

$$i(t) = \frac{100}{\pi}\left[-2 \sin \omega_0 t + \frac{4\pi}{3000} \sin\left(2\omega_0 t + \frac{\pi}{2}\right) + \cdots\right]$$

where $\omega_0 = 2\pi(10^3)$ radians/second.

Signal Symmetry

The symmetry possessed by a signal can be of help in determining whether certain terms in its Fourier series will have zero coefficients. We define a few basic types of symmetry and show that certain groups of terms are linked to these types of symmetry. A function $x(t)$ has *even* symmetry if it satisfies the condition

$$x(-t) = x(t) \tag{8.24}$$

and $f(t)$ has *odd* symmetry if

$$x(-t) = -x(t) \tag{8.25}$$

The reader should verify that a (zero-phase) cosine has even symmetry and a (zero-phase) sine has odd symmetry.

An odd symmetric function must have zero average value. Thus, by removing the constant average value of some functions, its basic odd symmetry is made clear. This is the case for the signal $v(t)$ used in Example 8.4. One additional symmetry that has proved useful is that referred to as *half-wave symmetry*. A function $x(t)$ has half-wave symmetry if

$$x(t) = -x\left(t + \frac{T}{2}\right) \tag{8.26}$$

Although this symmetry condition is more difficult to apply, the half-wave symmetry property can be visualized by taking half of a period, flipping this half period about the horizontal axis and sliding it over to overlap the other half period. If the two parts line up exactly, the function has half-wave symmetry.

The utility of considering symmetry is clear from the following remarks. The Fourier series shown in expression (8.18) will have

1. no sine terms if $x(t)$ has even symmetry
2. no cosine terms if $x(t)$ has odd symmetry
3. no even harmonic terms (sine or cosine) if $x(t)$ has half-wave symmetry.

These three statements follow easily from expressions (8.21) and (8.22). Note that the symmetric square wave of Example 8.3 had both half-wave symmetry and (after subtracting out its average value) odd symmetry. Consequently, the resulting series had no cosine terms and no even harmonics.

Beginning with the next section, we will consider a more general Fourier series expansion and introduce the important concepts of signal approximation. The exponential form of the Fourier series is presented formally in a later section. It is, however, exactly the series that could be written from a sine-cosine series with the use of the Euler equations for representing sines and cosines in terms of complex exponentials:

$$\cos\theta = \frac{1}{2}(e^{j\theta} + e^{-j\theta}) \tag{8.27}$$

and

$$\sin\theta = \frac{1}{j2}(e^{j\theta} - e^{-j\theta}) \tag{8.28}$$

In Chapter 9 we discuss the computational aspects of Fourier analysis and develop the efficient FFT algorithms.

8.2 SIGNAL APPROXIMATION

As a preliminary to developing the Fourier series representation, let us first consider the equally important concept of signal approximation. In the classical signal approximation problem, one wishes to approximate an arbitrary signal $x(t)$ by means of a linear combination of the n prespecified "basis signals," as follows:

$$\phi_1(t), \phi_2(t), \ldots, \phi_n(t)$$

We shall allow for the possibility of the signals $x(t)$ and $\phi_k(t)$ being complex functions of time. The basis signals are generally selected to be of a simple analytical structure (e.g., $\phi_k(t) = t^{k-1}$) and are specified a priori in any given application. It must be mentioned that there exists a variety of good selections for the basis signals (for example, polynomials, exponentials, and so forth), and it is for this very reason that we use the general notation $\phi_k(t)$ in order to encompass all these possibilities.

It is now desired to approximate the arbitrary signal $x(t)$ by means of a linear combination of basis signals, that is,

$$x_n(t) = x_1\phi_1(t) + x_2\phi_2(t) + \cdots + x_n\phi_n(t) \tag{8.29}$$

where the complex-valued scalars $x_1, x_2, \ldots, x_n$ identify the approximating signal $x_n(t)$. These coefficients are to be selected so that the approximation error signal

$$e(t) = x(t) - x_n(t) = x(t) - \sum_{k=1}^{n} x_k\phi_k(t) \tag{8.30}$$

is minimized in some sense. There are numerous criteria one may use for measuring the goodness of this approximation. Without a doubt, the single most widely used such measure is the mean squared error criterion as defined by

$$\int_{t_o}^{t_1} |e(t)|^2\,dt = \int_{t_o}^{t_1} e(t)e^*(t)\,dt \tag{8.31}$$

where $e^*(t)$ denotes the complex conjugate of $e(t)$ and $t_o \leq t \leq t_1$ gives the interval over which the comparison of $x(t)$ and $x_n(t)$ is to be made. This criterion is seen to measure the integral of the error's magnitude squared. If this mean squared error criterion is minimized by a proper selection of the $x_1, x_2, \ldots, x_n$ coefficients that characterize the approximating signal, it follows that we are at the same time tending to make $x_n(t)$ closely approximate $x(t)$. What makes the mean squared error criterion of such relevance and interest is that its minimization yields a convenient analytical closed-form solution for the optimum coefficients $x_1^o, x_2^o, \ldots, x_n^o$.

Example 8.5

To illustrate the basic concepts of signal approximation, let it be desired to approximate the signal

$$x(t) = \begin{cases} 1 & \text{for } 0 \leq t < 1/2 \\ 2 & \text{for } 1/2 \leq t \leq 1 \end{cases}$$

over the time interval $0 \le t \le 1$. The basis signals in this case will be chosen to be

$$\phi_1(t) = 1, \quad \phi_2(t) = t, \quad \text{and} \quad \phi_3(t) = t^2$$

so that the approximating signal is given by

$$x_n(t) = x_1 + x_2 t + x_3 t^2$$

It is now desired to select values for the coefficients x_1, x_2, and x_3, in order that this approximating signal best matches the given signal $x(t)$ over $0 \le t \le 1$ in the sense of minimizing the mean squared error criterion (8.32).

With this problem serving as a motivation, we shall now evolve a general solution procedure.

Minimization of Mean Squared Error Criterion

We now develop a simple and systematic procedure for selecting the x_k coefficients that characterize the approximating signal (8.29), so that the mean squared error criterion (8.31) is minimized. This first entails putting this criterion into an appropriate mathematical format. For notational purposes, the mean squared error criterion is denoted by the symbol $I(x)$, in order to emphasize its dependency on the $n \times 1$ coefficient vector $x = (x_1, x_2, \ldots, x_n)$. This dependency is made clear when the expression for the error signal (8.30) is substituted into the functional (8.31), thereby giving

$$\begin{aligned} I(x) &= \int_{t_0}^{t_1} \left[x(t) - \sum_{k=1}^{n} x_k \phi_k(t) \right] \left[x^*(t) - \sum_{m=1}^{n} x_m^* \phi_m^*(t) \right] dt \\ &= \int_{t_0}^{t_1} x(t)x^*(t)\, dt - \sum_{m=1}^{n} x_m^* \int_{t_0}^{t_1} x(t)\phi_m^*(t)\, dt \\ &\quad - \sum_{k=1}^{n} x_k \int_{t_0}^{t_1} dt\, x^*(t)\phi_k(t) + \sum_{k=1}^{n} \sum_{m=1}^{n} x_k x_m^* \int_{t_0}^{t_1} dt\, \phi_k(t)\phi_m^*(t) \end{aligned} \tag{8.32}$$

where the operations of integration and summation have been interchanged in arriving at the last expression. Once the signal $x(t)$ becomes known (the $\phi_k(t)$ have already been specified), it is possible to evaluate each of the integrals in this expression. It is notationally convenient to denote these integral values by the symbols

$$a_{mk} = \int_{t_0}^{t_1} \phi_k(t)\phi_m^*(t)\, dt \qquad \text{for } m, k = 1, 2, \ldots, n \tag{8.33}$$

$$b_k = \int_{t_0}^{t_1} x(t)\phi_k^*(t)\, dt \qquad \text{for } k = 1, 2, \ldots, n \tag{8.34}$$

From expression (8.33) it is apparent that $a_{km} = a_{mk}^*$. If the entities b_k and a_{mk} are now substituted into the mean squared error criterion, the following equivalent representation is obtained:

$$I(x) = \int_{t_0}^{t_1} x(t)x^*(t)\, dt - \sum_{m=1}^{n} b_m x_m^* - \sum_{k=1}^{n} b_k^* x_k + \sum_{k=1}^{n} \sum_{m=1}^{n} a_{mk} x_k x_n^* \tag{8.35}$$

Recall that our ultimate objective is that of selecting values for the coefficients $x_1, x_2, \ldots, x_n$ in order to minimize $I(x)$. To accomplish this goal, it is expedient to first express the generally complex coefficient x_k in terms of its rectangular representation, as follows:

$$x_k = \alpha_k + j\beta_k \qquad \text{for } k = 1, 2, \ldots, n \tag{8.36}$$

where the elements α_k and β_k are real numbers. Substituting this rectangular representation into the mean squared error criterion (8.35), we obtain the following relationship:

$$\begin{aligned} I(x) = \int_{t_0}^{t_1} x(t)x^*(t)\,dt &- \sum_{m=1}^{n} b_m(\alpha_m - j\beta_m) \\ &- \sum_{k=1}^{n} b_k^*(\alpha_k + j\beta_k) + \sum_{k=1}^{n}\sum_{m=1}^{n} a_{mk}(\alpha_k + j\beta_k)(\alpha_m - j\beta_m) \end{aligned} \tag{8.37}$$

which is now a function of the $2n$ real variables α_k and β_k. This being the case, we may then use a well-known theorem from calculus which states that a necessary condition for $I(x)$ to be a relative minimum is that

$$\frac{\partial I(x^o)}{\partial \alpha_p} = 0 \quad \text{and} \quad \frac{\partial I(x^o)}{\partial \beta_p} = 0 \qquad \text{for } p = 1, 2, \ldots, n$$

with x^o denoting an optimum choice for the coefficient vector. These required partial derivatives are readily obtained from expression (8.37) and are given by

$$\frac{\partial I(x)}{\partial \alpha_p} = -b_p - b_p^* + \sum_{k=1}^{n} a_{pk}(\alpha_k + j\beta_k) + \sum_{m=1}^{n} a_{mp}(\alpha_m - j\beta_m) \qquad \text{for } p = 1, 2, \ldots, n$$

$$\frac{\partial I(x)}{\partial \beta_p} = jb_p - jb_p^* - j\sum_{k=1}^{n} a_{pk}(\alpha_k + j\beta_k) + j\sum_{m=1}^{n} a_{mp}(\alpha_m - j\beta_m) \qquad \text{for } p = 1, 2, \ldots, n$$

The condition for the α_p and β_p to be minimized is then obtained by setting these $2n$ partial derivatives to zero. One may readily show that this is achieved only if these coefficients satisfy

$$\sum_{k=1}^{n} a_{pk}(\alpha_k^o + j\beta_k^o) = b_p \qquad \text{for } p = 1, 2, \ldots, n \tag{8.38}$$

This is recognized as being a system of n-linear equations in the n-unknowns $x_p^o = \alpha_p^o + j\beta_p^o$. It is now desirable to summarize our findings in the form of the following theorem.

Theorem 8.1. Fundamental Approximation Theorem. A signal composed of a linear combination of the n-basis signals $\phi_1(t), \phi_2(t), \ldots, \phi_n(t)$ as given by

$$x_n(t) = x_1\phi_1(t) + x_2\phi_2(t) + \cdots + x_n\phi_n(t) \tag{8.39}$$

best approximates an arbitrary signal $x(t)$ on the time interval $[t_0, t_1]$ in the sense of minimizing the mean squared error criterion

$$\int_{t_0}^{t_1} \left| x(t) - \sum_{k=1}^{n} x_k \phi_k(t) \right|^2 dt \tag{8.40}$$

only if the x_k coefficients satisfy the following consistent system of n linear equations in n unknowns:

$$\sum_{k=1}^{n} a_{mk} x_k^o = b_m \qquad \text{for } m = 1, 2, \ldots, n \tag{8.41}$$

where the elements b_m and a_{mk} are given by

$$b_m = \int_{t_0}^{t_1} x(t)\phi_k^*(t)\, dt \quad \text{and} \quad a_{mk} = \int_{t_0}^{t_1} \phi_k(t)\phi_m^*(t)\, dt \qquad \text{for } m, k = 1, 2, \ldots, n \tag{8.42}$$

To obtain the best approximating signal, one must then solve the consistent system of linear equations (8.41) for an optimum set of coefficients. This being the case, it is desirable to express this optimal condition in matrix algebra, that is,

$$Ax^o = b \tag{8.43}$$

where A is a $n \times n$ matrix with elements a_{mk}, and b and x^o are each $n \times 1$ vectors with elements b_k and x_k^o, respectively. The fundamental approximation theorem indicates that this system of equations always has at least one solution vector x^o. As a matter of fact, when the basis signals $\phi_k(t)$ are linearly independent, it can be shown that the matrix A is invertible.[1] Then there will be a unique best approximating signal whose coefficients are given by

$$x^o = A^{-1}b \tag{8.44}$$

where A^{-1} denotes the $n \times n$ inverse of matrix A. It is a well-known fact that the matrix A is invertible only when the basis signals are linearly independent. Fortunately, in most practical approximation problem formulations, the basis signals selected are almost always linearly independent.

To obtain a best approximation signal, one must of necessity solve a system of n linear equations. As the next example illustrates, even in very simple situations (i.e., small values of n), this can involve a considerable amount of hand calculation. For more practical situations in which n is very large (on the order of hundreds), even the incorporation of digital computer routines for solving large systems of linear equations is doomed to failure (an exceptionally large ordered square matrix is difficult to invert). Fortunately, we can circumvent this difficulty by requiring that

[1]The signals $\phi_1(t), \phi_2(t), \ldots, \phi_n(t)$ are said to be linearly independent on the time interval $[t_0, t_1]$ if the only manner in which the relationship

$$\alpha_1\phi_1(t) + \alpha_2\phi_2(t) + \cdots + \alpha_n\phi_n(t) = 0 \qquad \text{for } t_0 \le t \le t_1$$

is satisfied is for all the α_k coefficients to be zero.

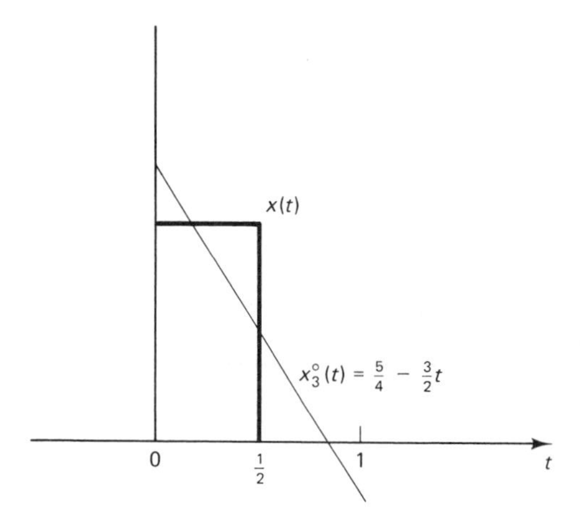

Figure 8.5 Approximation example.

the selected basis signals satisfy the additional property of being pairwise orthogonal. The next section is devoted to this most noteworthy concept.

Example 8.6

Let us now solve the problem presented in Example 8.5. This first requires the determination of the a_{kp} and b_k terms. Utilizing relationships (8.42), it is readily found that the system of equations (8.41) that is to be satisfied for the optimum coefficients is given by

$$\begin{bmatrix} 1 & 1/2 & 1/3 \\ 1/2 & 1/3 & 1/4 \\ 1/3 & 1/4 & 1/5 \end{bmatrix} \begin{bmatrix} x_1^o \\ x_2^o \\ x_3^o \end{bmatrix} = \begin{bmatrix} 1/2 \\ 1/8 \\ 1/24 \end{bmatrix}$$

Observe that all of the matrix terms are real, which is to be expected since in this example the signals $x(t)$, $\phi_1(t)$, $\phi_2(t)$, and $\phi_3(t)$ are all real. Solving this system of three linear equations, it is found that the unique solution is given by $x_1^o = \frac{5}{4}$, $x_2^o = -\frac{3}{2}$, and $x_3^o = 0$. Thus, the best approximation is given by

$$x^o(t) = \frac{5}{4} - \frac{3}{2}t \qquad \text{for } 0 \leq t \leq 1$$

A plot of the signal $x(t)$ and the best approximating signal is shown in Fig. 8.5. The approximation is rather poor, but it is the best possible under the existing choice of basis signals. One can achieve a significantly better approximation if one increases the number of basis signals to be used, e.g., $\phi_4(t) = t^3$, $\phi_5(t) = t^4$, $\phi_6(t) = t^5$, and so forth. This, of course, necessitates the solving of a larger system of linear equations to obtain the new optimum coefficients.

8.3 ORTHOGONAL BASIS SIGNALS

The concept of orthogonal basis signals provides a most significant simplification for the determination of the best approximating signal. Orthogonality connotes the notion of perpendicularity between vectors. In our case, the vectors are taken to be functions of continuous time. Many readers may find it difficult to envision a continuous-time signal as a vector. Nevertheless, it can be shown that continuous-time signals satisfy all of the axioms of a vector space and can therefore be interpreted as vectors. As a matter of fact, much of modern signal analysis theory uses this very interpretation. With these thoughts in mind, let us now define the concept of orthogonality between two continuous-time signals.

Definition 8.1. The continuous-time signals $x(t)$ and $y(t)$ are said to be *orthogonal* on the time interval $t_0 \le t \le t_1$ if

$$\int_{t_0}^{t_1} x(t)y^*(t)\,dt = 0 \tag{8.45}$$

where $y^*(t)$ denotes the complex conjugate of the signal $y(t)$.

It is important to note that orthogonality is dependent on a time interval $t_0 \le t \le t_1$. Two signals can be orthogonal on one choice of time interval and not on another choice. Furthermore, if the signals $x(t)$ and $y(t)$ are real, the conjugate operation * may be dropped.

Example 8.7

The signals $x(t) = e^{j(2\pi kt/T)}$ and $y(t) = e^{j(2\pi mt/T)}$ are orthogonal on the time interval $0 \le t \le T$ if the integer constants k and m are unequal. This is readily shown by substituting these signals into the relationship defining orthogonality, namely,

$$\int_0^T x(t)y^*(t)\,dt = \int_0^T e^{j(2\pi kt/T)}e^{-[j(2\pi mt/T)]}\,dt$$

$$= \left.\frac{e^{j[2\pi(k-m)t/T]}}{j\dfrac{2\pi(k-m)}{T}}\right|_0^T = 0 \qquad \text{for } k \ne m$$

It may be further shown that these two signals are not orthogonal on the time interval $0 \le t \le T/2$. As an added bit of information, the signals $x(t)$ and $y(t)$ can be shown to be orthogonal on *any* interval of T seconds' duration.

In the aforementioned signal approximation problem, let us consider the special case where the basis signals are pairwise orthogonal. The set of basis signals $\phi_1(t), \phi_2(t), \ldots, \phi_n(t)$ are said to be pairwise orthogonal on the time interval $t_0 \le t \le t_1$ if

$$\int_{t_0}^{t_1} \phi_k(t)\phi_m^*(t)\,dt = 0 \qquad \text{for } k \ne m \tag{8.46}$$

As we now show, if the selected basis signals satisfy this property, the task of computing the optimum set of coefficients characterizing the best approximating signal is significantly simplified. To see why this is so, it is first noted that the a_{km} parameters that characterize the system of equations (8.41) to be solved are zero for

$k \neq m$. This is a direct result of relationship (8.33) and the assumed orthogonality of the basis signals. With this in mind, the system of equations to be solved simplifies to

$$a_{mm}(\alpha_m^o + j\beta_m^o) = b_m \qquad \text{for } m = 1, 2, \ldots, n$$

The optimum coefficient selection is then obtained by dividing each side of this expression by a_{mm} to give

$$x^o = \alpha_m^o + j\beta_m^o = \frac{b_m}{a_{mm}} \qquad \text{for } m = 1, 2, \ldots, n$$

We next put this relationship into integral form using the identities for b_p and a_{pp} as given by expression (8.33). This results in

$$x_p^o = \frac{\int_{t_0}^{t_1} x(t)\phi_p^*(t)\, dt}{\int_{t_0}^{t_1} \phi_p(t)\phi_p^*(t)\, dt} \qquad \text{for } p = 1, 2, \ldots, n \tag{8.47}$$

Thus, when the basis signals are chosen to be orthogonal, one circumvents the task of having to solve a system of n-linear equations in n-unknowns. One need only evaluate the integrals involved in expression (8.47) to generate the optimum coefficient selection. Clearly, this is a most welcomed computational simplification, and the utilization of orthogonal basis signals is desirable in this sense.

One troublesome point remains to be dispensed with: If we restrict the basis signals to be pairwise orthogonal, do we lose the capability of obtaining "good" approximating signals? Fortunately, this is not the case since it may be shown that any set of linearly independent basis signals can be transformed into an equivalent set of orthogonal basis signals. These two sets of basis signals are said to be equivalent in the sense that use of either set yields the same best approximating signal $x_n^o(t)$ for a given $x(t)$. The process for generating the equivalent orthogonal set is commonly known as the *Gram-Schmidt orthogonalization method*, and its essential features are outlined in the problem section. With these thoughts in mind, we shall hereafter exclusively consider sets of orthogonal basis signals.

Individual signals $\phi_p(t)$ that constitute a basis signal set have "sizes" that are measured by the quantity a_{pp} as given in expression (8.33). If the value for the scalar a_{pp} happens to be one, that is,

$$a_{pp} = \int_{t_0}^{t_1} \phi_p(t)\phi_p^*(t)\, dt = 1 \tag{8.48}$$

the signal $\phi_p(t)$ is said to be *normalized*. If the signal $\phi_p(t)$ is not normalized, that is, relationship (8.48) is not satisfied, then an equivalent normalized signal can be substituted in its place by the simple scalar multiple $\hat{\phi}_p(t) = \phi_p(t)/\sqrt{a_{pp}}$. If a given set of orthogonal basis signals is made up of normalized signals exclusively (that is, $a_{pp} = 1$ for $p = 1, 2, \ldots, n$), the set is said to be *orthonormal*. The optimal coefficient selection for sets of orthonormal basis signals is particularly simple and is given by

$$x_p^o = b_p = \int_{t_0}^{t_1} x(t)\phi_p^*(t)\,dt \qquad \text{for } p = 1, 2, \ldots, n \tag{8.49}$$

since the dividing scalar a_{pp} is equal to one.

Frequently Used Sets of Orthogonal Basis Signals

We now consider specific sets of orthogonal basis signals. Most likely, it is apparent to many readers that there exists a large number (infinity) of different sets of orthogonal basis signals that can be used to approximate signals on any given time interval. The particular set to be used in any situation largely depends on the type (or class) of signals that are to be approximated. One set of orthogonal basis signals is generally preferred over others when the particular characteristics of this class are known a priori. The choice of which set to use is a very critical decision, and one must give its selection careful consideration. This selection procedure usually reflects the user's previous experience and knowledge. We now give a small sample of some of the most often used orthogonal basis signals.

Complex Exponentials Signals

The most widely used set of orthogonal basis signals for approximating a signal on any T-second interval $(t_0, t_0 + T)$ is the complex exponentials as defined by

$$\phi_k(t) = e^{j(2\pi kt/T)} \qquad \text{for } k = 0, \pm 1, \pm 2, \ldots \tag{8.50}$$

One should note that the integral index k is defined over negative as well as positive integer values. The orthogonality of these complex exponentials over *any* T-second interval is established by verifying the following integral relationship:

$$a_{km} = \int_{t_0}^{t_0+T} e^{j(2\pi kt/T)} e^{-[j(2\pi mt/T)]}\,dt = \begin{cases} 0 & \text{for } k \neq m \\ T & \text{for } k = m \end{cases}$$

This orthogonal property is independent of the initial time parameter t_0. When using the complex exponential signals to approximate any signal, the approximating signal is expressed in the following symmetrical manner:

$$x_n(t) = \sum_{k=-n}^{n} x_k\, e^{j(2\pi kt/T)} \tag{8.51}$$

As previously indicated, the complex exponential basis signals form one of the most important of all orthogonal sets. We shall shortly more fully discuss the salient features of this most essential set of orthogonal signals.

Rademacher Signals

In digital signal processing applications, one is often interested in approximating signals by simple two-valued signals. Let us generate what might be viewed as the simplest set of orthogonal signals that can take on the values of plus or minus one exclusively on the time interval [0, 1]. To begin, let us define the first such signal as

$$\phi_0(t) = \begin{cases} 1 & \text{for } 0 \le t < 1 \\ 0 & \text{otherwise} \end{cases}$$

The next signal of this set must be orthogonal to $\phi_0(t)$ and have as simple a

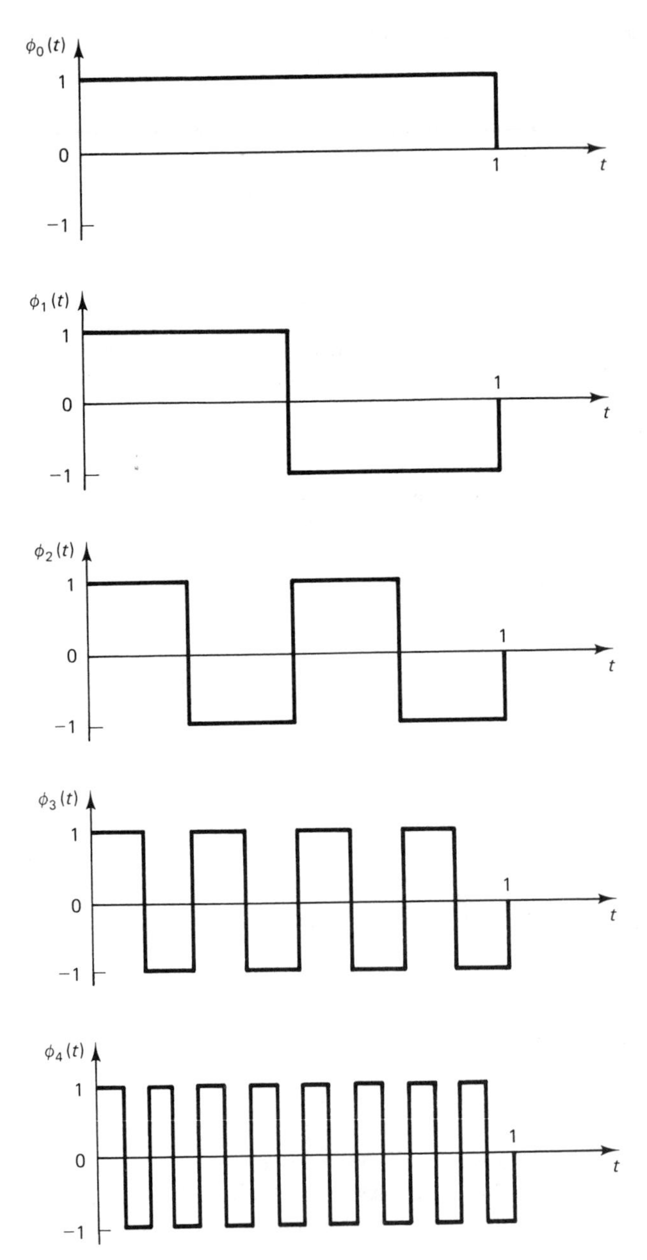

Figure 8.6 Rademacher signals.

structure as possible. An obvious choice would be

$$\phi_1(t) = \begin{cases} 1 & \text{for } 0 \le t < 1/2 \\ -1 & \text{for } 1/2 \le t < 1 \\ 0 & \text{otherwise} \end{cases}$$

The third signal of this set must be orthogonal to $\phi_0(t)$ and $\phi_1(t)$ and likewise be simple. A logical candidate might be

$$\phi_2(t) = \begin{matrix} 1 \\ -1 \\ 0 \end{matrix} \begin{cases} \text{for } 0 \le t < 1/2 \quad \text{and} \quad 1/2 \le t < 3/4 \\ \text{for } 1/4 \le t < 1/2 \quad \text{and} \quad 3/4 \le t < 1 \\ \text{otherwise} \end{cases}$$

which is readily found to satisfy the required orthogonality properties. It is observed that $\phi_2(t)$ is simply equal to two cycles of $\phi_1(t)$ compacted timewise to the interval $[0, 1]$.

If we continue this process, the set of orthonormal Rademacher signals would be generated. The formula for the general kth signal of this set is given by the iterative relationship

$$\phi_k(t) = \phi_{k-1}(2t) + \phi_{k-1}(2t - 1) \qquad \text{for } k = 2, 3, 4, \ldots \tag{8.52}$$

which reflects the fact that the kth basis signal is identical to two cycles of the $(k - 1)$st basis signal compacted in time by one-half. A sketch of the initial four Rademacher signals is shown on Fig. 8.6. It is observed that the kth Rademacher signal is a train of rectangular pulses consisting of 2^{k-1} cycles over the interval $0 \le t < 1$. The zero Rademacher signal is the lone exception to this pattern. As might be expected, Rademacher signals are useful in approximating signals that are pulselike in shape.

Legendre Signals

Undoubtedly, one has previously been made aware of the importance attached to the polynomial approximation of arbitrary signals. In such situations, it is desired to approximate a signal $x(t)$ by means of a linear combination of the form $x_o + x_1 t + \cdots + x_n t$ over the time interval $[-1, 1]$. The basis signals may then be taken to be $\phi_k(t) = t^k$ for $k = 0, 1, 2, \ldots, n$. Unfortunately, this set of basis signals is not orthogonal, thus a determination of the best approximation signal will involve the solving of a system of linear equations. This undesirable task may be avoided by transforming this simple set of basis signals into an equivalent set of orthonormal basis signals using the Gram-Schmidt procedure. This results in the following set of orthonormal signals on $[-1, 1]$:

$$\phi_0(t) = \frac{1}{2}, \qquad \phi_1(t) = \sqrt{\frac{3}{2}}\, t, \qquad \phi_2(t) = \sqrt{\frac{5}{2}}\left(\frac{3}{2}t^2 - \frac{1}{2}\right), \qquad \text{etc.}$$

The general nth term is of the form

$$\phi_n(t) = \sqrt{\frac{2n + 1}{2}}\, P_n(t) \qquad \text{for } n = 0, 1, 2, \ldots$$

where $P_n(t)$ are the so-called *Legendre polynomials* given by

$$P_n(t) = \frac{1}{2^n n!} \frac{d^n(t^2-1)^n}{dt^n} \qquad \text{for } n = 0, 1, 2, \ldots$$

The individual polynomials may be calculated using the iterative relationship

$$nP_n(t) = (2n-1)P_{n-1}(t) - (n-1)P_{n-2}(t)$$

8.4 CHANGING THE TIME INTERVAL OF APPROXIMATION

A practical consideration that frequently arises is that of approximating a signal $x(t)$ defined on the time interval $[t_0, t_1]$ by means of a set of orthogonal basis signals $\phi_k(t)$ defined on a different time interval $[t_2, t_3]$. Since the time intervals are different, there exists a basic incompatibility between the constituent signals. This is readily resolved by making a change in the time variable utilizing the following given orthogonality condition:

$$\int_{t_2}^{t_3} \phi_k(\tau)\phi_m^*(\tau)\, d\tau = 0 \qquad \text{for } k \neq m \tag{8.53}$$

so that the upper and lower limits become t_1 and t_0, respectively. It is readily shown that this is achieved by the change of variable substitution

$$\tau = \left(\frac{t_3 - t_2}{t_1 - t_0}\right)t + \left(\frac{t_2 t_1 - t_0 t_3}{t_1 - t_0}\right) \tag{8.54}$$

to yield the result

$$\int_{t_0}^{t_1} \hat{\phi}_k(\tau)\hat{\phi}_m^*(\tau)\, d\tau = 0 \qquad \text{for } k \neq m$$

The new orthogonal basis signals defined on the time interval $[t_0, t_1]$ are simply

$$\hat{\phi}_k(t) = \sqrt{\frac{t_3 - t_2}{t_1 - t_0}}\, \phi_k\left[\left(\frac{t_3 - t_2}{t_1 - t_0}\right)t + \left(\frac{t_2 t_1 - t_0 t_3}{t_1 - t_0}\right)\right] \qquad \text{for } t_0 \le t \le t_1 \tag{8.55}$$

It should also be noted that if the original basis signals are orthonormal on $[t_2, t_3]$, these transformed basis signals are likewise orthonormal on $[t_0, t_1]$. This follows from the fact that the value of integral (8.53) is not changed by the given change of variables operation.

Example 8.8

Approximate the time signal $x(t)$ considered in Example 8.8 by means of a linear combination of the first three Legendre signals. In this case, the signal $x(t)$ is defined on $[0, 1]$, whereas the Legendre basis signals $\phi_k(t)$ are defined on $[-1, 1]$. Letting $t_0 = 0$, $t_1 = 1$, $t_2 = -1$, and $t_3 = 1$, the Legendre basis signals are then transformed via relationship (8.55) into

$$\hat{\phi}_k(t) = \sqrt{2}\, \phi_k(2t-1) \qquad \text{for } 0 \le t \le 1$$

for $k = 0, 1, 2, \ldots$. The first three transformed Legendre basis signals are then given by

$$\hat{\phi}_0(t) = 1, \qquad \hat{\phi}_1(t) = \sqrt{3}\,(2t - 1), \qquad \hat{\phi}_2(t) = \sqrt{5}\,(6t^2 - 6t + 1)$$

It is readily shown that these signals are orthonormal on the time interval [0, 1].

The desired approximation signal is then given by

$$x_3^o(t) = \sum_{k=0}^{2} x_k^o \hat{\phi}_k(t)$$

where the x_k^o coefficients are evaluated according to

$$x_k^o = \int_0^1 x(t)\hat{\phi}_k(t)\,dt \qquad \text{for } k = 0, 1, 2$$

For the given signal $x(t)$, it is readily found that $b_0 = \frac{1}{2}$, $b_1 = \sqrt{\frac{3}{4}}$, and $b_2 = 0$, so that the best approximation is given by

$$x_3^o(t) = \frac{1}{2} - \frac{3}{4}(2t - 1) = \frac{5}{4} - \frac{3}{2}t \qquad \text{for } 0 \le t \le 1$$

The fact that this result is in agreement with that of Example 8.6 should not be surprising since the orthonormal Legendre signals $\hat{\phi}_k(t)$ used here are obtained from the signals t^k in Example 8.6 by the Gram-Schmidt orthonormalizing procedure.

8.5 GENERAL FOURIER SERIES

The desirability of using orthogonal basis signals approximating an arbitrary signal has been established in Section 8.3. Such a selection ensures a straightforward generation of the best approximation signal over a time interval $[t_0, t_1]$, that is,

$$x_n^o(t) = \sum_{k=1}^{n} \frac{b_k}{a_{kk}} \phi_k(t) \tag{8.56}$$

where the parameters a_{kk} and b_k are obtained by evaluating the integral expressions

$$a_{kk} = \int_{t_0}^{t_1} \phi_k(t)\phi_k^*(t)\,dt \quad \text{and} \quad b_k = \int_{t_0}^{t_1} x(t)\phi_k^*(t)\,dt \tag{8.57}$$

for $k = 1, 2, \ldots, n$. Thus, for a given set of orthogonal basis signals $\phi_1(t), \phi_2(t), \ldots, \phi_n(t)$, the specific linear combination (8.56) renders the underlying mean squared error criterion a "unique" minimum. Any other choice of the x_k coefficients causes a poorer quality of approximation.

Another dividend is accrued when orthogonal basis signals are used: The minimum value that the mean squared error criterion takes on has a convenient, closed-form expression. This is readily obtained by substituting the optimum coefficient selection $x_k^o = b_k/a_{kk}$ into expression (8.35) and using the orthogonal condition, which states that $a_{km} = 0$ for $k \neq m$, to obtain

$$I_n(x^o) = \int_{t_0}^{t_1} x(t)x^*(t)\,dt - \sum_{k=1}^{n} \frac{b_k b_k^*}{a_{kk}} \tag{8.58}$$

We have chosen to use the subscript n on $I(x)$ to emphasize that n basis signals are being used in the approximation signal representation. Note that the summand terms $b_k b_k^* / a_{kk}$ in this expression are always nonnegative real numbers.

A careful examination of this expression of the minimum mean squared error offers invaluable insight into the approximation problem. It is first noted that $I_n(x^o)$ is always nonnegative real due to its basic definition as given by expression (8.32). Thus, the smallest value that $I_n(x^o)$ can take on is zero; it will result only when the error signal is itself identically zero over the time interval $[t_0, t_1]$. Generally, the quantity $I_n(x^o)$ is positive, thereby implying the existence of a nonzero error signal $e(t) = x(t) - x_n^o(t)$.

The kth term of the best approximating signal as given by $(b_k/a_{kk})\phi_k(t)$ is seen to decrease the mean squared error criterion by the amount $b_k b_k^* / a_{kk}$. This quantity $b_k b_k^* / a_{kk}$ then provides a measure of the kth term's effectiveness in the approximation signal. The larger this quantity is, the more effective is the corresponding basis signal term $(b_k/a_{kk})\phi_k(t)$ in approximating the signal $x(t)$.

For a given set of orthogonal basis signals $\phi_1(t)$, $\phi_2(t)$, ..., $\phi_n(t)$, expression (8.58) gives the smallest value the mean squared error criterion can take on. If the quality of this approximation is not satisfactory, an improvement might be achieved by supplementing the given set of orthogonal basis signals by an additional orthogonal basis signal $\phi_{n+1}(t)$. The best approximation signal for this enlarged set of orthogonal basis signals would then be given by

$$\begin{aligned} x_{n+1}^o(t) &= \sum_{k=1}^{n+1} \frac{b_k}{a_{kk}} \phi_k(t) \\ &= x_n^o(t) + \frac{b_{n+1}}{a_{n+1\,n+1}} \phi_{n+1}(t) \end{aligned} \tag{8.59}$$

Thus, the best approximation signal $x_{n+1}^o(t)$ is seen to be simply generated by adding the term $(b_{n+1}/a_{n+1\,n+1})\phi_{n+1}(t)$ to the previous best approximation signal $x_n^o(t)$. This simple and convenient relationship is totally due to the orthogonality of the set of basis signals, and it does not hold for general sets of nonorthogonal basis signals (i.e., the coefficients change for the enlarged set of basis signals if the basis signals are not orthogonal).

Using this supplementation of the set of orthogonal basis signals, an improvement in the signal approximation is generally achieved. This may be readily shown by noting that for the optimum approximation signal (8.59), the mean squared error criterion becomes

$$I_{n+1}(x^o) = \int_{t_0}^{t_1} x(t)x^*(t)\,dt - \sum_{k=1}^{n+1} \frac{b_k b_k^*}{a_{kk}}$$

A comparison of this relationship with that of expression (8.55) is seen to give

$$I_{n+1}(x^o) = I_n(x^o) - \frac{b_{n+1} b_{n+1}^*}{a_{n+1\,n+1}}$$

Therefore, unless the elements $b_{n+1} = 0$, a decrease in the mean squared error criterion by the amount $b_{n+1}b^*_{n+1}/a_{n+1n+1}$ has been achieved by the supplementation process.

Conceivably, this supplementing procedure can be continued "indefinitely" with the mean square error criterion being decreased at each stage (or, at worst, not being increased), thereby rendering an improved approximation. If this criterion is in fact made equal to zero "in the limit," the signal $x^o(t)$ is said to "converge in the mean" to signal $x(t)$, that is,

$$x(t) \doteq x^o(t) = \sum_{k=1}^{\infty} \frac{b_k}{a_{kk}} \phi_k(t) \tag{8.60}$$

The dot over the equality implies that the error signal $e(t) = x(t) - x^o(t)$ is "almost everywhere" zero on the time interval $[t_0, t_1]$. We use the term "almost everywhere" zero to depict the fact that

$$\int_{t_0}^{t_1} e(t)e^*(t)\, dt$$

is zero. One must be careful to note that this does not necessarily imply $x^o(t) = x(t)$ for all values of t in $[t_0, t_1]$. Fortunately, this condition does imply that $x^o(t)$ and $x(t)$ are identically equal to each other for virtually all t in $[t_0, t_1]$. The meaning of this mathematical concept is illustrated in future sections.

Complete Set of Orthogonal Basis Signals

It might be conjectured that if one very intelligently chose the set of infinite basis signals, then the minimum mean squared error criterion (8.35) would be zero for an interesting class of signals to be approximated. Fortunately, this is in fact the case so long as (a) the signal $x(t)$ satisfies the so-called Dirichlet conditions, and (b) the set of countably infinite orthogonal basis signals $\{\phi_k(t)\}$ is "complete." It is beyond the scope of this text to prove this conjecture, but it is worthwhile to discuss these concepts.

The first requirement for the minimum mean squared error criterion to be zero is that the signal being approximated must satisfy the Dirichlet conditions. The signal $x(t)$ is said to satisfy the Dirichlet conditions if

1. it is absolutely integrable on $[t_0, t_1]$, that is, if

$$\int_{t_0}^{t_1} |x(t)|\, dt \tag{8.61}$$

   is finite.
2. it contains a finite number of relative maximum and minimum on $[t_0, t_1]$ and if it has a finite number of discontinuities on $[t_0, t_1]$.

It should be apparent that these conditions are satisfied by virtually all real-world signals and that the requirement that $x(t)$ satisfies them imposes no real restriction in the approximation problem.

The concept of the complete set of orthogonal basis signals is most critical to our discussion and will now be formally described.

Definition 8.2. The infinite orthogonal basis signal set $\{\phi_1(t), \phi_2(t), \phi_3(t), \ldots\}$ on the time interval $[t_0, t_1]$ is said to be complete if there exists no nontrivial signal $x(t)$ that is orthogonal to each of the $\phi_k(t)$, that is,

$$\int_{t_0}^{t_1} x(t)\phi_k^*(t)\, dt = 0 \qquad \text{for } k = 1, 2, 3, \ldots$$

A little thought should convince the reader that this is a most natural requirement to impose if one wishes to cause the mean squared error criterion to always be zero. For instance, suppose there does exist a nontrivial signal $x(t)$ which is orthogonal to each of the orthogonal basis signals. If we approximate this signal by means of a Fourier series, each of the b_k coefficients in the series would be zero, thereby rendering the approximation signal $x^o(t) = 0$. However, the error signal $e(t) = x(t) - x^o(t) = x(t)$ in this case would not be zero "almost everywhere" due to the assumption that $x(t)$ is nontrivial. The mean squared error criterion would then be nonzero, thereby contradicting the original premise.

It is generally a difficult task to show whether a given countably infinite set of orthogonal basis signals is complete or not. It can be shown, however, that the complex exponential signals and the Legendre signals considered in Section 8.3 are complete. On the other hand, the Rademacher signals are not complete and their use in a Fourier series representation, therefore, does not ensure that $x^o(t) = x(t)$.

8.6 EXPONENTIAL FOURIER SERIES

Without a doubt, the most widely used and studied of all the complete orthogonal basis signals are the exponentials. In using this particular choice of basis signals for representing a signal on the time interval $[t_0, t_1]$, it is beneficial to denote the time interval's length by the symbol T, where

$$T = t_1 - t_0$$

The corresponding exponential basis signals may then be expressed as

$$\phi_k(t) = e^{j(2\pi kt/T)} \qquad \text{for } k = 0, \pm 1, \pm 2, \ldots \tag{8.62}$$

in which it is observed that the integer k encompasses nonpositive as well as positive values. It has previously been shown that this set of basis signals is pairwise orthogonal on the time interval $[t_0, t_0 + T]$. What is so remarkable about exponential basis signals is that they remain pairwise orthogonal on any time interval of T seconds width (i.e., orthogonality is independent of the choice of t_0).

In using exponential signals for the basis set, the entity $2\pi/T$ will occur over and over again. This combination of three terms can be replaced by the single symbol

$$\omega_0 = \frac{2\pi}{T} \tag{8.63}$$

This enables us to express the exponential basis signals in the more desirable, concise notation

$$\phi_k(t) = e^{jk\omega_0 t} \qquad \text{for } k = 0, \pm 1, \pm 2, \ldots \tag{8.64}$$

In using this notational convenience, it is essential that the user always keep in mind that ω_0 is equal to $2\pi/T$. The parameter ω_0 can be thought of as a radian frequency constant as is made clear when using the Euler identity $e^{jk\omega_0 t} = \cos k\omega_0 t + j \sin k\omega_0 t$. With this in mind, we shall refer to ω_0 as the "fundamental" radian frequency.

The exponential Fourier series representation of the signal $x(t)$ is formally given by the following linear combination:

$$x^o(t) = \sum_{k=-\infty}^{\infty} X_k e^{jk\omega_0 t} \tag{8.65}$$

where it is noted that the summation index k has been expanded to include the nonpositive, indexed basis signals. The Fourier coefficients have here been denoted by X_k (instead of x_k) for reasons that will shortly be made clear. The appropriate values for these coefficients are generated according to relationships (8.56) and (8.57), which for the exponential basis signals become

$$X_k = \frac{1}{T} \int_{t_0}^{t_0+T} x(t)e^{-jk\omega_0 t}\, dt \qquad \text{for } k = 0, \pm 1, \pm 2, \ldots \tag{8.66}$$

where use of the readily derived fact that $a_{kk} = T$ has been made. Since this integral possesses an integrand which is generally complex-valued, it follows that the Fourier coefficients are normally complex numbers.

Dirichlet Conditions

It has been stated that the exponential basis signals are complete. This then implies that the error signal

$$e(t) = x(t) - x^o(t)$$

is almost everywhere zero on the time interval of approximation $[t_0, t_0 + T]$ for a given class of signals, that is, a special Hilbert space. As a matter of fact, it is well-known that a signal $x(t)$ can be equivalently represented by an exponential Fourier series as long as $x(t)$ satisfies the Dirichlet conditions. (See Section 9.5.)

Definition 8.3. The signal $x(t)$ is said to satisfy the Dirichlet conditions on the interval $[t_0, t_0 + T]$ if (a) it is absolutely integrable on $[t_0, t_0 + T]$, that is,

$$\int_{t_0}^{t_0+T} |x(t)|\, dt < \infty \tag{8.67}$$

and (b) it contains a finite number of relative maximum and minimum on $[t_0, t_0 + T]$, and it has a finite number of discontinuities on $[t_0, t_0 + T]$.

It should be apparent that these conditions are satisfied by virtually all real-world signals defined on any finite time interval, $T < \infty$. One may conclude that almost all relevant signals may be equivalently represented by a complex exponential Fourier series over a finite time interval.

Fourier Series Representation

Let us now assume that it is desired to equivalently represent a signal $x(t)$ by the complex exponential Fourier series (8.65). The signal $x(t)$ is said to have a Fourier series representation $x^o(t)$ as given by relationship (8.65) *only if*

$$\int_{t_0}^{t_0+T} | x(t) - x^o(t) | \, dt = 0 \tag{8.68}$$

One might infer from this fact that the signals $x(t)$ and $x^o(t)$ are identical on the time interval $[t_0, t_0 + T]$. This is not the case, however, unless the signal $x(t)$ being represented is everywhere continuous on the time interval of interest. In fact, the relationship between $x(t)$ and $x(t^o)$ has been found by mathematicians to be

$$x^o(t) = \begin{cases} \dfrac{x(t^+) + x(t^-)}{2} & \text{for } t_0 < t < t_0 + T \\ \\ \dfrac{x(t_0) + x(t_0 + T)}{2} & \text{for } t_0, t_0 + T \end{cases} \tag{8.69}$$

Thus, if $x(t)$ is continuous at time t, it is evident that $x^o(t) = x(t)$. On the other hand, if $x(t)$ happens to be discontinuous at time t, then $x^o(t)$ is simply equal to the average value of $x(t)$ at the discontinuity.

From the arguments above, it is clear that $x(t)$ and its Fourier series differ only at points of discontinuity of $x(t)$. Moreover, if the signal $x(t)$ satisfies the Dirichlet conditions, it can only have a finite number of discontinuities in the time interval $[t_0, t_0 + T]$. It then follows that for signals satisfying the Dirichlet conditions, the equivalent Fourier series will be such as to cause the error signal

$$e(t) = x(t) - x^o(t)$$

to be zero for all times t in $[t_0, t_0 + T]$ except at the finite (in number) points of discontinuity of $x(t)$. With this in mind, let us use the notation

$$x^o(t) \equiv x(t) \tag{8.70}$$

to indicate that $x(t)$ and $x^o(t)$ are equal to each other almost everywhere on the time interval $[t_0, t_0 + T]$. It is in this sense that the integral of the error's magnitude (8.68) is zero.

The Fourier series composed of complex exponential basis signals is seen to generate a representation signal $x^o(t)$ that equals $x(t)$ at its points of continuity. It should not be surprising that the complex exponential Fourier series does not exactly represent a signal at its points of discontinuity since the series itself is composed of a linear combination of everywhere-continuous basis signals $e^{jk\omega_0 t}$. Let us summarize our findings in the following theorem.

Theorem 8.2. Let the signal $x(t)$ have the complex exponential Fourier series representation on the time interval $[t_0, t_0 + T]$ as given by

$$x^o(t) = \sum_{k=-\infty}^{\infty} X_k e^{jk\omega_0 t} \tag{8.71}$$

with $\omega_0 = 2\pi/T$ and

$$X_k = \frac{1}{T}\int_{t_0}^{t_0+T} x(t)e^{-jk\omega_0 t}\, dt \tag{8.72}$$

The signals $x(t)$ and $x^o(t)$ are then related to each other by

$$x^o(t) = \begin{cases} \dfrac{x(t^+) + x(t^-)}{2} & \text{for } t_0 < t < t_0 + t \\ \\ \dfrac{x(t_0) + x(t_0 + T)}{2} & \text{for } t = t_0,\ t_0 + T \end{cases} \tag{8.73}$$

It is important to realize that any signal that satisfies the Dirichlet conditions will have a Fourier series representation in the sense of relationship (8.68). This is indeed fortunate since it is a simple task to determine whether a given signal satisfies the Dirichlet conditions. Virtually all real-world signals do! On the other hand, there do exist signals (generally, mathematically contrived signals), that have a Fourier series representation but do not satisfy the Dirichlet conditions.

Computation of Fourier Coefficients

As a computational aid in evaluating the Fourier coefficients X_k, it is usually convenient to evaluate the integral quantity

$$X_T(j\omega) = \int_{t_0}^{t_0+T} x(t)e^{-j\omega t}\, dt \tag{8.74}$$

It is then observed that when the variable ω is set equal to $k\omega_0$ in $X_T(j\omega)$, the formula for TX_k is obtained. We have therefore established the fact that

$$X_k = \frac{1}{T} X_T(jk\omega_0) \tag{8.75}$$

The quantity $X_T(j\omega)$ is referred to as the Fourier transform of the signal $x(t)$ which is assumed to be zero outside the time interval $[t_0, t_1]$.

Example 8.9

Determine the complex Fourier series representation of the signal consideration in Example 8.5, that is,

$$x(t) = \begin{cases} 1 & \text{for } 0 \le t < 1/2 \\ 0 & \text{for } 1/2 \le t \le 1 \end{cases}$$

This signal is seen to satisfy the Dirichlet conditions and it therefore can be represented by a complex exponential series with $t_0 = 0$, $t_1 = 1$, and $T = 1$. The exponential orthogonal basis signals are then given by

$$\phi_k(t) = e^{jk\omega_0 t} \qquad \text{for } k = 0, \pm 1, \pm 2, \ldots$$

with $\omega_0 = 2\pi$. The corresponding Fourier transform function is

$$X_T(j\omega) = \int_0^1 x(t)e^{-j\omega t}\,dt = \int_0^{1/2} e^{-j\omega t}\,dt$$

$$= \frac{1 - e^{-j\omega/2}}{-j\omega}$$

The general Fourier coefficient X_k is now obtained by making the substitution of $k\omega_0 = 2\pi k$ for ω everywhere it appears in $X_T(j\omega)$. A frequently occurring difficulty arises in evaluating the X_0 coefficient, however, since there is an indeterminacy in $X_T(j\omega)$ at $\omega = 0$, i.e., zero divided by zero. This may be easily resolved by either using l'Hospital's rule or evaluating integral expression (8.64) at $k = 0$. Using either approach, it is found that $X_0 = \frac{1}{2}$. For $k \neq 0$, no similar difficulty arises, so that

$$X_k = X(j2\pi k) = \frac{1 - e^{-j\pi k}}{-j2\pi k} = \frac{1 - (-1)^k}{-j2\pi k} \qquad \text{for } k \neq 0$$

It is observed that for k even and nonzero, the Fourier coefficients are identically zero. In summary, the resultant Fourier coefficients are given by

$$X_k = \begin{cases} 1/2 & \text{for } k = 0 \\ j/\pi k & \text{for } k = \pm 1, \pm 3, \pm 5, \ldots \\ 0 & \text{for } k = \pm 2, \pm 4, \pm 6, \ldots \end{cases}$$

If these Fourier coefficients are now substituted into relationship (8.71), the desired Fourier series representation of $x(t)$ results, that is,

$$x^o(t) = \frac{1}{2} + \sum_{k=\pm 1, \pm 3, \ldots}^{\infty} \left(\frac{j}{\pi k}\right) e^{j2\pi k t}$$

$$= \frac{1}{2} + \sum_{k=1,3,5}^{\infty} \left(\frac{j}{\pi k}\right)(e^{j2\pi k t} - e^{-j2\pi k t})$$

$$= \frac{1}{2} - \frac{2}{\pi} \sum_{k=1,3,5,\ldots}^{\infty} \left(\frac{\sin 2\pi k t}{k}\right)$$

This Fourier series exactly equals the given signal $x(t)$ for all time $0 \leq t \leq 1$ except at its single point of discontinuity located at $t = \frac{1}{2}$ and at the end points located at $t_0 = 0$ and $t_1 = 1$. Using expression (8.65), it is seen that $x^o(0) = x^o(1/2)$, so that the error signal for this Fourier series representation is

$$e(t) = x(t) - x^o(t) = \begin{cases} 1/2 & \text{for } t = 0, 1 \\ -1/2 & \text{for } t = 1/2 \\ 0 & \text{everywhere else in } [0, 1] \end{cases}$$

It is in this sense that the error signal is said to be "almost everywhere zero" in making a Fourier series representation of any signal that satisfies the Dirichlet conditions.

8.7 PERIODIC SIGNAL REPRESENTATION

If a signal $x(t)$ satisfies the Dirichlet conditions on the time interval $[t_0, t_0 + T]$, it is possible to equivalently represent that signal by means of the exponential Fourier series

$$x^o(t) \doteq \sum_{k=-\infty}^{\infty} X(k)e^{j(2\pi kt/T)} \qquad \text{for } t_0 \leq t \leq t_0 + T \tag{8.76}$$

This Fourier series exactly represents the signal on the time interval $[t_0, t_0 + T]$ at all points of continuity of $x(t)$ on that interval. If this Fourier series expression is examined for values of time outside the basic time interval $[t_0, t_0 + T]$ it is noted that

$$x^o(t + T) \doteq \sum_{k=-\infty}^{\infty} X(k)e^{j[2\pi k(t+T)/T]} = x^o(t) \tag{8.77}$$

for any value of time t. Recall that any signal that satisfies the relationship $x^o(t) = x^o(t + T)$ is said to be *periodic*. Thus, the Fourier series signal $x^o(t)$ is periodic, and it therefore repeats its time behavior on every T-second time interval. This periodicity is a direct consequence of the fact that the constituent complex exponential basis signals each satisfy the periodic relationship $\phi_k(t + T) = \phi_k(t)$.

Periodic Extension of $x(t)$

With the thoughts presented above in mind, we can then interpret Fourier series representation (8.76) as being the "periodic extension" of the signal $x(t)$ defined in the time interval $[t_0, t_0 + T]$. Namely, if $x(t)$ is taken to be zero outside the time interval $[t_0, t_0 + T]$, then the periodic extension of $x(t)$ is formally defined by[2]

$$x_p(t) = \sum_{k=-\infty}^{\infty} x(t - kT) \tag{8.78}$$

The concept of periodic extension is illustrated in Fig. 8.7, where the periodicity of $x_p(t)$ is apparent. It then follows that the Fourier series signal $x^o(t)$ represents the periodic extension $x_p(t)$ for all values of time t. In particular, these two signals are related to each other according to

$$x^o(t) = \frac{x_p(t^+) + x_p(t^-)}{2} \qquad \text{for all time } t \tag{8.79}$$

[2]In order to avoid possible difficulties, the time interval over which $x(t)$ is being represented should be given by $t_0 \leq t < t_0 + T$ or $t_0 < t \leq t_0 + T$.

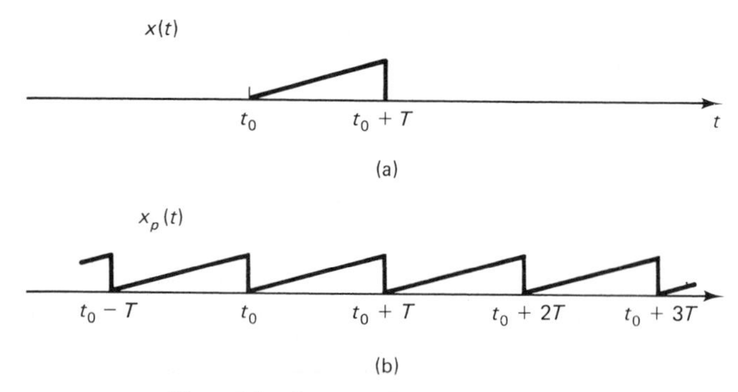

Figure 8.7 Concept of periodic extension.

This relationship implies that $x^o(t) = x_p(t)$ for all time t for which $x_p(t)$ is continuous and $x^o(t)$ takes on the intermediate value at points of discontinuity of $x(t)$.

Representation of Periodic Signals

The periodicity of the Fourier series representation offers an interesting and most important possibility. Namely, suppose that a given signal $x(t)$ is itself periodic and the time interval $[t_0, t_0 + T]$ is chosen so that it contains exactly one period of $x(t)$. It then follows that the Fourier series signal $x^o(t)$ not only represents the periodic signal $x(t)$ for t in the interval $[t_0, t_0 + T]$, but it also represents $x(t)$ for all values of t. These two signals would then be related to each other in the standard fashion, that is,

$$x^o(t) = \frac{x(t^+) + x(t^-)}{2} \qquad \text{for all } t \tag{8.80}$$

Let us illustrate this representation procedure by means of a specific example.

Example 8.10

Obtain the Fourier series representation for the pulse waveform shown in Fig. 8.8. It is first noted that this waveform is periodic with a basic period of T seconds. To generate the required Fourier series representation, one then selects the time instants t_0 and t_1 to cover exactly one period of $x(t)$. Any such selection will do and will result in the same representation. For example, if these parameters are selected to be $t_0 = -T/2$ and $t_0 + T = T/2$, the associated Fourier transform is given by[3]

$$X_T(j\omega) = \int_{-T/2}^{T/2} x(t)e^{-j\omega t}\,dt = \int_{-\Delta}^{\Delta} Ae^{-j\omega t}\,dt$$

$$= 2A\,\frac{\sin \Delta\omega}{\omega}$$

[3]This is the Fourier transform of the truncated signal $x(t)[u(t + T/2) - u(t - T/2)]$.

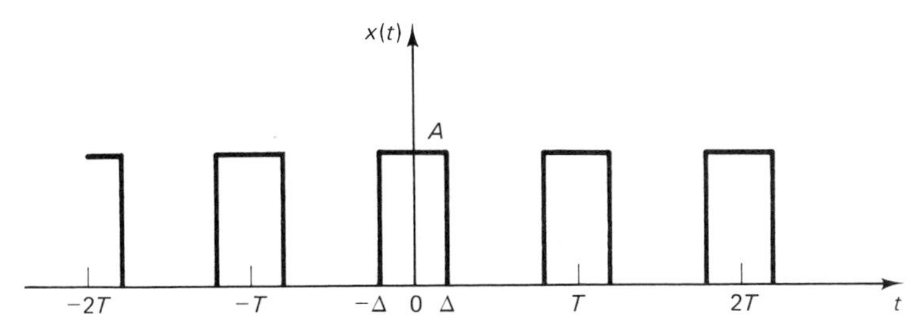

Figure 8.8 Periodic train of pulses with pulse width $\Delta < T$.

The required Fourier coefficients are then given by expression (8.68), that is,

$$X_k = \frac{1}{T} X_T\left(j\,\frac{2\pi k}{T}\right)$$
$$= \frac{A}{\pi k}\sin\frac{2\pi k\Delta}{T} \qquad \text{for } k = 0,\ \pm 1,\ \pm 2,\ \ldots$$

It is again observed that the $k = 0$ coefficient is ill-defined, but an evaluation of $X_T(j\omega)$ at $\omega = 0$ indicates that $X(0) = 2\Delta$. The Fourier series representation for the given periodic pulse train is then given by

$$x^o(t) = \frac{2\,\Delta A}{T} + \sum_{\substack{k=-\infty \\ \neq 0}}^{\infty} \frac{A}{\pi k}\sin\frac{2\pi k\Delta}{T}\,e^{j2\pi kt/T}$$

$$= \frac{2\,\Delta A}{T} + \frac{A}{\pi}\sum_{k=1}^{\infty}\frac{\sin\dfrac{2\pi k\Delta}{T}}{k}\left(e^{j2\pi k\Delta/T} + e^{-j2\pi kt/T}\right)$$

$$= \frac{2\,\Delta A}{T} + \frac{2A}{\pi}\sum_{k=1}^{\infty}\frac{\sin\dfrac{2\pi k}{T}}{k}\cos\frac{2\pi kt}{T}$$

This Fourier series exactly represents the given periodic pulse train for all values of time except at its points of discontinuity (i.e., $t = \Delta + kT$, or $t = -\Delta - kT$ for k as an integer) where the series expression takes on the intermediate value $A/2$.

From the development noted above, it is apparent that a complex exponential Fourier series can be used either to represent an aperiodic (nonperiodic) signal over a given time interval or a periodic signal for all values of time. One must realize that in the former case, the nonperiodic representation is valid only in the specified time interval $[t_0, t_0 + T]$ and that the two signals $x(t)$ and $x^o(t)$ generally differ for values of time outside that time interval. This Fourier series representation of aperiodic signals is mainly of mathematical interest, and it plays a prominent role in the establishment of the Fourier integral transformation. The primary utilization of the complex exponential Fourier series is almost exclusively restricted to the representation of periodic signals.

8.8 SPECTRAL CONTENT OF PERIODIC SIGNALS

As we have seen, any periodic signal that satisfies the Dirichlet conditions can be equivalently represented by a complex exponential Fourier series (hereafter abbreviated as Fourier series). This Fourier series is composed of an infinite sum of weighted, complex, exponential basis signals. The weights used are, in fact, equal to the Fourier coefficients, which are formally given by

$$X_k = \frac{1}{T}\int_{t_0}^{t_0+T} x(t)e^{-j2\pi kt/T}\,dt \qquad \text{for } k = 0, \pm 1, \pm 2, \ldots \tag{8.81}$$

with T denoting the basic period of the periodic signal being represented. Once these coefficients are evaluated, they are substituted into the relationship

$$x(t) \doteq \sum_{k=-\infty}^{\infty} X_k e^{j2\pi kt/T} \tag{8.82}$$

to form the desired Fourier series representation.

We now give this procedure for obtaining a Fourier series representation a most important interpretation, that is, that the periodic signal $x(t)$ can in fact be equivalently represented by the complex number sequence consisting of the Fourier coefficients

$$\ldots, X_{-2}, X_{-1}, X_0, X_1, X_2, \ldots \tag{8.83}$$

and vice versa. This is an equivalency in the sense that the original periodic signal $x(t)$ can be recovered from the Fourier coefficient sequence $\{X_k\}$ by simply forming the Fourier series (8.82). Moreover, the Fourier coefficient sequence $\{X_k\}$ can be generated from periodic signal $x(t)$ by evaluating the integral expression (8.81).

It is apparent that the continuous-time periodic signal $x(t)$ and the corresponding discrete-time Fourier coefficient sequence $\{X_k\}$ contain the same information based on the argument above. The kth element of this sequence is seen to contribute the component

$$X_k e^{j2\pi kt/T} = X_k\left[\cos\frac{2\pi kt}{T} + j\sin\frac{2\pi kt}{T}\right] \tag{8.84a}$$

to the periodic signal $x(t)$ in the Fourier series representation. This is recognized as a complex-valued, continuous-time sinusoidal signal with radian frequency $2\pi k/T$. Note also that the $-k$th element of the sequence contributes the component

$$X_{-k} e^{-j2\pi kt/T} = X_{-k}\left[\cos\frac{2\pi kt}{T} - j\sin\frac{2\pi kt}{T}\right] \tag{8.84b}$$

which is also a complex-valued, continuous-time signal with radian frequency $2\pi kt/T$. As such, the Fourier coefficient sequence $\{X_k\}$ is said to contain the spectral information (sinusoidal content) of the periodic signal $x(t)$.

We can display this sinusoidal content by making a plot of X_k versus $\omega_k = 2\pi k/T$. This depicts the so-called *line spectrum* information relative to the periodic signal $x(t)$. Since the Fourier coefficients X_k are generally complex-valued, this line spectrum is conveniently displayed by first expressing X_k in polar form

$$X_k = |X_k| e^{j\theta_k} \tag{8.85}$$

where $|X_k|$ and θ_k denote the magnitude and phase angle, respectively, of X_k. One may then make separate plots of the *magnitude spectrum* $|X_k|$ and *phase spectrum* θ_k versus the discrete set of radian frequencies $\omega_k = 2\pi kt/T$ for $k = 0, \pm 1, \pm 2, \ldots$. These plots may be used to ascertain the sinusoidal content of the periodic signal being represented. Namely, those values of $2\pi k/T$ that have a large (small) value of $|X_k|$ indicate that $x(t)$ has a correspondingly large (small) sinusoidal content at that radian frequency.

Example 8.11

Determine and plot the magnitude and phase spectrums of the periodic pulse train considered in Example 8.10, where it was shown that the required Fourier coefficients are given by

$$X_k = \frac{2\Delta A}{T} \frac{\sin \dfrac{2\pi k\Delta}{T}}{\dfrac{2\pi k\Delta}{T}}$$

$$= \frac{2\Delta A}{T} \operatorname{sinc} \frac{2\pi k\Delta}{T}$$

where sinc $x = (\sin x)/x$. This representation of X_k in terms of the sinc function was

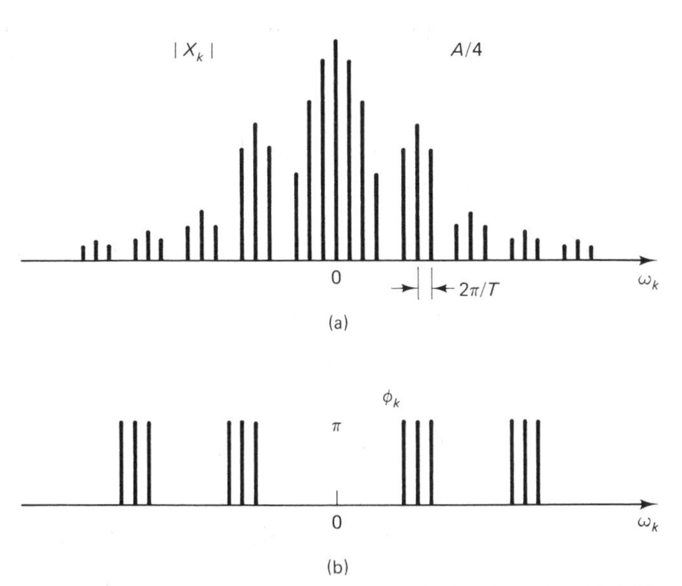

Figure 8.9 Magnitude and phase spectrum of periodic pulse train with $\Delta/T = 1/8$.

made for notational purposes.[4] The magnitude and phase angle components that correspond to X_k are found to be

$$|X_k| = \frac{2\Delta A}{T}\left|\text{sinc}\,\frac{2\pi k\Delta}{T}\right|$$

$$\theta_k = \begin{cases} 0 & \text{if sinc}\,\dfrac{2\pi k\Delta}{T} \geq 0 \\ \pi & \text{otherwise} \end{cases}$$

A plot of these two sequences versus $2\pi k/T$ for the case $\Delta T = \frac{1}{4}$ is shown in Fig. 8.9. It is clear from this sketch of the magnitude spectrum that the lower frequency sinusoids play a more significant role in representing the periodic signal than the higher frequency sinusoids.

Real Periodic Signals

In virtually all practical applications of the Fourier series, the periodic signal $x(t)$ being represented is a real-valued function of time. In such cases, there exists a desirable symmetry between the positive and negative Fourier coefficients, which is readily seen by taking the complex conjugate of the general expression for the kth coefficient (8.81). Taking complex conjugates of each side of this expression and noting that the complex conjugate of the right-side integrand is $x(t)e^{j2\pi kt/T}\,dt$, it follows that

$$X_k^* = \frac{1}{T}\int_{t_0}^{t_1} x(t)e^{j2\pi kt/T}\,dt$$

This integral expression is recognized as being equal to the Fourier coefficient X_{-k}, (for example, let $k = 5$ and observe that $X_5^* = X_{-5}$). We have therefore shown that when the periodic signal $x(t)$ is real, the Fourier series coefficients possess the following complex conjugate symmetry:

$$X_{-k} = X_k^* \qquad \text{for all integers } k \tag{8.86}$$

In such cases, one then only evaluates the values of the nonnegative (or nonpositive) Fourier coefficients to obtain the desired Fourier series representation.

When $x(t)$ is a real-valued signal, it is also always possible to extensively simplify the Fourier series representation (8.82). This is achieved by using the symmetric relationship between X_k and X_{-k} described above, which when incorporated with the polar representation of X_k indicates that

$$\begin{aligned} X_k &= |X_k|\,e^{j\theta_k} \\ X_{-k} &= |X_k|\,e^{-j\theta_k} \end{aligned} \qquad \text{for } k = 0, 1, 2, \ldots \tag{8.87}$$

If these polar representations are substituted into the Fourier series relationship (8.82) and the two-sided infinite summation is broken up into the $k = 0$ term and the

[4]It is noted that sinc 0 = 1 so that $X(0) = (2\,\Delta A)/T$ as was found in Example 8.10. The reader is cautioned that some authors define sinc x as $(\sin \pi x)/(\pi x)$.

positive and negative sum terms, one obtains

$$x(t) \doteq X_0 + \sum_{k=1}^{\infty} |X_k| e^{j\theta_k} e^{j2\pi kt/T} + \sum_{k=1}^{\infty} |X_k| e^{-j\theta_k} e^{-j2\pi kt/T}$$

where the first summation corresponds to the positive Fourier coefficient terms and the second summation to the negative Fourier coefficient terms (i.e., $-\infty < k < -1$). Combining the exponential products and the two summations over the same values of k, the desired representation is given by

$$\begin{aligned} x(t) &\doteq X_0 + \sum_{k=1}^{\infty} |X_k| \{e^{j(2\pi kt/T + \theta_k)} + e^{-j(2\pi kt/T + \theta_k)}\} \\ &= X_0 + 2\sum_{k=1}^{\infty} |X_k| \cos\left(\frac{2\pi kt}{T} + \theta_k\right) \end{aligned} \tag{8.88}$$

In summary, when the periodic signal being represented is real, the Fourier series expression is composed of an infinite sum of sinusoidal signals. The amplitude and phase angle of these individual terms are seen to be directly related to the magnitude and phase angle of the associated Fourier coefficient. With this in mind, the periodic signal $x(t)$ is then said to have a kth harmonic component as given by

$$2|X_k| \cos\left(\frac{2\pi kt}{T} + \theta_k\right) \tag{8.89}$$

which has an amplitude of $2|X_k|$ and a phase of θ_k. The $k = 1$ harmonic is commonly referred to as the *fundamental harmonic*, since its basic period of T seconds is identical to the period of the periodic signal $x(t)$ being represented. Similarly, the real term X_0 is usually called the *average value* or the *d.c. component* of $x(t)$.

With the interpretations above in mind, the spectral plots of magnitude and phase take on an even greater significance. In effect, the magnitude spectrum $|X_k|$ versus $\omega_k = 2\pi k/T$ depicts the sinusoidal content of the periodic signal being represented. A signal of basic period T seconds is then seen to have sinusoidal content at the discrete radian frequencies $\omega_k = 2\pi k/T$ for $k = 0, 1, 2, \ldots$ and nowhere else. The amount of sinusoidal energy of these discrete radian frequencies is measured by the Fourier coefficient's magnitude $|X_k|$. Similar statements can be made concerning the phase spectrum information.

8.9 PARSEVAL'S THEOREM AND SIGNAL POWER

In many applications, one frequently wishes to measure the size of a signal. There are a variety of such measures, among which one of the most widely used is that of signal "power." The average power contained in a signal $x(t)$ over the T-second time interval $[t_0, t_0 + T]$ is defined to be

$$P_x = \frac{1}{T}\int_{t_0}^{t_0+T} x(t)x^*(t)\, dt \tag{8.90}$$

where the integrand is observed to be $|x(t)|^2$. Based on this measure, one is able to compare the power content of various signals. Undoubtedly, the importance of this measure is appreciated by the consumer of electric power.

If the signal $x(t)$ has a Fourier representation in the time interval $[t_0, t_0 + T]$, it is always possible to express that signal's power content in terms of its Fourier coefficients. This may be readily shown by the basic procedure of substituting the equivalent Fourier series representation of $x(t)$ into the power measure relationship and then simplifying. To demonstrate this process, let us consider the case whereby the signal under examination can be represented by the complex exponential Fourier series

$$x(t) \doteq \sum_{k=-\infty}^{\infty} X_k e^{j(2\pi kt/T)} \tag{8.91}$$

The power contained in this signal over the time interval $[t_0, t_0 + T]$ may be obtained by inserting representation (8.91) into expression (8.90), to yield

$$P_x = \frac{1}{T} \int_{t_0}^{t_0+T} \left(\sum_{k=-\infty}^{\infty} X_k e^{j(2\pi kt/T)} \right) \left(\sum_{m=-\infty}^{\infty} X_m^* e^{-[j(2\pi mt/T)]} \right) dt$$

with the sum on m being recognized as $x^*(t)$. If $x(t)$ is in fact a periodic signal, this expression for P_x is seen to give the average power of one period of that signal.

To evaluate this complex expression for P_x, it is advisable to now interchange the order of integration and summation. After extracting all multiplicative terms not involving the integration variable t, the result is

$$P_x = \frac{1}{T} \sum_{k=-\infty}^{\infty} \sum_{m=-\infty}^{\infty} X_k X_m^* \int_{t_0}^{t_0+T} e^{j(2\pi kt/T)} e^{-[j(2\pi kt/T)]} \, dt$$

The two multiplicative exponential signals in this integrand are recognized as being complex exponential basis signals that are known to be orthogonal on the time interval $[t_0, t_0 + T]$ for $k \neq m$. This integral then equals zero for all summation variables $k \neq m$ and equals T whenever $k = m$. With this in mind, the integral expression for P_x may be equivalently replaced by $T\delta(k - m)$ where δ denotes the unit-impulse sequence. Thus, the expression for power has been simplified to

$$P_x = \frac{1}{T} \sum_{k=-\infty}^{\infty} \sum_{m=-\infty}^{\infty} X_k X_m^* T\delta(k - m)$$

This power expression may be further simplified by noting that a double summation is evaluated by first fixing the outside summation variable k at a given integer value and letting the inside summation variable vary through its infinite range $-\infty < m < \infty$. Once this has been accomplished, the outside variable is then incremented by one and m is again varied through its infinite range. This procedure is continued until all values of the outside summation variable have been taken into account. With this in mind, let us perform one step of this procedure with k being assigned a fixed integer value. As m varies throughout its infinite range, the only value of m that causes the summand term $\delta(k - m)$ to be other than zero is $m = k$

where $\delta(k - m) = 1$. It must then follow that

$$P_x = \sum_{k=-\infty}^{\infty} X_k X_k^*$$
$$= \sum_{k=-\infty}^{\infty} |X_k|^2$$

This is a most profound result, and it is generally credited to the eminent mathematician Parseval. To emphasize its fundamental importance, we will express it as a theorem.

Theorem 8.3. Parseval's Theorem. If the signal $x(t)$ has a complex exponential Fourier series representation on the time interval $[t_0, t_0 + T]$, that signal's power on that time interval is given by

$$\frac{1}{T}\int_{t_0}^{t_0+T} |x(t)|^2\, dt = \sum_{k=-\infty}^{\infty} |X_k|^2 \tag{8.92}$$

where X_k denotes the associated Fourier coefficients as given by

$$X_k = \frac{1}{T}\int_{t_0}^{t_0+T} x(t)e^{-[j(2\pi kt/T)]}\, dt \qquad \text{for } k = 0, \pm 1, \pm 2, \ldots$$

This particular form of Parseval's theorem is of most value when investigating the power content of periodic signals. If $x(t)$ is a periodic signal that has a complex exponential Fourier series, that signal may be equivalently represented by a Fourier series for *all* values of time. Expression (8.92) is seen to give the power contained in one period of that periodic signal. Based on the comments of the previous section when $x(t)$ is a real-valued signal, the quantity $|X_k|^2$, i.e., $(1/2)\{|X_k|^2 + |X_{-k}|^2\}$, is seen to measure the power contained in the kth harmonic component

$$2|X_k| \cos\left(\frac{2\pi kt}{T} + \theta_k\right)$$

of the series representation. Thus, a plot of the behavior of $|X_k|$ versus k is seen to give us a measure of the power content of a real periodic signal as a function of the discrete frequency value $2\pi k/T$.

Example 8.12

Examine the power content of the periodic signal shown in Fig. 8.9a. This signal is seen to have a fundamental period of 1 second, and we shall arbitrarily select $t_0 = -\frac{1}{2}$ and $t_0 + T = \frac{1}{2}$. In this case, the power contained over one period is readily evaluated by the time-domain integral expression

$$P_x = \frac{1}{T}\int_{-T/2}^{T/2} |x(t)|^2\, dt = \int_{-1/4}^{1/4} 1\, dt = \frac{1}{2}$$

On the other hand, Parseval's theorem indicates that this power content is equivalently given by the infinite sum of the Fourier coefficients' magnitudes squared. The Fourier coefficients for this signal were previously determined in Example 8.10 with $T = 1$, $\Delta = \frac{1}{4}$, and $A = 1$, and they resulted in

$$X_k = \frac{1}{\pi k} \sin \frac{\pi k}{2} \qquad \text{for } k = \pm 1, \pm 2, \pm 3, \ldots$$

$$X_0 = \frac{1}{2}$$

From Parseval's theorem

$$P_x = \frac{1}{4} + 2 \sum_{k=1}^{\infty} \left(\frac{1}{\pi k} \sin \frac{\pi k}{2} \right)^2$$

$$P_x = \frac{1}{4} + \frac{2}{\pi^2} \sum_{k=0}^{\infty} \frac{1}{(2k+1)^2}$$

The value of P_x from this last expression is not obvious, but we can expand a few terms to show that P_x is 1/2.

$$P_x \cong \frac{1}{4} + \frac{2}{\pi^2} \left(1 + \frac{1}{9} + \frac{1}{25} + \frac{1}{49} + \frac{1}{81} + \frac{1}{121} \right) = 0.492$$

Thus, the calculation of signal power can often be greatly simplified by Parseval's theorem since it is generally easier to sum the squared Fourier coefficients than to evaluate the integral of the squared magnitude.

Theorem 8.4. General Parseval's Theorem. Let the signal $x(t)$ have a Fourier series representation in the orthogonal basis signal set $\phi_1(t)$, $\phi_2(t)$, $\phi_3(t)$, ... on the time interval $[t_0, t_0 + T]$ as given by

$$x(t) \doteq \sum_{k=1}^{\infty} \frac{b_k}{a_{kk}} \phi_k(t)$$

where

$$a_{kk} = \int_{t_0}^{t_0+T} \phi_k(t)\phi_k^*(t)\, dt \quad \text{and} \quad b_k = \int_{t_0}^{t_0+T} x(t)\phi_k^*(t)\, dt$$

The signal's power on that time interval is then given by

$$\frac{1}{T} \int_{t_0}^{t_0+T} |x(t)|^2\, dt = \frac{1}{T} \sum_{k=1}^{\infty} \frac{|b_k|^2}{a_{kk}} \tag{8.93}$$

One may readily show that this general Parseval's relationship simplifies to that given by expression (8.92) when the orthogonal basis signals used are the complex exponentials. This, of course, requires a reordering of the indexing of the basis signals, i.e., ..., $\phi_{-1}(t)$, $\phi_2(t)$, $\phi_1(t)$, ..., and utilization of the identity that $b_k = X_k$ and $a_{kk} = 1/T$ when complex, exponential basis signals are used.

8.10 FOURIER SERIES REPRESENTATION OF IMPULSE TRAIN

In studies related to the exponential Fourier series representation of periodic pulse-like signals or in analog-to-digital conversion, the so-called impulse train signal arises in a natural manner. As its name implies, the impulse train signal consists of a train of unit-impulse signals separated by T-second intervals as defined by

$$\delta_T(t) = \sum_{k=-\infty}^{\infty} \delta(t - kT) \tag{8.94}$$

The time behavior of this signal is shown in Fig. 8.10, where it is apparent that the impulse train is a periodic signal with period T.

When determining the exponential Fourier series representation of the impulse train, one must be careful to select the time instance t_0 to avoid coinciding with the time of application of one of the unit impulses (i.e., $t_0 \neq kT$). This is necessary since by the integral property of impulse signals as treated in Chapter 2, the limits of integration must always properly enclose the unit-impulse signal's time of application. With this in mind, let us then select t_0 to be $-T/2$, so that the Fourier coefficients corresponding to the impulse train signal is given by

$$\begin{aligned} X_k &= \frac{1}{T}\int_{-T/2}^{T/2} \delta_T(t) e^{-jk\omega_0 t}\, dt \\ &= \frac{1}{T}\int_{-T/2}^{T/2} \delta(t) e^{-jk\omega_0 t}\, dt = \frac{1}{T} \end{aligned} \tag{8.95}$$

The exponential Fourier series representation of the impulse train is then given by

$$\begin{aligned} \delta_T(t) &\doteq \sum_{k=-\infty}^{\infty} \frac{1}{T} e^{jk\omega_0 t} \\ &= \frac{1}{T} + \frac{2}{T}\sum_{k=1}^{\infty} \cos k\omega_0 t \end{aligned} \tag{8.96}$$

where $\omega_0 = 2\pi/T$.

The impulse train signal is seen to have Fourier coefficients all equal to the reciprocal of the period (i.e., $X_k = 1/T$). It then follows from the comments of Section 8.8 that the impulse train has a uniform spectral content in the sense that all of its harmonics have equal amplitude and phase. It is interesting to observe that when the Fourier series representation (8.96) is evaluated at time $= kT$, it results in

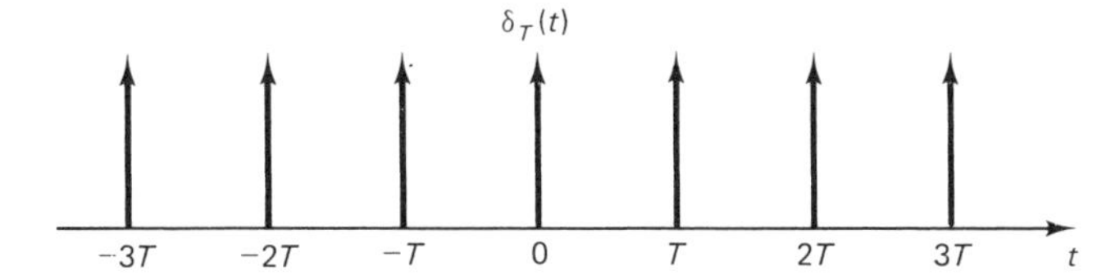

Figure 8.10 Impulse train.

$\delta_T(kT)$ being infinite as is desired. Moreover, since $\delta_T(t)$ is continuous for $t \neq kT$, it must also follow that this Fourier series representation will be zero for $t \neq kT$ as is desired.

Example 8.13

It is possible to investigate other periodic signals that consist of unit impulses in a similar manner. To demonstrate this point, let us consider the derivative of the periodic train of pulses shown in Fig. 8.8. This derivative is depicted in Fig. 8.11 and is observed to be a periodic signal consisting of unit impulses. Selecting $t_0 = 0$, it follows that the corresponding Fourier coefficients are given by

$$X_k = \frac{A}{T}\int_0^T [-\delta(t-\Delta) + \delta(t - T + \Delta)]e^{-jk\omega_0 t}\, dt$$

$$= \frac{A}{T}[-e^{-jk\omega_0\Delta} + e^{jk\omega_0(T-\Delta)}]$$

The Fourier series representation of the derivative of the pulse train is then

$$\frac{dx(t)}{dt} \doteq \sum_{k=-\infty}^{\infty} A\left[\frac{-e^{-jk\omega_0\Delta} + e^{-jk\omega_0(T-\Delta)}}{T}\right]e^{jk\omega_0 t}$$

$$= \frac{A}{T}\sum_{k=-\infty}^{\infty}[e^{jk\omega_0(t-T+\Delta)} - e^{jk\omega_0(t-\Delta)}]$$

$$= \frac{2A}{T}\sum_{k=1}^{\infty}\cos[k\omega_0(t - T + \Delta)] - \cos[k\omega_0(t-\Delta)]$$

Since $\omega_0 = 2\pi/T$, it follows that the first term in this summation can be replaced by $\cos[k\omega_0(t - T + \Delta)] = \cos[k\omega_0(t + \Delta)]$, that is,

$$\frac{dx(t)}{dt} = \frac{2A}{T}\sum_{k=1}^{\infty}\cos[k\omega_0(t+\Delta)] - \cos[k\omega_0(t-\Delta)]$$

$$= \frac{2A}{T}\sum_{k=1}^{\infty} -2\sin k\omega_0\Delta \sin k\omega_0 t$$

$$= -\frac{4A}{T}\sum_{k=1}^{\infty}\sin\frac{2\pi k\Delta}{T}\sin\frac{2\pi kt}{T}$$

We follow this idea in the next section.

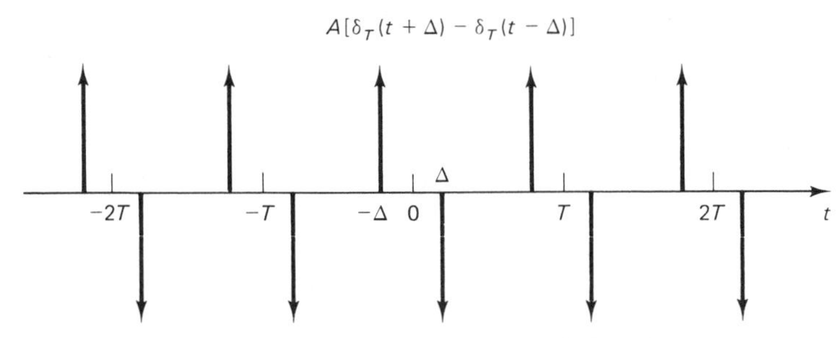

Figure 8.11 Derivative of pulse train of Figure 8.8.

8.11 DIFFERENTIATION OF FOURIER SERIES

Suppose that one is given a periodic signal $x(t)$ which has an exponential Fourier series representation as given by

$$x(t) \doteq \sum_{k=-\infty}^{\infty} X_k e^{jk\omega_0 t} \tag{8.97}$$

where $\omega_0 = 2\pi/T$. Furthermore, let it be assumed that the derivative of this periodic signal (*note:* $dx(t)/dt$ is also periodic) is such as to also have an exponential Fourier series representation, that is,

$$\frac{dx(t)}{dt} \doteq \sum_{k=\infty}^{\infty} X'_k e^{jk\omega_0 t} \tag{8.98}$$

where X'_k designates the Fourier coefficient of the periodic signal $dx(t)/dt$.

Since the periodic signals $x(t)$ and $dx(t)/dt$ are related to each other through the simple differentiation operation, it is intuitively appealing to expect that their respective Fourier coefficients X_k and X'_k would also be related in some obvious manner. This is in fact the case, as can be shown by differentiating the Fourier series representation (8.97), to give

$$\frac{dx(t)}{dt} \doteq \sum_{k=-\infty}^{\infty} jk\omega_0 X_k e^{jk\omega_0 t}$$

A comparison of this relationship with the Fourier series representation for the periodic signal $dx(t)/dt$ as given by expression (8.98) indicates that

$$X'_k = jk\omega_0 X_k \tag{8.99}$$

Thus, as anticipated, the Fourier coefficients of a periodic signal and its derivative signal are related to one another in a simple manner.

This interesting relationship between the Fourier coefficients X_k and X'_k is most often used when determining the Fourier series of a pulselike periodic signal $x(t)$. This first involves finding the Fourier coefficients X'_k of the derivative signal $dx(t)/dt$ and then using relationship (8.99) to generate the X_k. It should be observed that there is an ambiguity for the $k = 0$ term, however, which may be resolved by recalling that X_0 is equal to the d.c. level of the signal $x(t)$.

Example 8.14

Determine the exponential Fourier series representation of the periodic sawtooth signal $x(t)$ shown in Fig. 8.12 using the process of differentiation. The derivative of the given periodic signal is readily found to be

$$\begin{aligned}\frac{dx(t)}{dt} &= -\frac{1}{T} + \sum_{k=-\infty}^{\infty} \delta(t - kT)\\ &= -\frac{1}{T} + \delta_T(t)\end{aligned}$$

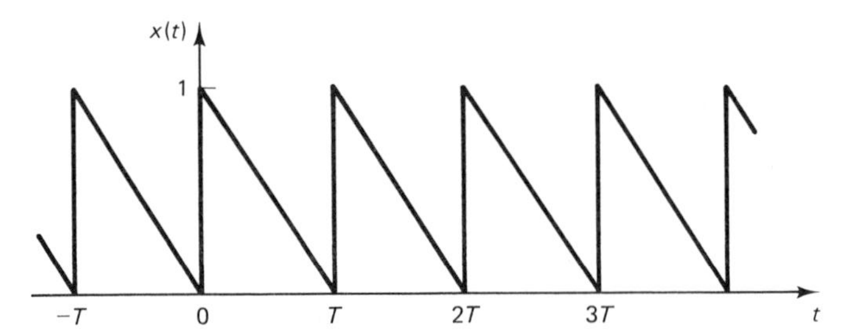

Figure 8.12 Periodic sawtooth signal.

The exponential Fourier series for the impulse train signal $\delta_T(t)$ was found in the last section and when substituted into the relationship above, yields

$$\frac{dx(t)}{dt} = -\frac{1}{T} + \sum_{k=-\infty}^{\infty} \frac{1}{T} e^{jk\omega_0 t}$$

$$= \sum_{\substack{k=-\infty \\ \neq 0}}^{\infty} \frac{1}{T} e^{jk\omega_0 t}$$

where $\omega_0 = 2\pi/T$. It is then concluded that the Fourier coefficients for the periodic signal $dx(t)/dt$ are given by

$$X'_k = \begin{cases} 1/T & \text{for } k = \pm 1, \pm 2, \ldots \\ 0 & \text{for } k = 0 \end{cases}$$

Using relationship (8.99), it is found that the Fourier coefficients for the periodic signal $x(t)$ are then

$$X_k = \frac{1}{jk\omega_0} \frac{1}{T} = \frac{1}{j2\pi k} \qquad \text{for } k = \pm 1, \pm 2, \ldots$$

while the d.c. level is seen to be $\frac{1}{2}$, which implies that

$$X_0 = \frac{1}{2}$$

The desired Fourier series representation of the periodic sawtooth signal is then

$$x(t) = \frac{1}{2} + \sum_{\substack{k=-\infty \\ \neq 0}}^{\infty} \frac{1}{j2\pi k} e^{jk\omega_0 t}$$

$$= \frac{1}{2} + \frac{1}{\pi} \sum_{k=1}^{\infty} \frac{\sin k\omega_0 t}{k}$$

where $\omega_0 = 2\pi/T$.

8.12 RESPONSE OF LINEAR SYSTEMS TO PERIODIC INPUTS

One may use the theory of Fourier series representation to straightforwardly determine the response of a time-invariant, linear continuous-time operator to a periodic input signal. Recall that such an operator is characterized by

$$y(t) = \int_{-\infty}^{\infty} h(\tau)x(t-\tau)\,d\tau \tag{8.100}$$

with $h(t)$ denoting the operator's unit-impulse response. If an exponential input signal is applied to this operator, it was also shown that the following input-response signal pair is generated:

$$x(t) = e^{st} \quad \text{and} \quad y(t) = H(s)e^{st} \tag{8.101}$$

where $H(s)$ is referred to as the *exponential transfer function*, which is formally given by

$$H(s) = \int_{-\infty}^{\infty} h(t)e^{-st}\,dt \tag{8.102}$$

Based on this response characteristic, let us now find the response of linear operator (8.100) to a periodic input signal. Since the input signal is periodic, it may be equivalently represented by the exponential Fourier series.

$$x(t) \doteq \sum_{k=-\infty}^{\infty} X_k e^{j(2\pi kt/T)} \tag{8.103}$$

where T denotes the period of the periodic input signal. Let us now substitute this representation into the governing operator relationship (8.100) to determine the corresponding response, that is,

$$y(t) = \int_{-\infty}^{\infty} h(\tau)\left[\sum_{k=-\infty}^{\infty} X_k e^{j[2\pi k(t-\tau)/T]}\right] d\tau \tag{8.104}$$

In order to evaluate this complex expression, we will now interchange the operations of summation and integration. This is predicated on an extension of a well-known integration property.[5] Using the fact that the Fourier coefficients X_k and the multiplicative exponential $e^{j(2\pi kt/T)}$ terms are constants with respect to the integration variable τ, it follows that this interchange of operations yields

$$y(t) = \sum_{k=-\infty}^{\infty} X_k e^{j(2\pi kt/T)} \int_{-\infty}^{\infty} h(\tau)e^{-[j(2\pi k\tau/T)]}\,d\tau$$

[5] $$\int_{-\infty}^{\infty} [a_1x_1(t) + a_2x_2(t)]\,dt = a_1\int_{-\infty}^{\infty} x_1(t)\,dt + a_2\int_{-\infty}^{\infty} x_2(t)\,dt$$

where a_1 and a_2 are constants.

A comparison of the integral term involved in this response expression with the exponential transfer function (8.102) indicates that the integral is equal to $H(s)$ evaluated at $s = j2\pi k/T$. With this in mind, the desired signal is then given by

$$y(t) = \sum_{k=-\infty}^{\infty} H\left(j\frac{2\pi k}{T}\right) X_k e^{j(2\pi kt/T)} \tag{8.105}$$

Upon examination of expression (8.105), which gives the response of a linear operator to a periodic input signal, it is noted that the response may be expressed as

$$y(t) = \sum_{k=-\infty}^{\infty} Y_k e^{j(2\pi kt/T)} \tag{8.106a}$$

where the coefficients Y_k are given by

$$Y_k = H\left(j\frac{2\pi k}{T}\right) X_k \qquad \text{for } k = 0, \pm 1, \pm 2, \ldots \tag{8.106b}$$

This response expression is in the format of an exponential Fourier series, which implies that the response signal itself must be periodic with a period of T seconds. We then conclude that whenever a time-invariant linear operator is excited by a periodic input signal there results a periodic response signal of the same period. Perhaps of even greater significance, the linear operation has changed the spectrum content of the input signal with the kth harmonic (of radian frequency $2\pi k/T$) coefficient being changed from X_k to $X_k H(j2\pi k/T)$. To reflect the importance of these results, let us state them in the following theorem:

Theorem 8.5. If a time-invariant linear operator with exponential transfer function $H(s)$ is excited by a periodic input signal of period T whose Fourier coefficients are X_k, then the corresponding response signal will also be periodic with period T. Its Fourier coefficients are given by

$$Y_k = H\left(j\frac{2\pi k}{T}\right) X_k \qquad \text{for } k = 0, \pm 1, \pm 2, \ldots \tag{8.107}$$

The reason the results noted above are so significant is that they indicate that time-invariant linear operators may be used in applications requiring frequency (spectrum) discrimination capabilities. Namely, by judiciously selecting the exponential transfer function $H(s)$, one may change the spectral content of the input signal in some desirable manner. The next example points out one such possibility.

Example 8.15

Investigate the periodic signal transmission characteristics of the RC network shown in Fig. 8.13. It has already been shown that the exponential transfer function that corresponds to this network is given by

$$H(s) = \frac{1}{1 + RCs}$$

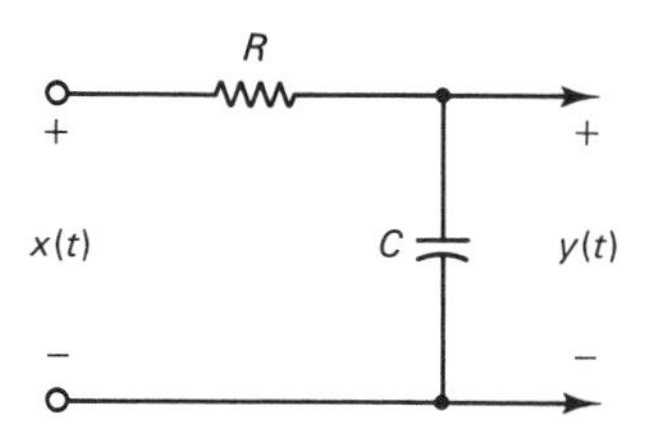

Figure 8.13 Simple RC network.

Thus, this RC network will alter the kth harmonic component of a periodic input as

$$Y_k = \frac{1}{1 + \dfrac{j2\pi kRC}{T}} X_k$$

The magnitude of the response signal's kth coefficient is then given by

$$|Y_k| = \frac{1}{\sqrt{1 + \dfrac{4\pi^2k^2R^2C^2}{T^2}}} |X_k| \qquad \text{for } k = 0, \pm 1, \pm 2, \ldots$$

Clearly, the factor multiplying $|X_k|$ in this expression decreases uniformly as k grows in size independent of the values of R, C, and T. One then concludes that the given RC network more readily transmits the low-frequency harmonics of the periodic input signal than it does the higher harmonics. Due to this dynamical property, the given network is often referred to as a *low-pass filter*, that is, it passes low-frequency sinusoids but tends to reject high-frequency sinusoids.

8.13 PROBLEMS

8.1. Find the transfer function $T(s) = [E_2(s)]/[E_1(s)]$ from the network of Fig. P8.1.

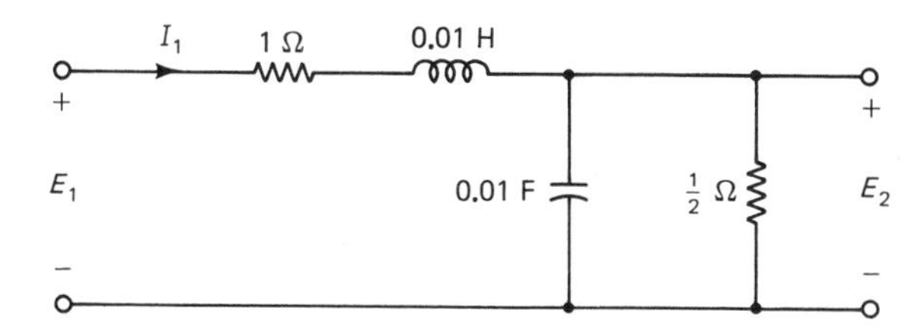

Figure P8.1

8.2. Determine the impedance $Z(s) = [E_1(s)]/[I_1(s)]$ from the network of Fig. P8.1 when E_2 is open.

8.3. Repeat Problem 8.1 for the network in Fig. P8.3.

8.4. Repeat Problem 8.2 for the network in Fig. P8.3.

8.5. Plot the magnitude and phase functions versus frequency ω for the transfer function developed in

(a) Problem 8.1

(b) Problem 8.2

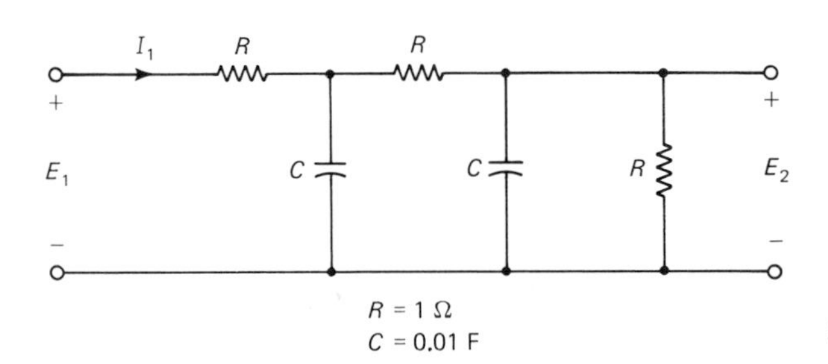

Figure P8.3

(c) Problem 8.3
(d) Problem 8.4

8.6. Determine the sinusoidal response using the results of Problem 8.5 to an input of the form

$$10 + 5 \cos 100t + 3 \sin 200t$$

for
(a) Problem 8.1
(b) Problem 8.2
(c) Problem 8.3
(d) Problem 8.4

8.7. A periodic voltage source consists of two superimposed sinusoids:

$$v(t) = 5 \sin t + 10 \cos \frac{4t}{3} \qquad \text{volts}$$

If $v(t)$ is applied to the circuit of Fig. P8.7, determine the resulting sinusoidal current $i(t)$.

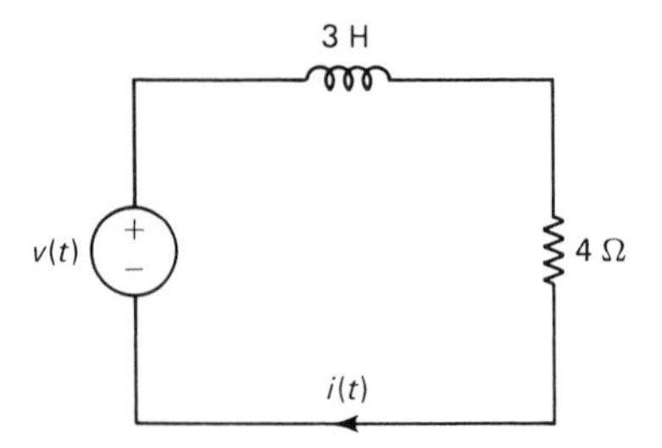

Figure P8.7

8.8. Find the d.c. and fundamental components of a half-wave–rectified sinusoid having period 2π and unit amplitude using direct integration.

8.9. Determine which of the functions in Fig. P8.9 have even, odd, or half-wave symmetry.

8.10. The frequency response of a digital filter can be found from its z-domain–transfer function by substituting $z = e^{j\omega T}$, where T is the sampling period. The discrete-time function $H(z) = z/z - 1$ approximates a continuous-time integrator, $1/s$.

(a) Evaluate the magnitude and angle functions of $1/s$ as a function of ω. Hint: Write $1/j\omega$ in polar form.
(b) Evaluate the magnitude and angle functions of $H(e^{j\omega T})$ for $T = 1$ second. Since $e^{j\omega}$ is periodic with period 2π, you may restrict the calculation to $0 < \omega < \pi$.
(c) Compare the magnitude curves of parts (a) and (b). For what frequency range is $H(z)$ a reasonably good approximation for an integrator?

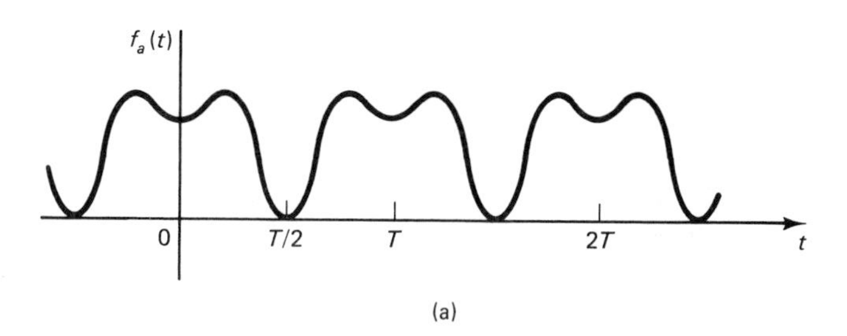

(a)

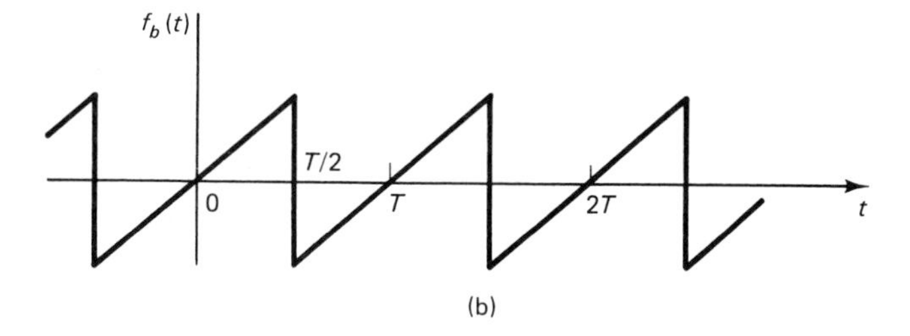

(b)

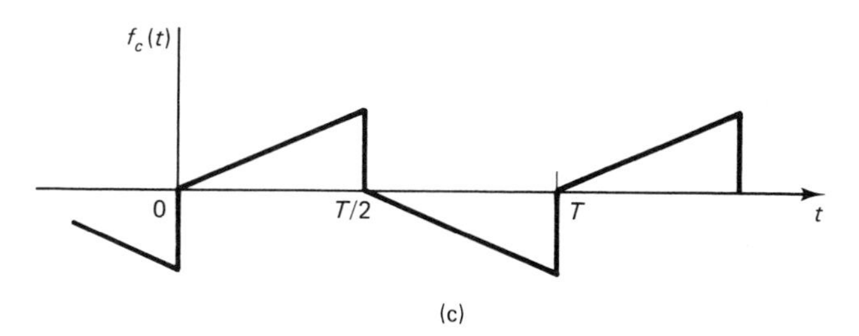

(c)

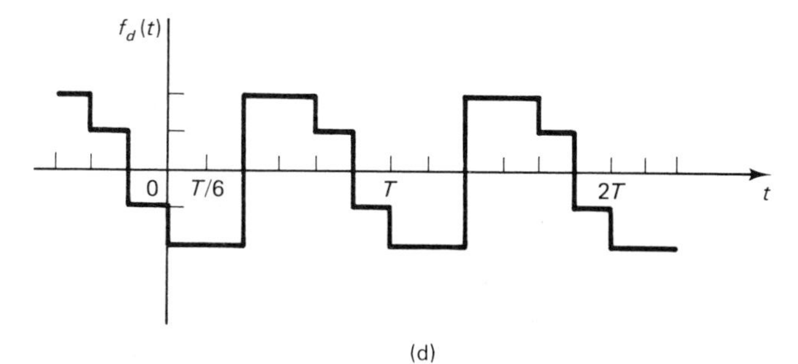

(d)

Figure P8.9

8.11. Determine if the signals $\phi_1(t) = t - 1$, $\phi_2(t) = t^2 - t + 1$, and $\phi_3(t) = t^3 - t^2 + t - 1$ are linearly independent over the interval $0 \leq t \leq 1$.

8.12. Write the equations corresponding to expression (8.41) for obtaining the approximation of $x(t) = u(t - 1/2)$ in terms of the functions in Problem 8.11 over the interval $0 \leq t \leq 1$.

8.13. Show that the signals $\{\cos n2\pi t$, for $n = 0, 1, 2, \ldots\}$ are orthogonal over the interval $0 \leq t \leq 1$.

8.14. Repeat Problem 8.13 for the signals $\{\sin n2\pi t, \text{ for } n = 0, 1, 2, \ldots\}$.

8.15. Make a sketch of the first three Rademacher signals. Show that $\phi_2(t)$ and $\phi_3(t)$ are orthogonal.

8.16. Determine if the signals $\{1, t, t^2, t^3\}$ are orthogonal over the interval $0 \le t \le 1$. If not, use the Gram-Schmidt procedure to derive a set of orthonormal signals.

8.17. Convert the signals given in Problem 8.13 to a set of orthogonal signals over $0 \le t \le 2$.

8.18. In making an analogy between vectors and signals we have already used the idea of an inner (or dot) product in defining orthogonality, that is, the inner product of two vectors $\boldsymbol{x} \cdot \boldsymbol{y}$ is analogous to

$$\int_a^b x(t)y^*(t)\, dt$$

for two (possibly complex-valued) signals. Continuing the analogy, we can say that the component of signal $x(t)$ "in the direction" of signal $y(t)$ is

$$\frac{\displaystyle\int_a^b x(t)y^*(t)\, dt}{\left(\displaystyle\int_a^b |y(t)|^2\, dt\right)^{1/2}}$$

where the denominator vanishes if the signal $y(t)$ is normalized.

(a) Determine the components of $x(t) = (4/\pi) \sin 2\pi t$ in terms of the first three Rademacher signals defined over the interval $0 \le t \le 1$.

(b) Determine the components of

$$x(t) = u(t) - 2u\left(t - \frac{1}{2}\right)$$

in the direction of $\cos 2\pi t$ and $\cos 4\pi t$ defined over the interval $0 \le t \le 1$.

8.19. **(a)** Show that the set of complex exponential functions given by expression (8.62) forms an orthogonal set.

(b) Derive the normalization constant that makes the set orthonormal.

8.20. **(a)** Evaluate the Fourier coefficients for the signal x if

$$x(t) = \begin{cases} 1 & \text{for } |t| < \tau \\ 0 & \text{for } \tau \le |t| \le 1 \end{cases}$$

where $0 < \tau < 1$.

(b) Expand your result into a cosine series.

8.21. **(a)** Sketch the periodic extension of $x(t)$ in Problem 8.20.

(b) Determine the Fourier cosine series for $x(t)$.

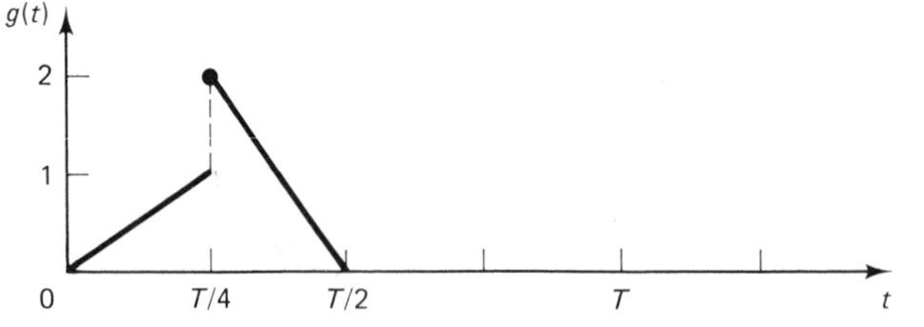

Figure P8.22

8.22. **(a)** Complete $g(t)$ in Fig. P8.22 so that $g(t)$ is periodic with period T.

(b) Set up the integral for finding the Fourier coefficients of $g(t)$.

8.23. Determine and plot the line spectrum for $x(t)$ given in Problem 8.20.

8.24. Sketch and find the d.c. component of the signal y given by

$$y(t) = \begin{cases} \sin t & \text{for } 0 \le t < \delta \\ 0 & \text{for } \delta \le x < 2\pi \end{cases}$$

if y is periodic with period 2π.

8.25. Given that f in Fig. P8.25a has the Fourier series

$$f(x) = \frac{8}{\pi^2}\left(\sin x - \frac{1}{9}\sin 3x + \frac{1}{25}\sin 5x - \cdots\right) \qquad \text{for } 0 \le x \le 2$$

Determine the cosine (Fourier) series for $v(t)$ in Fig. P8.25b.

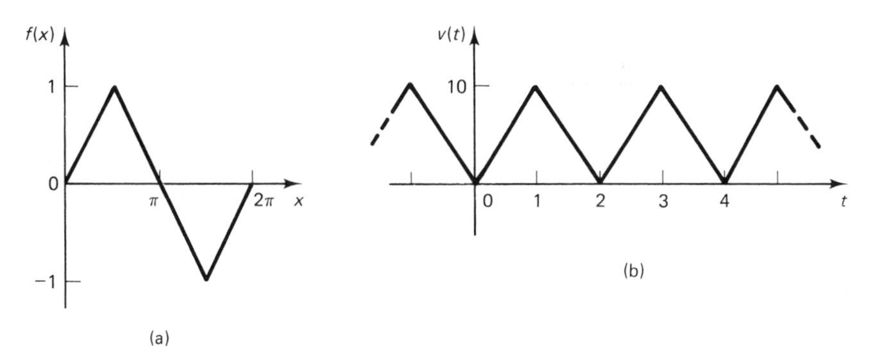

Figure P8.25

8.26. **(a)** Find the Fourier series representation for the "full-wave rectified" sine wave of Fig. P8.26.

(b) Calculate the total average power in this signal.

(c) Check your answer in part (b) by applying Parseval's theorem.

(d) Plot the spectrum.

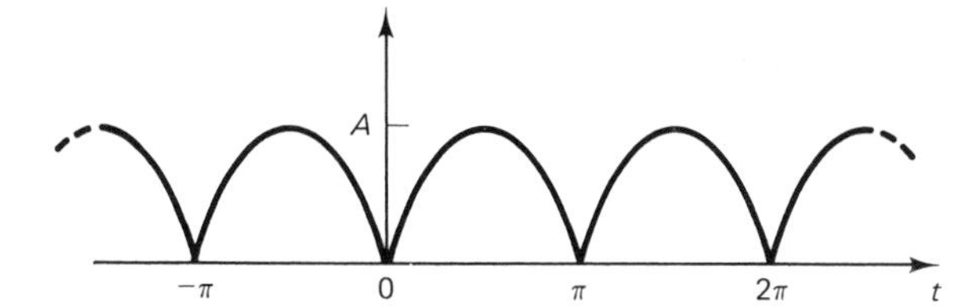

Figure P8.26

8.27. **(a)** Calculate a particular solution to the system shown in Fig. P8.27 if $x(t)$ is the periodic extension of f in Fig. P8.25a and $RC = 1$ second.

(b) Compare the power input and the power output of the network for the first three terms of the corresponding Fourier series.

(c) Express the total solution $y(t)$ if $y(0) = 1$ volt.

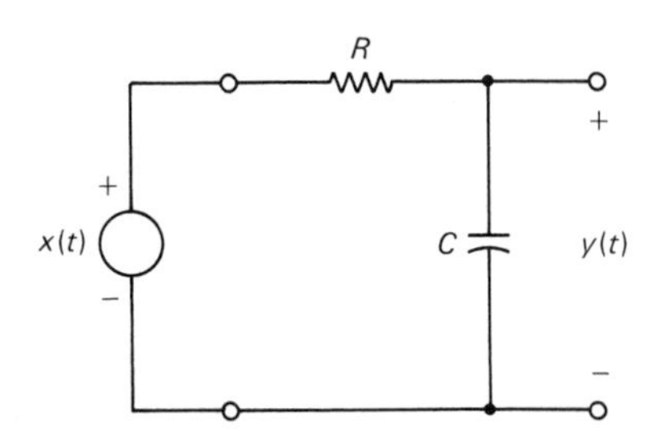

Figure P8.27

8.28. Evaluate the Fourier series of $x(t)$ given in Problem 8.20 by first finding the Fourier series of dx/dt.

8.29. **(a)** Find the steady-state output $y(t)$ in Fig. P8.29 if the input

$$x(t) = 10 + 5\cos 3t - \sin 6t$$

(b) Find the total output $y(t)$ when $v_C(0) = 1$ and $i_L(0) = 1$.

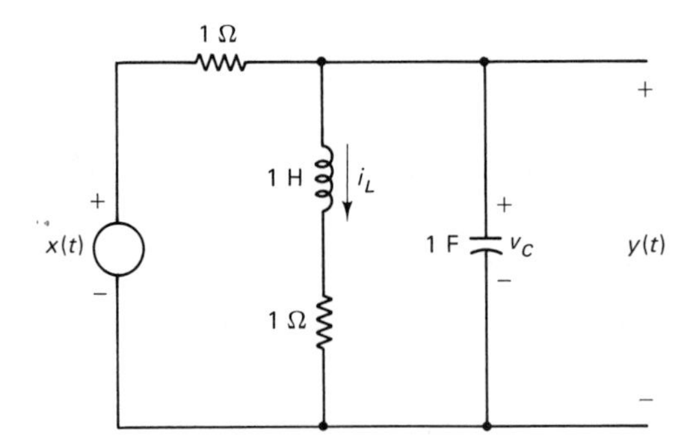

Figure P8.29

8.30. If the input voltage source is a "distorted" sinusoid given by

$$x(t) = 10 + \sin\frac{t}{10} + \frac{1}{10}\cos 10t$$

find the total response $y(t)$ in Fig. P8.30 after the switch is closed at $t = 0$. The capacitor is initially charged to 1 volt as shown.

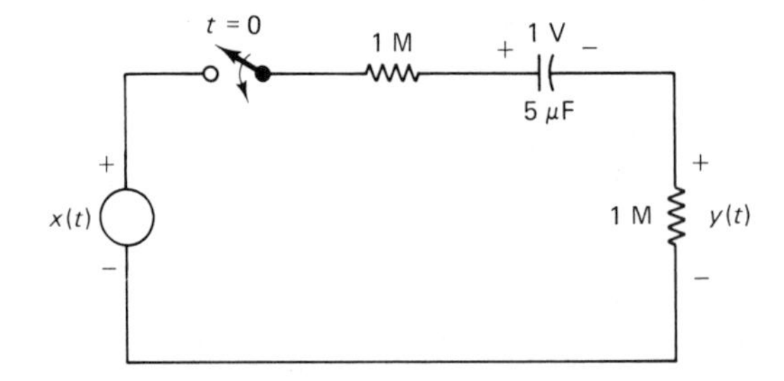

Figure P8.30

8.31. Given that the Fourier transform of $x(t)$ in Fig. P8.31a is

$$X(j\omega) = \frac{2}{\omega}\sin\frac{\omega}{2}$$

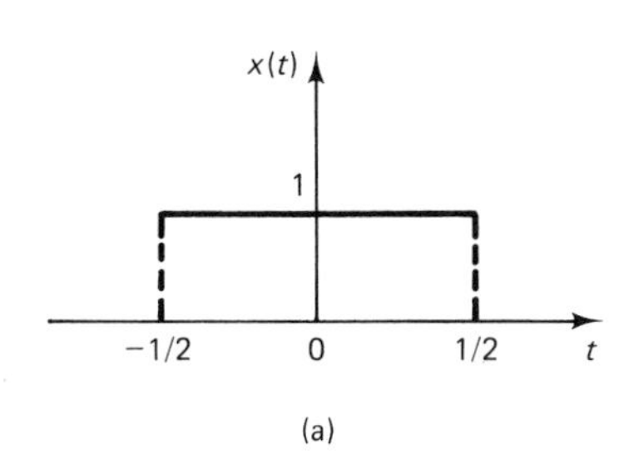

(a)

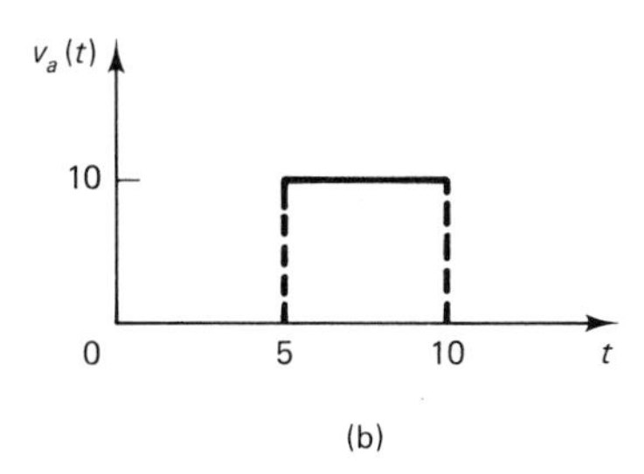

(b)

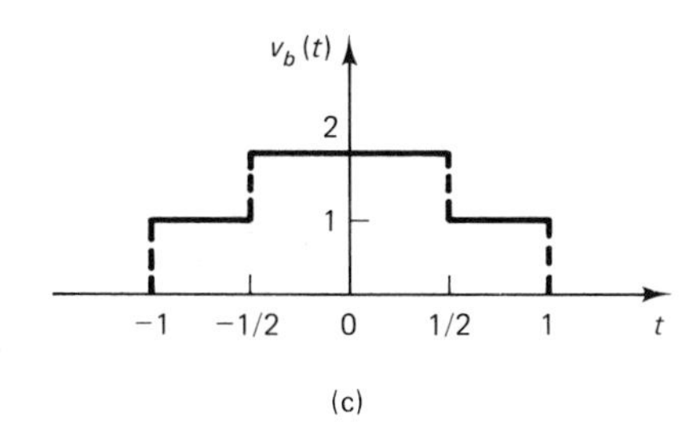

(c)

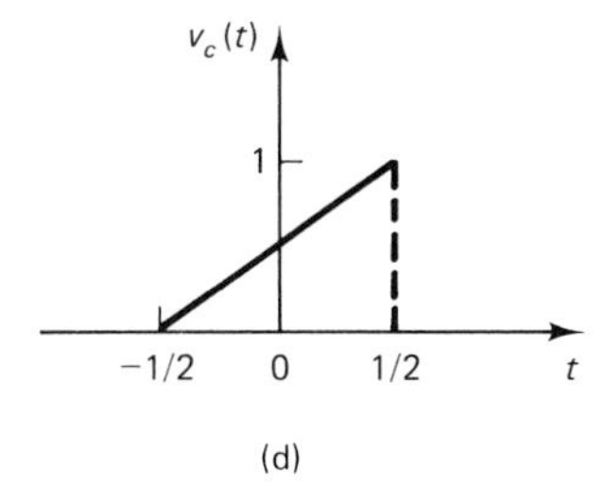

(d)

Figure P8.31

use the basic properties of symmetry, scaling, and so forth to find the Fourier transform of v_a, v_b, and v_c in the figure.

8.32. Find the percentage of the energy of $x(t)$ in Problem 8.31 contained in the frequency band $\omega \geq 10$ radians/second. You may leave any integrals unevaluated.

8.33. An ideal filter is specified as shown in Fig. P8.33. Calculate the filter output if its input $x(t)$ is the signal given in Fig. P8.25b.

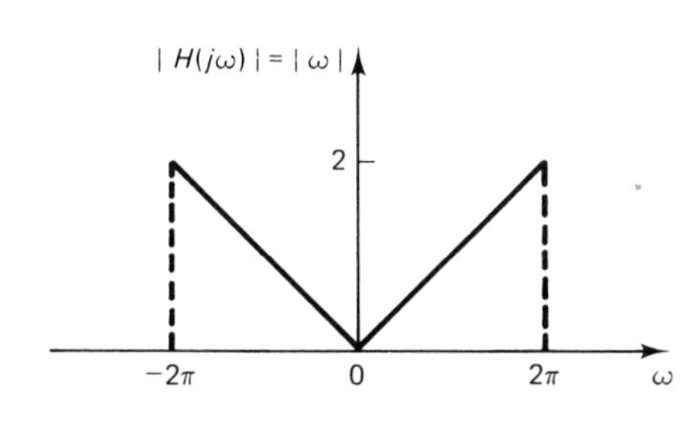

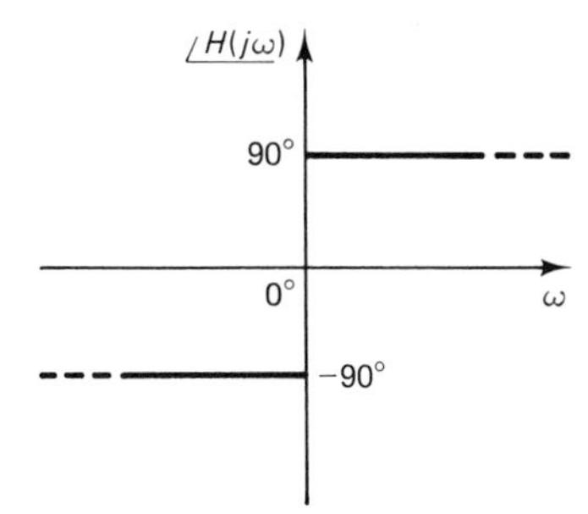

Figure P8.33

8.34. For a system with input $x(t)$ and output $y(t)$ described by

$$\ddot{y} + 4\dot{y} + 3y = 3x$$

(a) Sketch $|H(j\omega)|$ versus ω by calculating a few points along the graph. $H(s)$ is the transfer function $Y(s)/X(s)$.

(b) If $x(t)$ is sinusoidal with amplitude 0.1 volts at a frequency 3 radians/second, determine the amplitude of the sinusoidal part of the response.

(c) Determine the complete solution $y(t)$ if

$$x(t) = (\cos 3t)u(t) \quad \text{and} \quad y(0) = \dot{y}(0) = 0$$

8.35. For the filter whose frequency response is described in Fig. P8.35, find the forced response when the input

$$x(t) = 5 + 2 \cos 0.1w_c t + \sin 0.5w_c t$$

Assume that $t = 0.1$ second.

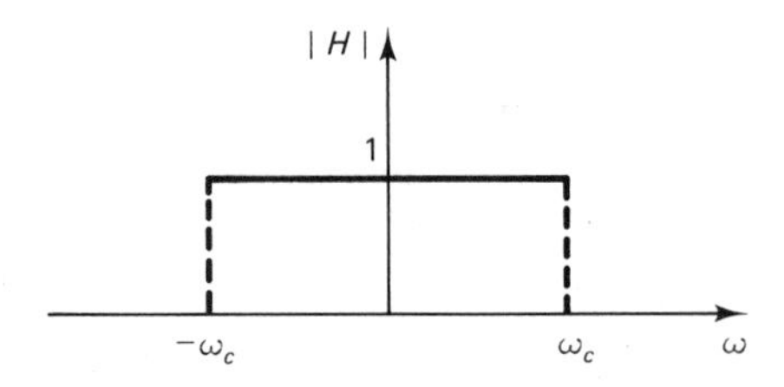

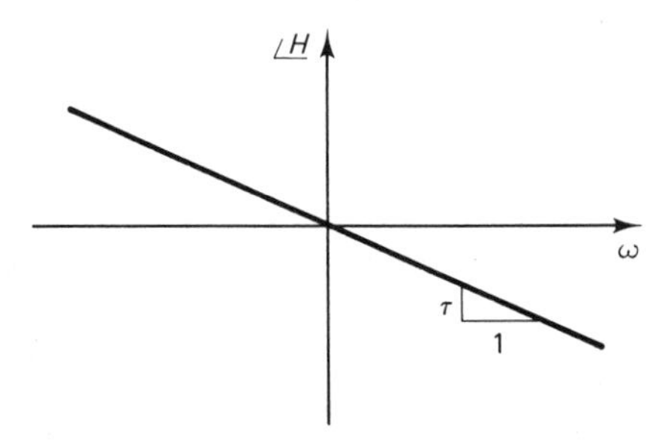

Figure P8.35

8.36. Sketch the magnitude and phase characteristics for a system with transfer function given by

(a) $\dfrac{1}{s(s+1)}$

(b) $\dfrac{(s^2+1)}{(s+1)(s^2+4)}$

9

The Discrete Fourier Transform and the Fast Fourier Transform Algorithm

9.1 INTRODUCTION

In previous chapters we have seen the power of the transform method in dealing with linear processing of signals. For instance, we know well the process of calculating a response from a linear system given a description of the system and the corresponding input signal. However, until now we have stressed only the theory aspect, permitting us to manipulate signals by means of algebra, transform-pair tables, and so forth. Utilizing the ever-growing capabilities of the digital computer is the remaining aspect of transform theory to be covered in this chapter.

The basic Fourier transform is introduced as a subclass of the (double-sided) Laplace transform. From this starting point the discrete Fourier transform (DFT) is developed with special care to relate it to the ordinary Fourier transform. One of the main algorithms used to calculate the DFT in an efficient manner is then studied.

9.2 THE FOURIER TRANSFORM

The reader will be relieved to find out that the Fourier transform of a signal is directly related to the Laplace transform, which was investigated in detail in Chapter 5. In this section we briefly state in which way our previous work applies to the Fourier transform.

The defining relation (5.1) is repeated here for convenience.

$$X(s) = \int_{-\infty}^{\infty} x(t)e^{-st}\,dt \tag{9.1}$$

where s, in general, is a complex variable $s = \sigma + j\omega$. If we now restrict s to be purely imaginary, namely, $s = j\omega$; then we have the classical *Fourier transform.*

$$X(j\omega) = \mathscr{F}[x(t)] \tag{9.2}$$

or

$$x(t) \leftrightarrow X(j\omega)$$

to denote the transform relationship of $x(t)$ and $X(j\omega)$.

Also, some authors prefer writing the transformed signal as $X(\omega)$, thereby suppressing the $j = \sqrt{-1}$ factor. The reader should be ready for either notation.

In light of having studied the material in Chapter 5, we simply state that the Fourier transform (9.2) exists (converges) if and only if the region of convergence of the corresponding expression (9.1) includes the $\text{Re}(s) = 0$ line ($j\omega$-axis). And, in keeping with this direct correspondence, Table 5.1 may be used as a table for Fourier transform pairs by simply changing the Laplace transform variable s to $j\omega$.

Example 9.1

The Fourier transform of the signal $x(t) = e^{-at}u(t)$, $a > 0$ is given by entry 1 in Table 5.1

$$X(j\omega) = \frac{1}{j\omega + a}$$

Note that the corresponding region of convergence contains $\text{Re}(s) = 0$. However, for $a = 0$, the Fourier transform (9.2) does not converge. This also shows up in entry 5 of Table 5.1, where it is seen that $\text{Re}(s) = 0$ is *not* included in the given region of absolute convergence.

The convergence problem is a mathematical constraint on using (9.2) that is easily overcome by *defining* the *Fourier transform* (in the limit) as

$$X(j\omega) = \lim_{\sigma \to 0} \int_{-\infty}^{\infty} x(t)e^{-(\sigma + j\omega)t}\, dt \tag{9.3}$$

even though the actual limit may not formally exist. By this definition for the Fourier transform we extend the use of Table 5.1 for Fourier transform pairs to include all those whose convergence region either includes the line $\text{Re}(s) = 0$ or is bounded by the $\text{Re}(s) = 0$ line on one side. Thus, for example, the Fourier transform of $u(t)$ is $1/j\omega$. Before looking at more Fourier transform pairs, let us consider the properties of the Fourier transform.

Clearly, the properties of the Laplace transform also apply to the Fourier transform wherever convergence is satisfied. Thus, the Fourier transform is linear, the transform of the time derivative of x is $j\omega$ times the transform of x, and so on. We summarize these properties in Table 9.1.

In addition to the basic properties listed in Table 9.1 there is an interesting symmetry between $x(t)$ and $X(j\omega)$. Consider that every real signal $f(t)$ can be written as a sum of an *even* part and an *odd* part.

$$f(t) = f_e(t) + f_o(t) \tag{9.4}$$

TABLE 9.1 FOURIER TRANSFORM PROPERTIES

	Property	Time Domain	Frequency Domain
1.	Linearity	$a_1x_1(t) + a_2x_2(t)$	$a_1X_1(j\omega) + a_2X_2(j\omega)$
2.	Time differentiation	$\dfrac{d^nx(t)}{dt^n}$	$(j\omega)^nX(j\omega)$
3.	Time shift	$x(t - t_0)$	$e^{-j\omega t_0}X(j\omega)$
4.	Scaling	$x(at)$	$\dfrac{1}{\lvert a\rvert}X\left(j\dfrac{\omega}{a}\right)$
5.	Modulation	$e^{j\omega_0 t}x(t)$	$X(j\omega - j\omega_0)$
6.	Time multiplication	$x_1(t)x_2(t)$	$\dfrac{1}{2\pi}\displaystyle\int_{-\infty}^{\infty} X_1(ju)X_2(j\omega - ju)\, du$
7.	Time integration	$\displaystyle\int_{-\infty}^{t} x(\tau)\, d\tau$	$\dfrac{1}{j\omega}X(j\omega) + \pi X(0)\delta(\omega)$
8.	Frequency differentiation	$-jtx(t)$	$\dfrac{d}{d\omega}X(\omega)$
9.	Convolution	$x_1(t) * x_2(t)$	$X_1(j\omega)X_2(j\omega)$
10.	Time transpose	$x(-t)$	$X(-j\omega) = X^*(j\omega)$

where

$$f_e(t) = f_e(-t) \quad \text{and} \quad f_o(t) = -f_o(-t)$$

To show this we write

$$f(t) = f(t) + \frac{1}{2}f(-t) - \frac{1}{2}f(-t)$$

$$f(t) = \left[\frac{f(t) + f(-t)}{2}\right] + \left[\frac{f(t) - f(-t)}{2}\right] \tag{9.5}$$

where the first term on the right is the even part of $f(t)$ and the second term, the odd part of $f(t)$.

By considering the transform (9.2) and using Euler's relations,

$$X(j\omega) = \int_{-\infty}^{\infty} [x_e(t) + x_o(t)][\cos(\omega t) - j\sin(\omega t)]\, dt$$

$$X(j\omega) = \int_{-\infty}^{\infty} x_e(t)\cos\omega t\, dt - j\int_{-\infty}^{\infty} x_o(t)\sin\omega t\, dt$$

$$X(j\omega) = 2\int_{0}^{\infty} x_e(t)\cos\omega t\, dt - 2j\int_{0}^{\infty} x_o(t)\sin\omega t\, dt \tag{9.6}$$

From (9.6) we see that a real, even signal $x_e(t)$ has a real and even transform $X(j\omega)$. This follows since $\cos\omega t$ is an even symmetric function. Other symmetry relations between transform pairs can be stated. For instance, assuming that $x(t)$ is a real-valued signal, its transform has the following symmetry:

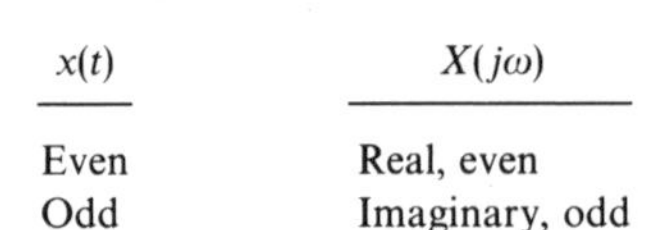

$x(t)$	$X(j\omega)$
Even	Real, even
Odd	Imaginary, odd

1. Time domain $x(t)$ — Frequency domain $X(f)$

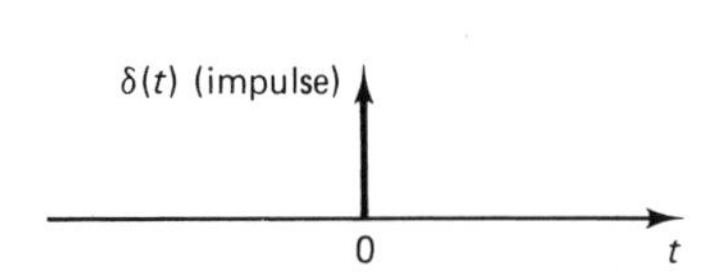

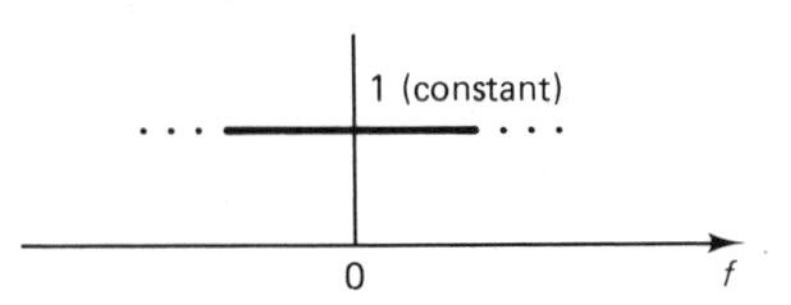

2.

cos πt
1
0
−1
t

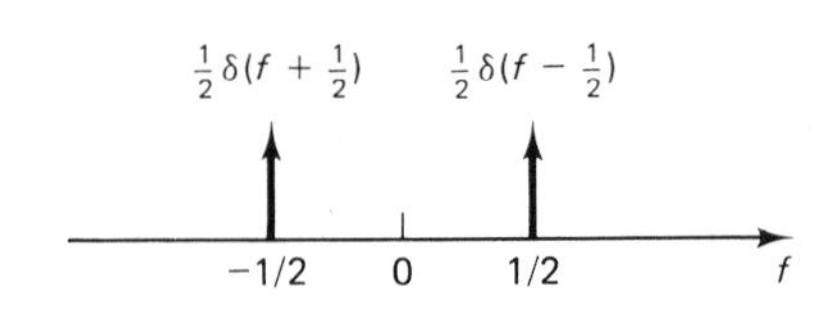

3.

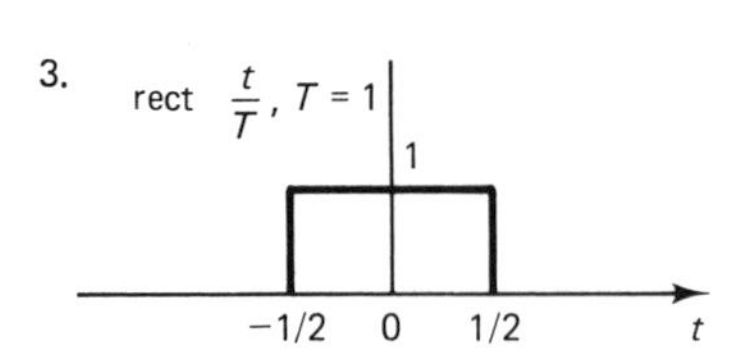

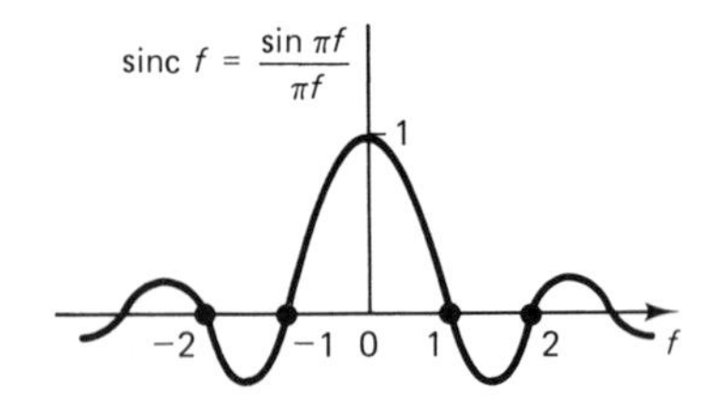

4.

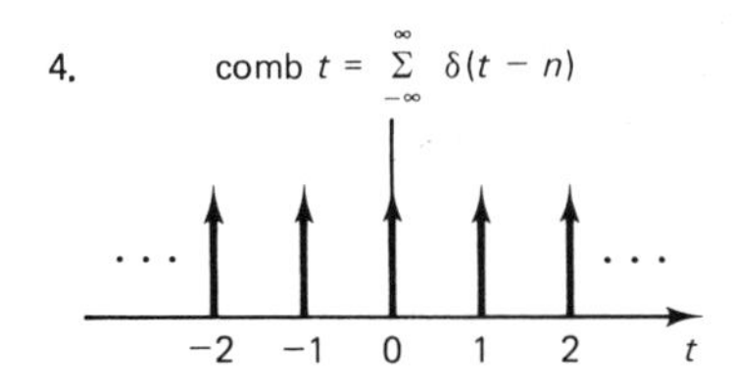

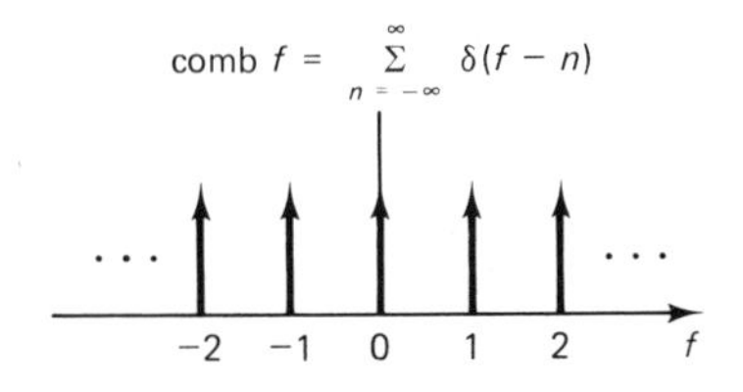

5.

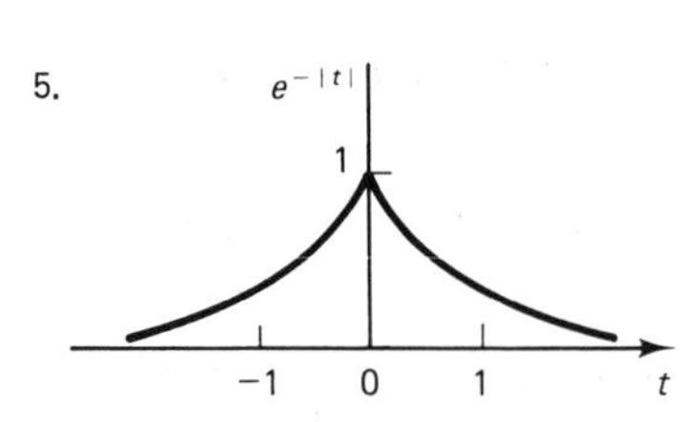

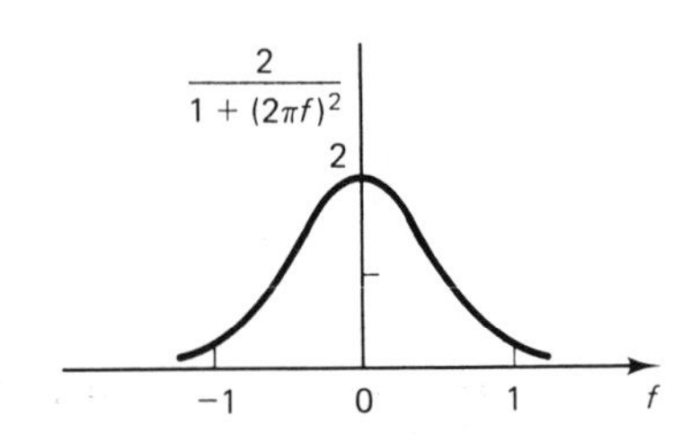

Figure 9.1 Some useful Fourier transform pairs.

We conclude this subsection with Fig. 9.1, which illustrates some useful transform pairs. Note that the transform of a real and even signal is real and can be depicted on a single graph, whereas, more generally, illustrating a complex-valued $X(j\omega)$ would require two graphs (either real and imaginary parts or magnitude and angle). To emphasize the symmetry, we use the real frequency f rather than the radian frequency ω. Thus, the plots are for[1]

$$X(f) = \int_{-\infty}^{\infty} x(t)e^{-j2\pi ft}\, dt$$

and

$$x(t) = \int_{-\infty}^{\infty} X(f)e^{j2\pi ft}\, df \qquad (9.7)$$

Note that for even signals the time-frequency roles may be reversed, that is column 1 in Figure 9.1 could be taken as frequency domain, in which case column 2 would be the corresponding time domain.

9.3 THE DISCRETE FOURIER TRANSFORM

The purpose of this section is to introduce the discrete Fourier transform (DFT). As the name implies, the DFT is closely related to the (continuous) Fourier transform. We develop and explain the DFT in a way that emphasizes this relation, by showing how the DFT approximates the actual Fourier transform. The DFT, being discrete in time and frequency, is naturally structured for implementation on a digital computer, and it is this property that motivates our study. Without computational aid, the practical use of the transform method of analysis is severely limited. As we shall see, the DFT and special algorithms for calculating the DFT, called *fast Fourier transform (FFT) algorithms*, are needed to efficiently calculate close approximations of Fourier transforms from data sources, i.e., samples of actual signals. We will also come to understand how the linear processing corresponding to the signal operation is done computationally.

To begin the development, we assume that a finite set of samples is available from a signal x, namely,

$$\{x_k\} = \{x_0, x_1, \ldots, x_{N-1}\} \qquad (9.8)$$

The DFT will transform this time sequence into a frequency sequence

$$\{X_n\} = \{X_0, X_1, \ldots, X_{N-1}\} \qquad (9.9)$$

[1]Corresponding to (9.2), the inverse transform is given by

$$x(t) = \frac{1}{2\pi}\int_{-\infty}^{\infty} X(j\omega)e^{j\omega t}\, d\omega$$

The fact that X is a function of f or $j\omega$ is determined from context.

through the defining relation DFT $\{x_k\} = \{X_n\}$, where

$$X_n = \sum_{k=0}^{N-1} x_k e^{-jkn2\pi/N} \tag{9.10}$$

for $n = 0, 1, \ldots, N-1$. In fact, we could establish from this definition all the properties of this transform in its own right without regard to the Fourier transform; but since this does not serve our purpose here, let us go back to see how the actual Fourier transform ties in.

Given a continuous-time signal $x(t)$, defined for all time, let us assume that we obtain the samples of (9.8). To describe these samples mathematically, we take each sample to represent a pulse having some signal energy. This may be done ideally, that is, for an infinitely narrow pulse, by the use of impulse functions. Figure 9.2 shows the mathematical operations of multiplying $x(t)$ by a comb function to extract the ideal samples of $x(t)$ and the following truncation of the infinite sample train into the finite set of N-samples to work with. The origin has been taken arbitrarily to be at the center of the data, and for graphical convenience the height of the impulses are used to indicate their relative weight. The N-samples obtained are relabeled, so that $x(-k_0) = x_0$ consecutively until $x(k_0) = x_{N-1}$.

As is typically done with a finite length of data, it may be thought of as one period of a periodic signal. By extending the data $x_d(t)$ of Fig. 9.2 periodically, we know that the frequency domain equivalent is discrete (Fourier series of a periodic signal). The selected components of this frequency-domain signal then become the set $\{X_n\}$, which is achieved by the DFT operation. To illustrate this result, we consider both the time and the frequency descriptions after each operation, shown in Fig. 9.3 for $x(t) = \exp(-|t|)$. In Fig. 9.3a we have entry 5 of Fig. 9.1. The first operation of sampling is a multiplication of signals in time and, therefore, a convolution of their transforms. Thus, $X(f)$ is convolved with

$$\mathscr{F}\left(\frac{1}{\Delta t} \operatorname{comb} \frac{t}{\Delta t}\right) = \operatorname{comb}(\Delta t\ f) \tag{9.11}$$

which, if Δt is the spacing in time, $1/\Delta t$ is the spacing in frequency. Recalling that convolution of a continuous function with an impulse function simply replicates the continuous function at the location of the impulse, we obtain the periodic frequency description indicated in Fig. 9.3b. Note that overlap occurs in the regions where the periodic segments fit together. This is already introducing error into the basic shape of $X(f)$! The next operation of truncation is again a multiplication in the time domain and, therefore, a convolution of our periodic frequency description of Fig. 9.3b with[2]

$$\mathscr{F}\left(\operatorname{rect} \frac{t}{T}\right) = T \operatorname{sinc} Tf \tag{9.12}$$

[2]In this chapter the sinc function is defined as

$$\operatorname{sinc} x = \frac{\sin \pi x}{\pi x}$$

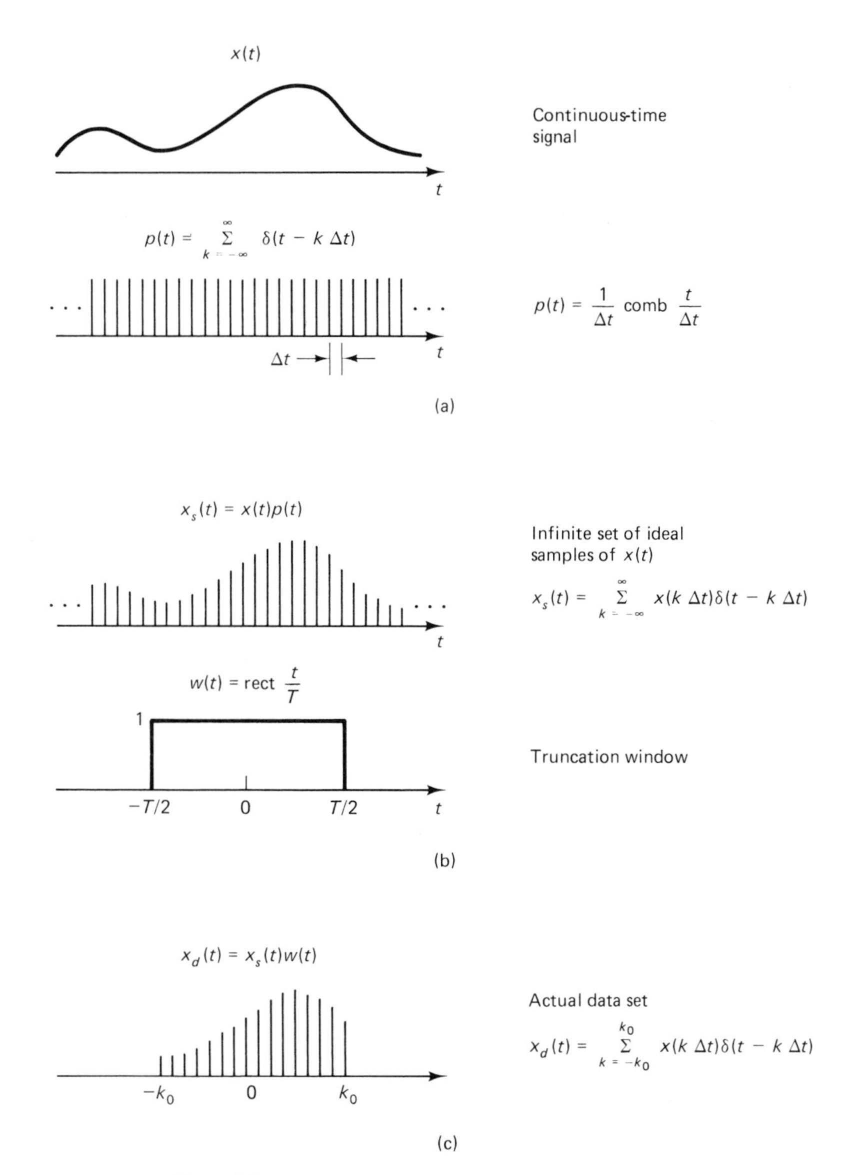

Figure 9.2 Mathematical operations to obtain time data.

Qualitatively, T is large so that the zeros of sinc Tf occurring at $f = n/T$ for $n = \pm 1, \pm 2, \ldots$ are close together; that is to say, the function T sinc Tf is a narrow pulse with ripples tailing off on either side (see Fig. 9.1). When the convolution is made, the result is to maintain the basic periodic shape as in Fig. 9.3b, but the ripples of the sinc function introduce further distortion on the basic shape of $X(f)$, as shown in Fig. 9.3c.

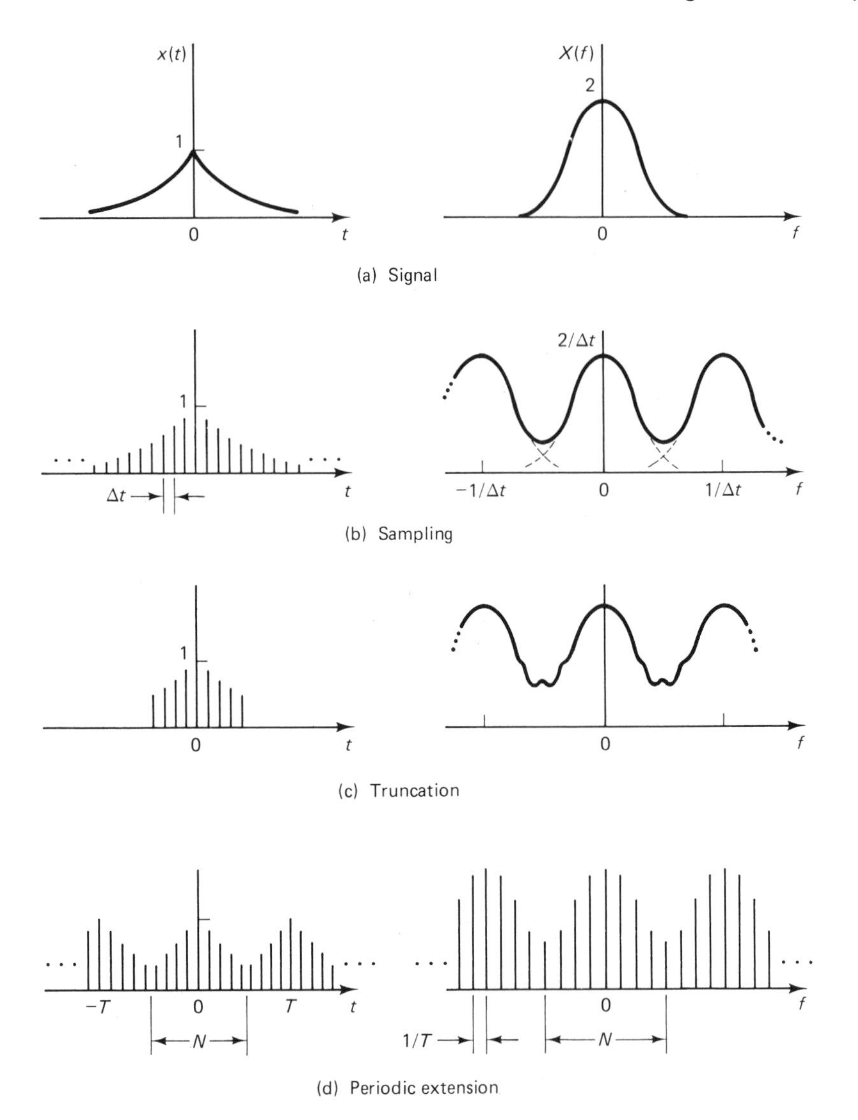

Figure 9.3 Graphical relation of DFT to Fourier transform.

The last operation of extending the time samples periodically amounts to convolving the time-domain signal with comb t/T, where T is the truncation interval. And, since

$$\mathscr{F}\left(\text{comb}\,\frac{t}{T}\right) = T\text{ comb } Tf \tag{9.13}$$

the frequency signal is multiplied by T comb Tf, which is equivalent to sampling in frequency with the samples $1/T$ apart. Finally, the DFT calculation can be shown to transform the N-samples in time to the corresponding N-samples in frequency as shown in Fig. 9.3d. The interpretation is that from a finite set of samples of $x(t)$ the DFT approximates a corresponding set of samples from $X(f)$. Unfortunately, errors have been introduced so that the frequency samples shown in Fig. 9.3d are not the exact samples of $X(f)$ as desired. However, from the example, samples near the origin ($f = 0$) appear to be more nearly the correct values. This offers us some hope to increase our accuracy, particularly at low frequencies, by taking a larger amount of data. The reader should note that only half of the frequency samples give us information since we already know that the negative frequency components are the complex-conjugate values of the corresponding positive frequency components. These observations are elaborated on subsequently. Motivated by the previous development, we define the DFT and its inverse operation as follows.

Definition 9.1

$$X\left(\frac{n}{N\,\Delta t}\right) = \text{DFT}\,\{x(k\,\Delta t)\} = \sum_{k=0}^{N-1} x(k\,\Delta t)e^{-j2\pi nk/N} \qquad \text{for } n = 0, 1, 2, \ldots, N-1$$

Definition 9.2

$$x(k\,\Delta t) = \text{DFT}^{-1}\left\{X\left(\frac{n}{N\,\Delta t}\right)\right\}$$

$$= \frac{1}{N}\sum_{n=0}^{N-1} X\left(\frac{n}{N\,\Delta t}\right)e^{j2\pi nk/N} \qquad \text{for } k = 0, 1, 2, \ldots, N-1$$

To show that Definitions 9.1 and 9.2 are compatible definitions, let us use the simpler notations x_k for $x(k\,\Delta t)$ and X_n for $X[n/(N\,\Delta T)]$ and show that $\{x_k\}$ and $\{X_n\}$ form a transform pair.

From Definition 9.1,

$$X_n = \sum_{k=0}^{N-1} x_k e^{-j2\pi nk/N}$$

Introducing Definition 9.2 into this expression,

$$X_n = \sum_{k=0}^{N-1}\left(\frac{1}{N}\sum_{m=0}^{N-1} X_m e^{j2\pi mk/N}\right)e^{-j2\pi nk/N}$$

$$X_n = \frac{1}{N}\sum_{m=0}^{N-1} X_m\left[\sum_{k=0}^{N-1} e^{j2\pi k(m-n)/N}\right] \tag{9.14}$$

But

$$\sum_{k=0}^{N-1} e^{j2\pi kr/N} = \begin{cases} N & \text{for } r = 0 \\ 0 & \text{for } r \neq 0 \end{cases}$$

(This is known as the *orthogonality property* for the discrete, complex exponential functions.) Thus, the right-hand side of (9.14) reduces to the left-hand side, completing the proof that $\{x_k\}$ and $\{X_n\}$ form a DFT pair.

With this brief development let us now work with sampling and finding the DFT of a more general signal. However, the methods of viewing the transformation as resulting from sampling, truncation, and periodic extension still apply. In the previous introductory work (see Fig. 9.3), we assumed an even symmetry for our signal, so that its Fourier transform was also real-valued. This is not generally the case, but a similar correspondence between the time and frequency domains may still be maintained by using magnitudes of the frequency-domain descriptions.

Before we tackle this practical problem, it is worthwhile to consider when the DFT gives us the *exact* samples of the Fourier transformed signal. First, from Fig. 9.3 to avoid aliasing error, that is, overlapping of periodic segments in the frequency domain (see Fig. 9.3b), the original signal $x(t)$ must be band-limited, and the sampling must be sufficiently fast (Δt sufficiently small) so that no aliasing occurs.[3] Second, if the signal x is already periodic and the truncation interval equals an integer number of periods, then the operations shown in Fig. 9.3c and d have no effect, so that the DFT sequence will be the same as the samples of $X(f)$, that is, the Fourier coefficients of the periodic signal (within a scale factor).[4]

DFT of a General Signal

Let us now consider that our signal x has no known special qualities such as being band-limited or time-limited. In fact, we may only have the discrete data and not the continuous-time x signal to work with. Since the data must originate at some point in time, let us assume that x is a single-sided signal with $x(t) = 0$ for $t < 0$.

Figure 9.4 illustrates the three operations on $x(t) = e^{-t}u(t)$ as were shown for an even symmetric function in Fig. 9.3. There are only minor differences between the two figures, but they are worth noting. The data $\{x_k\}$ need not be relabeled since the samples of x begin at $t = 0$. Again the errors of aliasing and those resulting from the truncation in time (finite set of data) occur. But, most importantly, the DFT sequence $\{X_n\}$ is a set of numbers as shown in Fig. 9.4d, indexed from 0 to $N - 1$ on the f-axis. The first half of these values represent approximate samples of $X(f)/\Delta t$. In general, they are complex numbers. The second half of the $\{X_n\}$ values are the same as the "negative" frequency values (shifted periodically). These give no new information and are simply complex conjugates of their corresponding "positive" frequency values. The midpoint $n = N/2$ is called the *folding frequency*. As we see in the next section, values near the folding frequency are of lesser importance.

In the next section we take a closer look at the computational aspects of the DFT in order to begin to develop an analytical feel for what a DFT algorithm can provide.

[3]Signal x is said to be *band-limited* if for some $B > 0$

$$|X(f)| = 0 \qquad \text{for } |f| > B$$

[4]For equivalence between the continuous Fourier transform and the DFT the DFT values must be scaled by a factor of Δt. For correspondence with Fourier series coefficients the DFT values must be scaled by a factor of $1/N$.

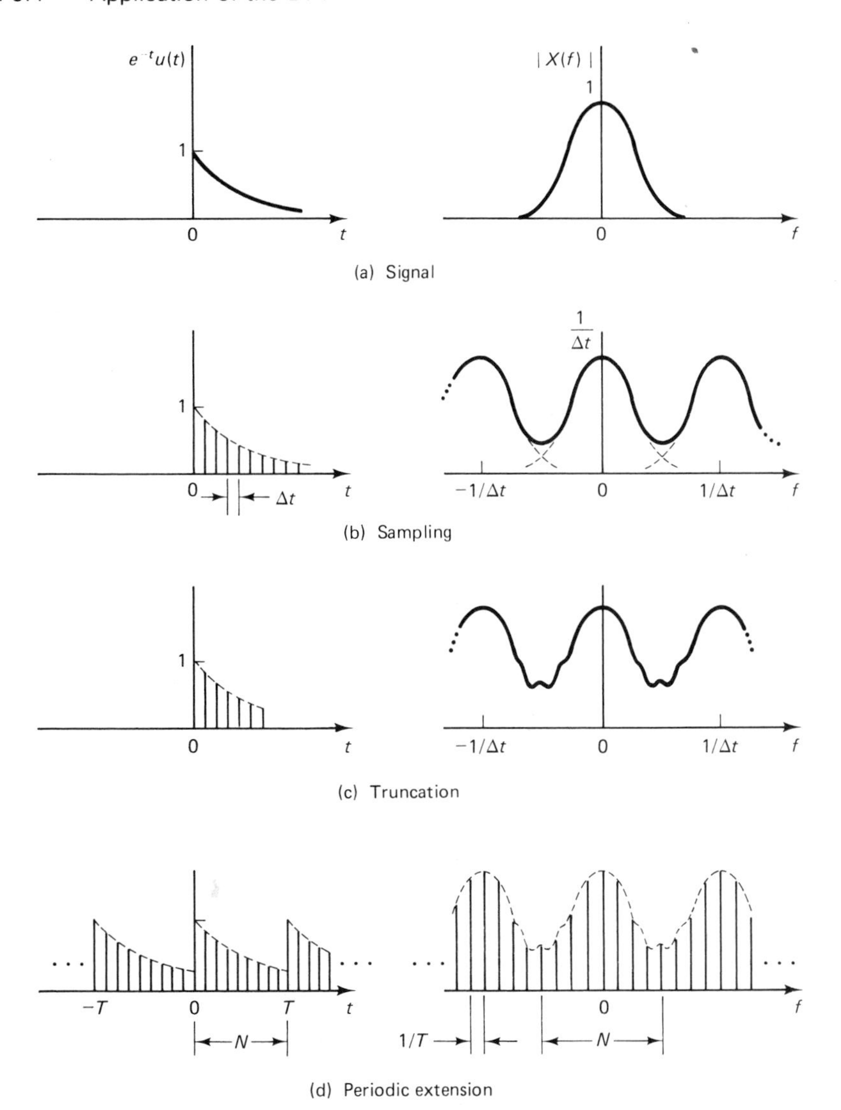

Figure 9.4 DFT process applied to a single-sided signal.

9.4 APPLICATION OF THE DFT

From our original definition of the DFT

$$X_n = \sum_{k=0}^{N-1} x_k W^{nk} \tag{9.15}$$

for $n = 0, 1, \ldots, N-1$ and where $W = e^{-j2\pi/N}$ has been defined to simplify the notation. Note that W depends upon N. When necessary to show this dependence explicitly, we write W_N; but most of the time it is understood from context. As a

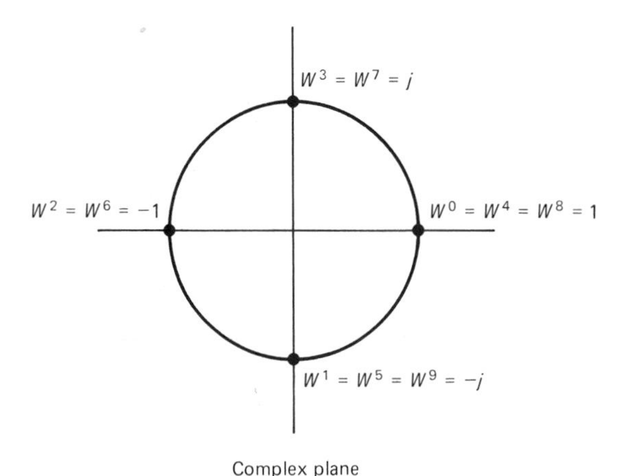

Figure 9.5 Exponential factors for a 4-point DFT.

simple example of the DFT in operation, we consider a "4-point" formulation, that is, one in which the sequence $\{x_k\}$ has length four. Since (9.15) represents four equations ($n = 0, 1, 2, 3$), we may combine them into a matrix form for the calculation.

$$\begin{bmatrix} X_0 \\ X_1 \\ X_2 \\ X_3 \end{bmatrix} = \begin{bmatrix} W^0 & W^0 & W^0 & W^0 \\ W^0 & W^1 & W^2 & W^3 \\ W^0 & W^2 & W^4 & W^6 \\ W^0 & W^3 & W^6 & W^9 \end{bmatrix} \begin{bmatrix} x_0 \\ x_1 \\ x_2 \\ x_3 \end{bmatrix} \tag{9.16}$$

This matrix operation will, in general, require N^2 (complex) multiplications and $N(N-1)$ (complex) additions. In the next section we see that much more efficient algorithms exist for computing DFTs.

From the definition of W in (9.15) we can depict its various powers as values on the unit circle of a complex plane. Figure 9.5 illustrates this for the 4-point case. Higher powers of W are found by proceeding around the circle in a clockwise manner; each higher integer power representing an additional $-90°$ ($e^{-j2\pi/N}$) rotation. Note that the power of W may be reduced by mod (N),[5] for instance, $W^9 = W^{1 \bmod(4)}$. It is by using the special properties of W^k that the efficient (FFT) algorithms were developed. In the 4-point case, (9.16) reduces to the matrix operation.

$$\boldsymbol{X} = \begin{bmatrix} 1 & 1 & 1 & 1 \\ 1 & -j & -1 & j \\ 1 & -1 & 1 & -1 \\ 1 & j & -1 & -j \end{bmatrix} \boldsymbol{x} \tag{9.17}$$

In the following examples we will be selecting various parameters to suit the problem at hand. First, the parameter N, the number of data points, determines the amount of calculation needed and, as we shall see, the corresponding accuracy

[5]An integer number K is said to equal the integer L (modulo M) if $K = nM + L$ where n is any integer.

obtained. Working from the time domain, the spacing between sample points is Δt seconds, and the total data record is of length T seconds, where $T = N\ \Delta t$. By the intrinsic properties of the Fourier transform, the resulting effects in the frequency domain are twofold: (1) The frequency sampling interval is given by $\Delta f = 1/T$ hertz (Hz))—notice that the total record length in time determines the frequency resolution (the smallest difference in frequency); and (2) the maximum useful frequency range is given by $N/2$ frequency samples, i.e., $\Delta f(N/2) = 1/(2\ \Delta t)$ Hz. Therefore, the sample width in time determines the maximum frequency of concern. The reader should be able to follow these parameter values in the subsequent examples.

Example 9.2 (Fourier Coefficients)

As a first example we will apply a 4-point DFT to a sinusoid (single frequency) with samples truncated to one period. In this case we should have the exact (single) frequency resulting from the DFT calculation.

From Fig. 9.6 we find that the appropriate data vector is

$$x = [1 \quad 0 \quad -1 \quad 0]^T$$

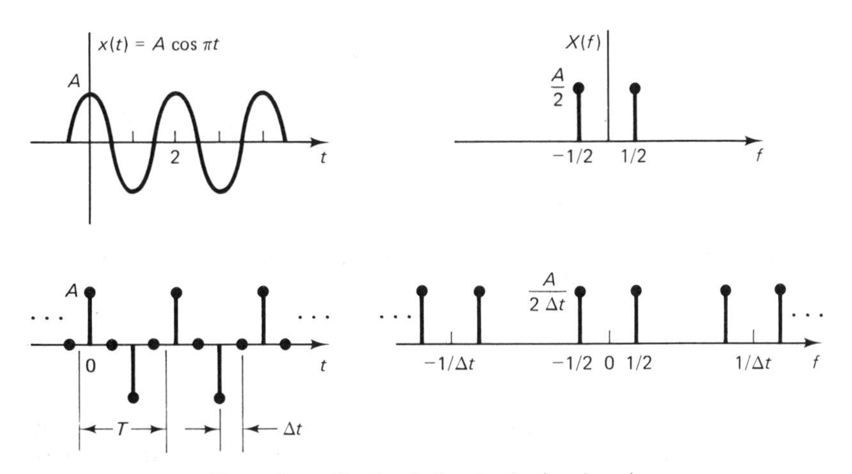

Truncation to T and periodic extension (no change)

(a)

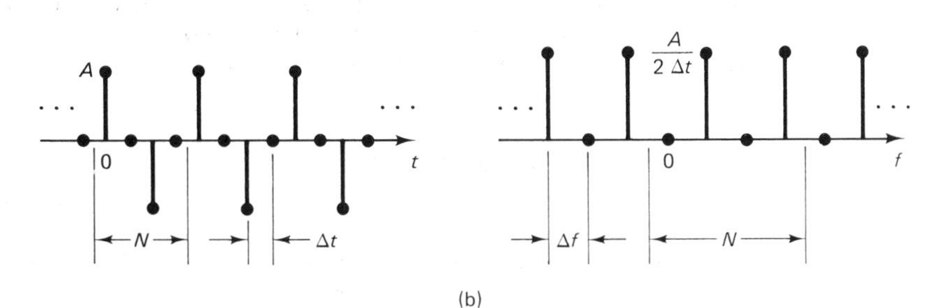

(b)

Figure 9.6 A 4-point DFT (Example 9.2).

Applying (9.17), the DFT output X is obtained:

$$X = [0 \quad 2 \quad 0 \quad 2]^T$$

The interpretation is that a single frequency is present (see Fig. 9.5), $X_1 = 2$ (X_3 is the "negative" frequency component of X_1). That there is no d.c. or average value to the data is given by $X_0 = 0$.

Since $N = 4$, if $x(t) = \cos \pi t$, then $T = 2$ (one period) and $\Delta t = 0.5$ (sample interval). From Definition 9.1 we note that the nonzero component X_1 corresponds to $X(1/2)$, i.e., $f = \frac{1}{2}$ (the correct frequency of the signal). Thus, if we interpret X as the exponential Fourier series of $x(t)$, we would have

$$x(t) = X_1 e^{j2\pi(1/2)t} + X_3 e^{-j2\pi(1/2)t} = 4 \cos \pi t$$

where a gain of $N = 4$ is observed. It is to compensate for this gain that Definition 9.2 has the factor $1/N$. Thus, to obtain the proper Fourier coefficients, one must divide the DFT result by N. Another way to see this result is to think of using a rectangular approximation to the integrals of (9.7), so that

$$\left.\begin{aligned} X_n &= \Delta t \sum x_k e^{-j2\pi nk/N} \\ &\text{and} \\ x_k &= \Delta f \sum X_n e^{j2\pi nk/N} \end{aligned}\right\} \tag{9.18}$$

showing us that the factor $\Delta t \; \Delta f = \Delta t/T = 1/N$ must be inserted. To complete the example, the Fourier coefficient is found to be $X_1/N = \frac{1}{2}$ as expected.

Example 9.3 (Fourier Series)

In this example we assume a periodic symmetric square wave in an attempt to find its Fourier series representation (harmonic analysis).

For $x(t) = 1$ if $0 \leq t < 4$ or $12 \leq t < 16$ and $x(t) = -1$ if $4 \leq t < 12$ as shown in Fig. 9.7, let us first work a 4-point DFT. Thus, $\Delta t = 4$ and

$$\boldsymbol{x} = [1 \quad 0 \quad -1 \quad 0]^T$$

We note that this small amount of data will give us d.c. and fundamental component information only. In fact, when we compare $\boldsymbol{x}$ with that of the previous example, we see that it is the same. Therefore, we expect that

$$X = [0 \quad 2 \quad 0 \quad 2]^T$$

where, on dividing by $N = 4$, our result approximates the fundamental component by 0.5. The exact fundamental component is given by $2/\pi = 0.637$.

Now, for a more realistic 8-point DFT, $\Delta t = 2$ and

$$\boldsymbol{x} = [1 \quad 1 \quad 0 \quad -1 \quad -1 \quad -1 \quad 0 \quad 1]^T$$

we find (using the matrix approach) that

$$X = [0 \quad 2 + 2\sqrt{2} \quad 0 \quad 2 - 2\sqrt{2} \quad 0 \quad 2 - 2\sqrt{2} \quad 0 \quad 2 + 2\sqrt{2}]^T$$

which agains says no d.c. component; fundamental component = 0.604 (after dividing by $N = 8$); no second-harmonic component; and the third-harmonic component is -0.104 (its actual value is -0.212). Notice that the half-wave symmetry is maintained

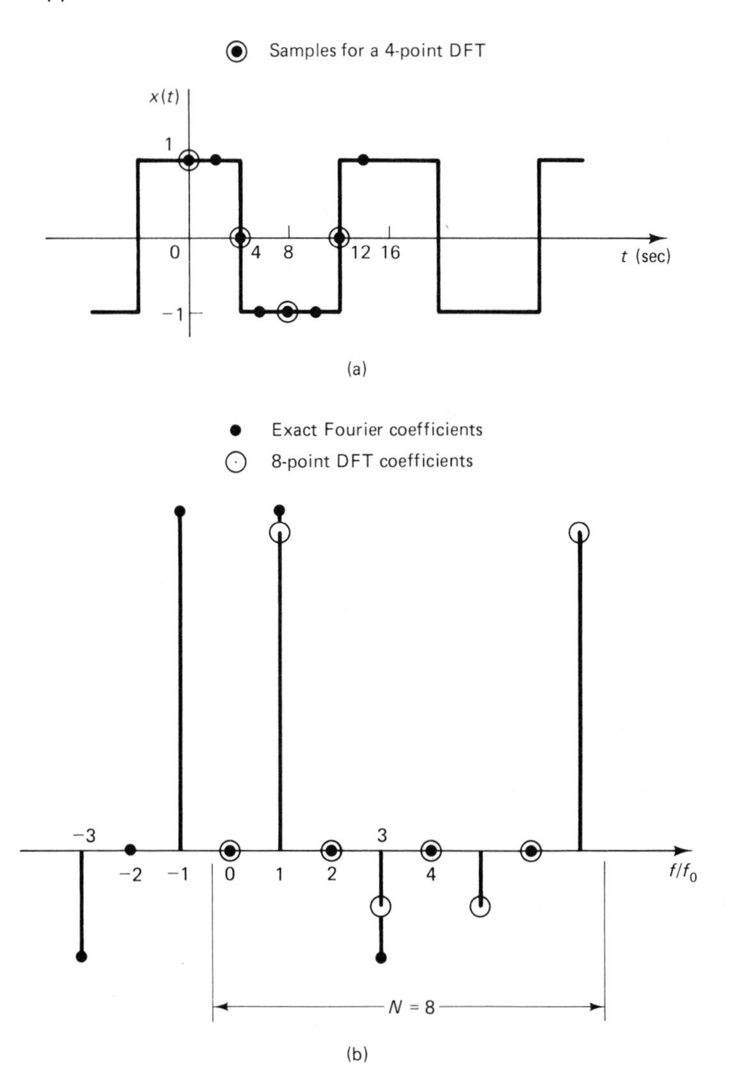

Figure 9.7 (a) Square-wave signal and (b) comparison of Fourier series coefficients and DFT output.

in the DFT, indicating that no even-harmonic terms are present. The approximation to the fundamental component has improved with the larger data set. The results are shown in Fig. 9.6.

As we go to more and more data, two things happen: (1) We begin to "see" more frequency terms, and (2) the lower frequency terms become increasingly more accurate. We conclude this example with a rule of thumb, which states that to obtain n-terms of a Fourier series to a reasonable accuracy, a minimum $8n$-point DFT must be used. The reader should note that at a discontinuity of a function, the value of the function is the midpoint between the two (right and left) limits. Thus, in this example the values at 4 and 12 were taken as zero.

The previous examples may be termed *harmonic analysis*, or finding the Fourier series of a (periodic) waveform. Harmonic synthesis corresponds to the inverse operation of sampling the Fourier transformed signal in frequency (at an integer multiple of the fundamental frequency) in order to synthesize the time-domain signal. In general, the frequency values are complex, and their conjugate images must be reflected about the $N/2$ folding frequency. The resulting operation produces a periodic time signal that corresponds to the truncated Fourier series developed from the original frequency sampling.

Example 9.4 (Fourier Transform)

If we now take a simple signal such as the exponential function $x(t) = e^{-t}u(t)$, which is not periodic, we can sample to obtain the data $\{x_k\}$ and then DFT to find approximate samples of $X(f)$. To illustrate with a 4-point DFT, consider $\Delta t = 1$, so that

$$\{X_n\} = \{0.500, 0.368, 0.135, 0.050\}$$

Note that because of the discontinuity at $t = 0$, the first sample is 0.5 and not 1.0. Applying the DFT, we obtain

$$\{X_n\} = \{1.053, 0.484\underline{/-41.1^\circ}, 0.218, 0.484\ \underline{/41.1^\circ}\}$$

To compare, the exact values should be the samples of $(1 + j2\pi f)^{-1}$ at $f = 0, 0.25, 0.50$ (since $N = 4$ and $\Delta t = 1$, $T = 4$ or $\Delta f = 1/T = 0.25$). From this we have

$$\{1.00, 0.537\ \underline{/-57.5^\circ}, 0.303\ \underline{/-72.3^\circ}\}$$

for which the first three entries of $\{X_n\}$ are rough approximations, particularly the $N/2$ term, X_2. In this case $\Delta t = 1$, but normally the DFT output $\{X_n\}$ would require an additional factor of Δt, as shown in (9.18).

Similarly, when using the DFT to approximate inverse Fourier transforms, one must remember to form the complex conjugate samples following the folding frequency $N/2$. Also, corresponding to (9.18), a factor of $\Delta f = 1/T$ must be included.

Example 9.5 (Inverse Fourier Transform)

In this example we will use the $(\text{DFT})^{-1}$ operation for the same basic signal of the previous example. The input samples are taken from

$$X(f) = \frac{1}{1 + j2\pi f} = \frac{1}{1 + (2\pi f)^2} + j\frac{-2\pi f}{1 + (2\pi f)^2}$$

Again we take $N = 4$ for convenience of illustrating the procedure. Also, taking $\Delta f = 0.5$, we obtain

$$X_0 = X(0) = 1.00$$

$$X_1 = X(0.5) = 0.092 - j0.289$$

$$X_2 = X(1) = 0.025 - j0.155$$

We must take care to represent the folding-frequency term $N/2 = 2$ correctly. Since the X_3 value is the conjugate of X_1, we take $X_2 = 0.025 + j0$. The imaginary part of zero simply recognizes that a discontinuity must occur at that point and, in keeping with proper sampling procedure, takes on the midpoint value between the two limits. Thus, the input data is given by

$$X = \{X_n\} = \{1.000, 0.092 - j0.289, 0.025, 0.092 + j0.289\}$$

The operation of $(\text{DFT})^{-1}$ is given by Definition 9.2, which can be reduced to the matrix operation

$$x = \frac{1}{2}\begin{bmatrix} 1 & 1 & 1 & 1 \\ 1 & j & -1 & -j \\ 1 & -1 & 1 & -1 \\ 1 & -j & -1 & j \end{bmatrix} X$$

which includes the additional factor $1/\Delta t$, making the scalar factor Δf. Performing the indicated calculations, we obtain

$$x = \{0.604, 0.777, 0.420, 0.199\}$$

which approximates the exact samples, that is, for $\Delta t = (N\,\Delta f)^{-1} = 1/2$,

$$\{0.500, 0.607, 0.368, 0.223\} = e^{-(k\,\Delta t)} \qquad \text{for } k = 0, 1, 2, 3.$$

Again we see only gross approximation.

Example 9.6 (Digital Filter Design)

In this example we explore the usefulness of the DFT to design a low-pass digital filter. We assume that the ideal specification is a magnitude frequency-response function as shown in Fig. 9.8a. And we are to design an approximating digital filter of eighth order.

Solution: The corresponding ideal frequency spectrum for the digital filter is given in Fig. 9.8b (since its spectrum must be periodic). Note the change of frequency scale between Fig. 9.8a and b. This was done in order to conveniently use the transform pairs:

$$\text{rect } \pi \hat{f} \leftrightarrow \frac{1}{\pi} \text{ sinc } \frac{t}{\pi}$$

and

$$|\hat{G}| = \text{rect } \pi \hat{f} * \text{comb } \hat{f} \leftrightarrow \frac{1}{\pi} \text{ sinc } \frac{t}{\pi} \cdot \text{comb } t$$

These relations are simply an aid to obtaining the (exponential) Fourier coefficients of the periodic frequency function $\hat{G}$ whose magnitude is given in Fig. 9.8b. Figure 9.8c presents the time-domain coefficients corresponding to $|\hat{G}|$. As indicated by the previous transform relationship, they are available as the "samples" of a sinc function.

To avoid Gibb's phenomenon, since we have a discontinuous ideal function, we multiply the coefficients shown in Fig. 9.8c by a "window" function, which has the effect of smoothing the eventual frequency response. Data windowing is discussed more fully in Section 9.7. The particular choice of window function is a *raised cosine.*

Corresponding to the coefficients -4 to $+4$, the raw coefficients are

$$\{-0.060, 0.015, 0.145, 0.268, 0.318, 0.268, 0.145, 0.015, -0.060\}$$

And the window coefficients are

$$\{0.096, 0.346, 0.655, 0.905, 1.00, 0.905, 0.655, 0.346, 0.096\}$$

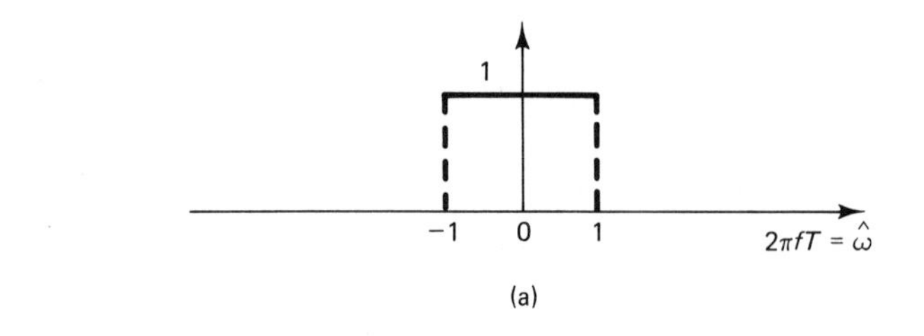

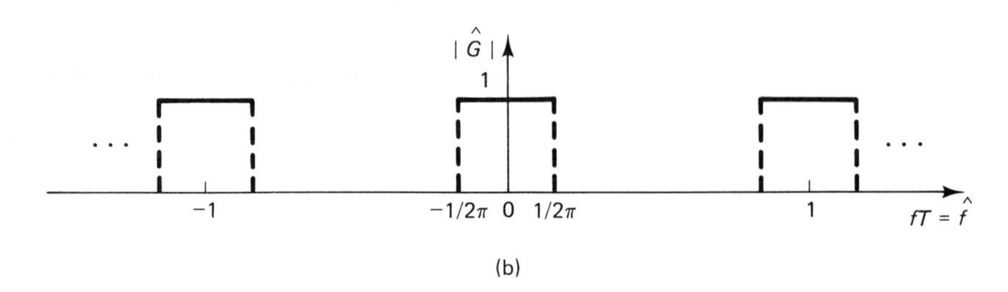

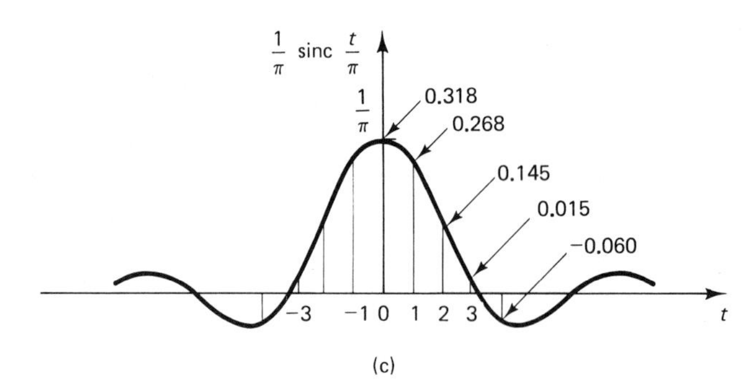

Figure 9.8 (a) Ideal filter response, (b) ideal digital filter response, and (c) Fourier coefficients of $|\hat{G}|$.

The procedure is to multiply point by point these two arrays. Doing this, we arrive at the unit-pulse–response sequence

$$\{h(k)\} = \{-0.006,\ 0.005,\ 0.095,\ 0.242,\ 0.318,\ 0.242,\ 0.095,\ 0.005,\ -0.006\}$$

We interpret this as a causal sequence so that the digital filter is given by

$$H(z) = \sum_{k=0}^{8} h(k)z^{-k}$$

Finally, we can calculate the resulting frequency response by evaluating

$$H(z)\Big|_{z=e^{j\hat{\omega}}}$$

Several points were plotted in Fig. 9.9 showing the actual (designed) frequency response

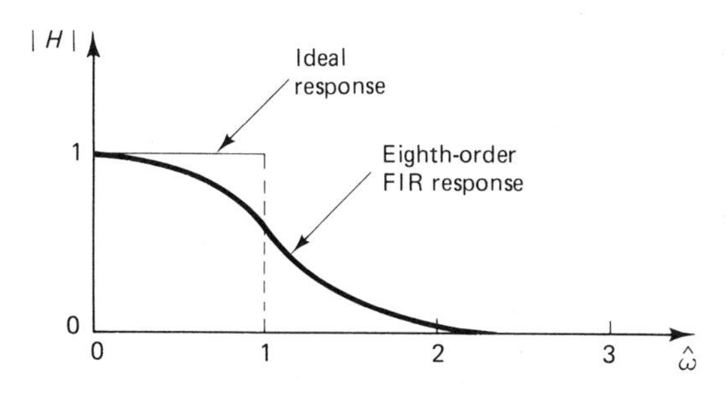

Figure 9.9 Comparison of designed and ideal filter-response functions.

in comparison with the original ideal response function. To achieve a better approximation, we would have to go to a higher-order filter; but, even at eighth order, the response exhibits a distinct low-pass structure. This type of filter is called a finite-impulse–response (FIR) filter since $\{h(k)\}$ is a finite sequence.

From these few examples it is clear that to work with practical problems much more than a 4-point algorithm is required. It is not unusual for a 2^{10}-point or even higher-order DFT to be needed. For jobs of this magnitude, machine calculations are called for. We are already aware that the simple matrix operation requires $N(N-1)$ complex multiplications (not counting multiplying by 1) and the same number of additions to compute N–output points. And since each complex multiplication involves four real multiplications and two real additions as shown below,

$$(a + jb)(c + jd) = (ab - bd) + j(ad + bc)$$

a total of $4N(N-1)$ real multiplications and additions are required. As we shall see, an N-point FFT algorithm (where N is an integer power of 2) requires conservatively $2N \log_2 N$ real multiplications and $3N \log_2 N$ real additions.[6]

Thus, we can see that for large N there can be a sizable computational saving to use one of the more efficient (FFT) algorithms for determining the DFT of an N-sequence. Comparing the multiplications required for each (since they are significantly more time-consuming than additions, in general), we find the following required numbers of real multiplications:

N	DFT	FFT
4	48	16
64	16,128	768
1024	4,190,208	20,480

Thus, while not absolutely necessary for small N, the FFT is mandatory for very large N. In the next section we develop the original FFT algorithm (Cooley-Tukey algorithm).

[6]Many of the multiplications are by ± 1 or $\pm j$, which are much simpler than a general multiplication.

9.5 THE FAST FOURIER TRANSFORM (FFT) ALGORITHMS

Computationally efficient algorithms for calculating the DFT are known collectively as FFTs, and as we have seen in the previous section, for an N-point DFT, where $N > 1000$, an FFT greatly relieves the burden of computation over the direct (matrix) calculation.

There have been two main approaches to the development of FFT algorithms: decimation in time and decimation in frequency. We emphasize the former approach. Recall from Definition 9.1 and expression (9.15) that the DFT of $\boldsymbol{x}$ can be written as

$$\boldsymbol{X} = \boldsymbol{W}\boldsymbol{x} \tag{9.19}$$

where W is a matrix of exponential factors as in (9.16).

To proceed with the development, we represent the indices n and k as binary numbers. For example, for $N = 4$ we require two bits to represent n, where $n = 0, 1, 2, 3$. Therefore, we let

$$k = k_1 k_0 \quad \text{and} \quad n = n_1 n_0 \tag{9.20}$$

where $k_1 k_0$ and $n_1 n_0$ are ordered pairs of binary numbers, i.e.,

$$n_1 n_0 = \{00, 01, 10, 11\} \qquad \text{for } n = \{0, 1, 2, 3\}$$

It can be shown that the matrix operation (9.16) can be factored into the "double" operation

$$\begin{bmatrix} X(00) \\ X(10) \\ X(01) \\ X(11) \end{bmatrix} = \begin{bmatrix} 1 & W^{00} & 0 & 0 \\ 1 & W^{10} & 0 & 0 \\ 0 & 0 & 1 & W^{01} \\ 0 & 0 & 1 & W^{11} \end{bmatrix} \begin{bmatrix} 1 & 0 & W^{00} & 0 \\ 0 & 1 & 0 & W^{00} \\ 1 & 0 & W^{10} & 0 \\ 0 & 1 & 0 & W^{10} \end{bmatrix} \begin{bmatrix} x(00) \\ x(01) \\ x(10) \\ x(11) \end{bmatrix} \tag{9.21}$$

where a binary notation has been used for the powers of W and also where we note that X has been written in the order $X(0)$, $X(2)$, $X(1)$, and $X(3)$, rather than the normal order. This is now a two-stage operation, and we modify the notation so that the input data $\boldsymbol{x}$ become $\boldsymbol{x}_0$, i.e., the original data are denoted with a subscript zero. The result of the first-stage operation we write as $\boldsymbol{x}_1$; thus,

$$\begin{bmatrix} x_1(00) \\ x_1(01) \\ x_1(10) \\ x_1(11) \end{bmatrix} = \begin{bmatrix} 1 & 0 & W^{00} & 0 \\ 0 & 1 & 0 & W^{00} \\ 1 & 0 & W^{10} & 0 \\ 0 & 1 & 0 & W^{10} \end{bmatrix} \begin{bmatrix} x_0(00) \\ x_0(01) \\ x_0(10) \\ x_0(11) \end{bmatrix} \tag{9.22}$$

And the second stage of the two-stage process is given by

$$\begin{bmatrix} X(00) \\ X(10) \\ X(01) \\ X(11) \end{bmatrix} = \begin{bmatrix} 1 & W^{00} & 0 & 0 \\ 1 & W^{10} & 0 & 0 \\ 0 & 0 & 1 & W^{01} \\ 0 & 0 & 1 & W^{11} \end{bmatrix} \begin{bmatrix} x_1(00) \\ x_1(01) \\ x_1(10) \\ x_1(11) \end{bmatrix} \tag{9.23}$$

The individual (scalar) operations in both stages are quite simple, e.g.,

$$x_1(00) = x_0(00) + W^{00}x_0(10) \tag{9.24}$$

or

$$X(00) = x_2(00) = x_1(00) + W^{00}x_1(01)$$

These same matrix operations may also be conveniently represented in a signal-flow graph where each arrow represents multiplication by a factor and each node represents a summation. In Fig. 9.10a we show expression (9.24) represented as a segment of a signal-flow graph. The subscript-zero elements act as inputs in formulating

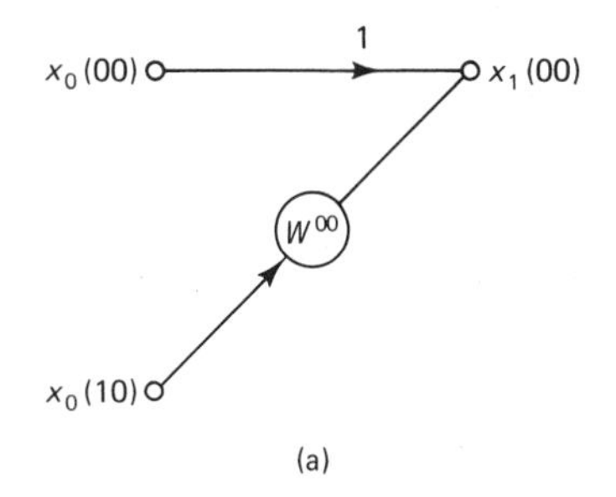

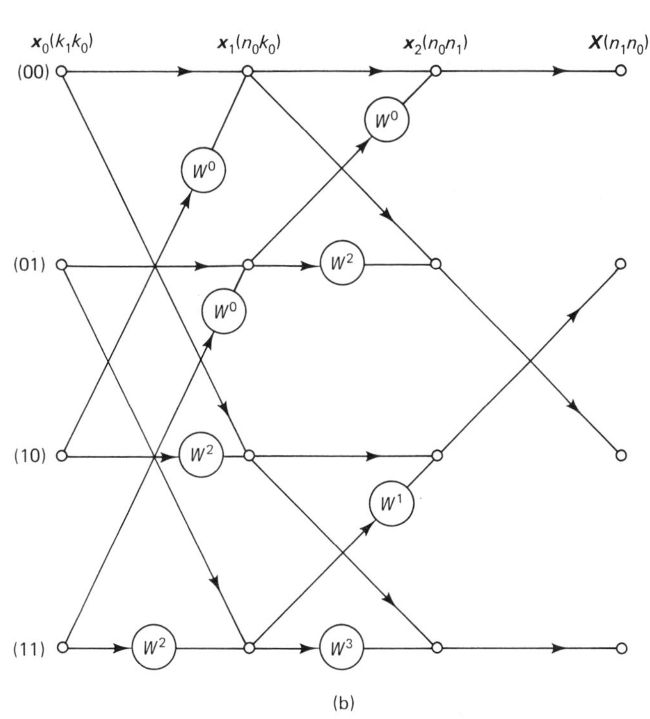

Figure 9.10 (a) Signal-flow graph representing (9.24) and (b) signal-flow graph for a 4-point FFT.

$x_1(00)$ according to (9.24). Combining all the operations indicated in (9.22) and (9.23), we obtain the signal-flow graph of Fig. 9.10b. For simplicity of notation, an unlabeled arrow is taken to be multiplication by unity. In addition, the final stage shows an unscrambling of the output vector, which fortunately only amounts to reversing the bits of the subscript on the components; thus, for instance, $1 = 01$ becomes $2 = 10$. The signal-flow diagram of Fig. 9.10b is a concise representation of the required operations shown in (9.21).

Formal Development

Beginning with (9.15) repeated here as

$$X(n) = \sum_{k=0}^{N-1} x_0(k) W^{nk} \tag{9.25}$$

we take the case where $N = 4$. Using the binary notation for n and k,

$$X(n_1 n_0) = \sum_{k_0=0}^{1} \sum_{k_1=0}^{1} x_0(k_1 k_0) W^{(2n_1+n_0)(2k_1+k_0)} \tag{9.26}$$

$$X(n_1 n_0) = \sum_{k_0=0}^{1} \left[\sum_{k_1=0}^{1} x_0(k_1 k_0) W^{2n_0 k_1} \right] W^{(2n_1+n_0)k_0} \tag{9.27}$$

where we have used the fact that $W^{4n_1k_1} = 1$ (since $N = 4$). The term in brackets is defined as $x_1(n_0\,k_0)$; it does not depend on k_1, the summation index. Finally, we write

$$X(n_1 n_0) = \sum_{k_0=0}^{1} x_1(n_0\,k_0) W^{(2n_1+n_0)k_0} = x_2(n_0\,n_1) \tag{9.28}$$

This is the original Cooley-Tukey formulation of the FFT algorithm, as shown in Fig. 9.10b. It is recommended that the reader trace through the signal-flow diagram

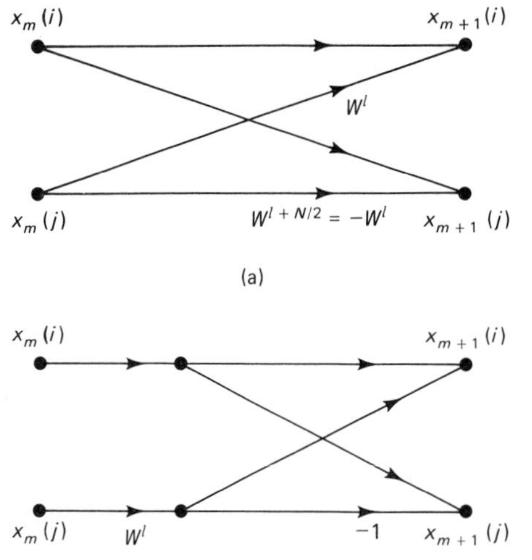

Figure 9.11 (a) Basic butterfly pattern and (b) simplified form.

of Fig. 9.10b using (9.28). This algorithm has been termed *decimation in time.* We say more about this later.

At each stage of the FFT algorithm there are a pair of nodes, called *dual nodes,* which involve a computation known as a "butterfly" because of the geometric pattern exhibited. This butterfly has the mathematical form

$$\begin{aligned} x_{m+1}(i) &= x_m(i) + W^l x_m(j) \\ x_{m+1}(j) &= x_m(i) + W^{l+N/2} x_m(j) \end{aligned} \tag{9.29}$$

where the integers i and j denote the sample numbers involved and l is an integer power of W. Figure 9.11 shows two forms of the basic butterfly computation. And Fig. 9.12 illustrates an 8-point FFT using the simplified butterfly of Fig. 9.11; see

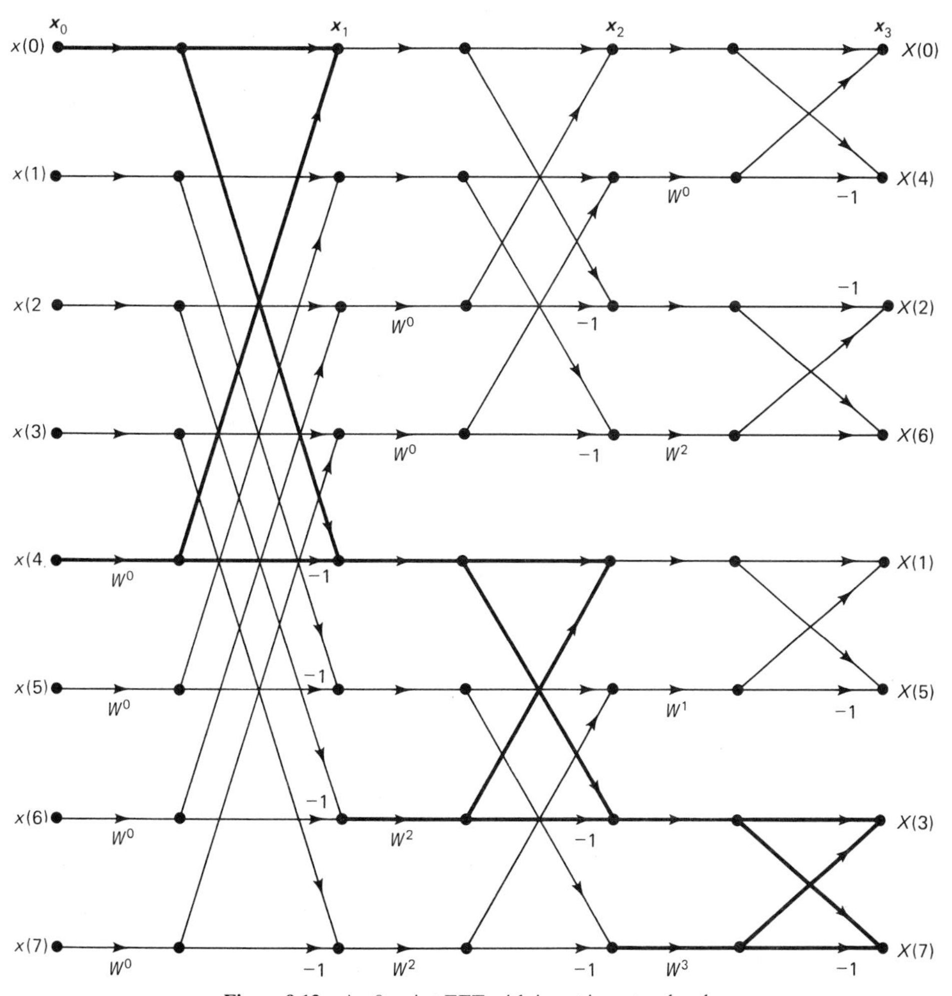

Figure 9.12 An 8-point FFT with input in natural order.

that nodes $x_1(0)$ and $x_1(4)$ are dual because they are calculated from $x_0(0)$ and $x_0(4)$ and no other nodes are concerned with $x_0(0)$ and $x_0(4)$. Similarly, the reader can visualize other butterfly computations shown in heavy lines on Fig. 9.12. The butterfly is important in that the entire FFT algorithm is a combination of appropriate butterfly computations.

The 8-point algorithm of Fig. 9.12 would require a 3-bit binary number to represent the indices. Consequently, the formal (binary) development would be slightly more involved, e.g., expression (9.26) would have 3 summations over the binary digits. We will not, however, take this approach to develop the general base-2 algorithm, as the reader already has a basic understanding of the algorithm from the previous 4-point development.

In order to generalize our results, let us show that the calculations for an N-point DFT can be reduced to calculations on two $N/2$-point DFTs. With this result we can see that each stage (see Fig. 9.12) is simply a decomposition into calculations on one-half of the previous stage elements until, at the final stage, a series of only 2-point DFTs are being made. To do this, we begin by splitting the summation of (9.15) into two parts.

$$X(n) = \sum_{k=0}^{(N/2)-1} x(k)W_N^{kn} + \sum_{k=N/2}^{N-1} x(k)W_N^{kn} \tag{9.30}$$

Introducing $m = k - N/2$ into the second sum,

$$X(n) = \sum_{k=0}^{(N/2)-1} x(k)W_N^{kn} + W_N^{(N/2)n} \sum_{m=0}^{(N/2)-1} x\left(m + \frac{N}{2}\right)W_N^{mn}$$

Since $W_N^{(N/2)n} = (-1)^n$, we obtain

$$X(n) = \sum_{k=0}^{(N/2)-1} \left[x(k) + (-1)^n x\left(k + \frac{N}{2}\right)\right]W_N^{nk} \tag{9.31}$$

We now consider even values and odd values of n separately. For this purpose we use $n = 2m$ and $n = 2m + 1$ to represent even and odd values of n, respectively (m is integer-valued). For even n

$$X(2m) = \sum_{k=0}^{(N/2)-1} \left[x(k) + x\left(k + \frac{N}{2}\right)\right]W_N^{2km} \tag{9.32}$$

and for odd n

$$X(2m + 1) = \sum_{k=0}^{(N/2)-1} \left[x(k) - x\left(k + \frac{N}{2}\right)\right]W_N^{k} W_N^{2km} \tag{9.33}$$

where $m = 0, 1, \ldots, (N/2) - 1$. But $W_N^{2km} = W_{N/2}^{km}$, so that (9.32) and (9.33) are seen to be two $(N/2)$-point DFTs. Figure 9.13 shows this result graphically. By this process, the two $(N/2)$-point DFTs can, in turn, be replaced by four $(N/4)$-point DFTs and so on until the computation involves only 2-point DFTs. Strictly speaking, this development is known as *decimation in frequency*, since the frequency-domain sequence X is decimated or divided into subsequently smaller parts. There

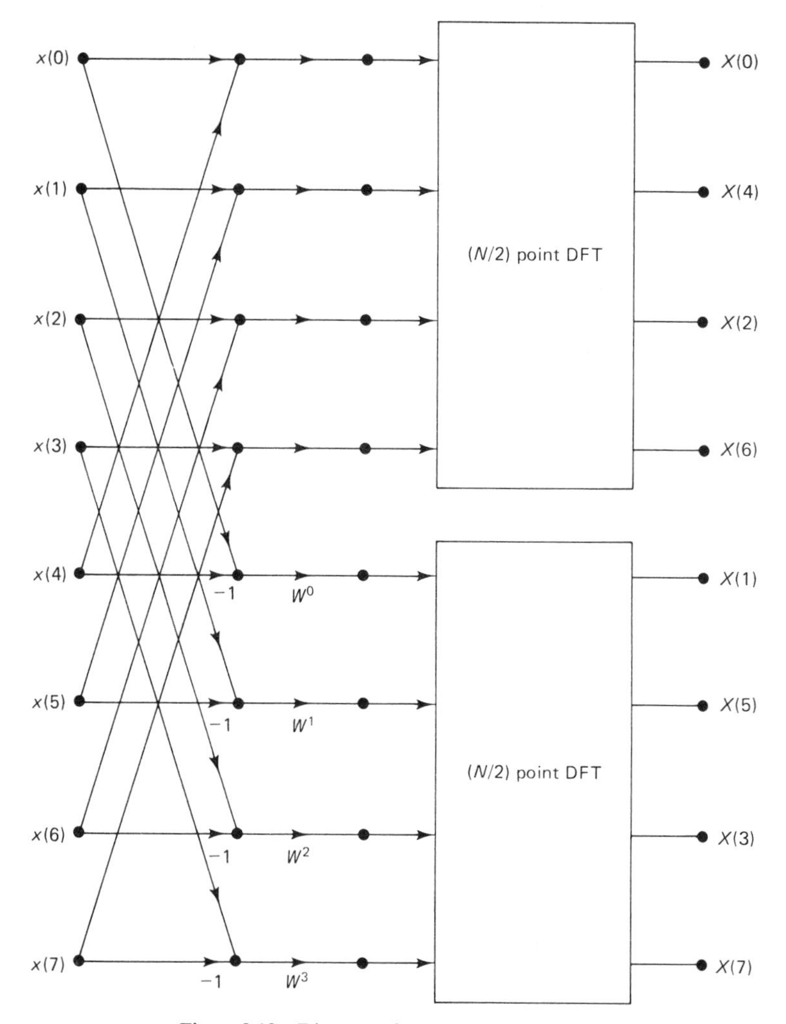

Figure 9.13 Diagram after a single decimation.

are, however, so many slight variations of FFT that referring to decimation in time or frequency is not very helpful. For instance, our binary development was basically decimation in time (see Fig. 9.10). But, by rearranging the diagram so that the output is in its natural order and the input is in bit-reversed order, the algorithm appears very different, as shown in Fig. 9.14.

Example 9.7 (FFT Calculation)

In this example we rework the DFT of Example 9.3 where the input data is given by

$$\mathbf{x}_0 = [1 \quad 1 \quad 0 \quad -1 \quad -1 \quad -1 \quad 0 \quad 1]^T$$

by following the signal-flow diagram of Fig. 9.12. Note that the butterfly computation

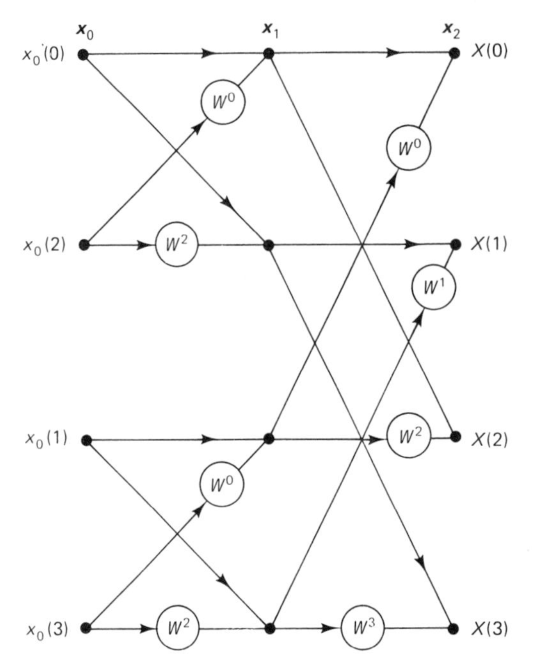

Figure 9.14 Signal-flow graph for a 4-point FFT, input in bit-reversed form.

may be used effectively. Thus, for example,

$$x_1(0) = 1 - 1 = 0$$

$$x_1(4) = 1 + 1 = 2$$

Completing the first stage, we have

$$\boldsymbol{x}_1 = [0 \quad 0 \quad 0 \quad 0 \quad 2 \quad 2 \quad 0 \quad -2]^T$$

Similarly, the results of computing the second and third stages are

$$\boldsymbol{x}_2 = [0 \quad 0 \quad 0 \quad 0 \quad 2 \quad 2+2j \quad 2 \quad 2-2j]^T$$

and

$$\boldsymbol{x}_3 = [0 \quad 0 \quad 0 \quad 0 \quad 2+2\sqrt{2} \quad 2-2\sqrt{2} \quad 2-2\sqrt{2} \quad 2+2\sqrt{2}]^T$$

where we have used the values of W_8^k as obtained from the unit-circle diagram of Fig. 9.15. Unscrambling the output, we obtain the final result:

$$\boldsymbol{X} = [0 \quad 2+2\sqrt{2} \quad 0 \quad 2-2\sqrt{2} \quad 0 \quad 2-2\sqrt{2} \quad 0 \quad 2+2\sqrt{2}]^T$$

which checks our result in Example 9.3.

Up to this point we have concentrated on the decimation-in-time approach; however, there is a dual approach called *decimation-in-frequency approach.* The developments follow in a manner similar to those we have studied. For reference, the signal-flow graph for an 8-point decimation-in-frequency algorithm is included

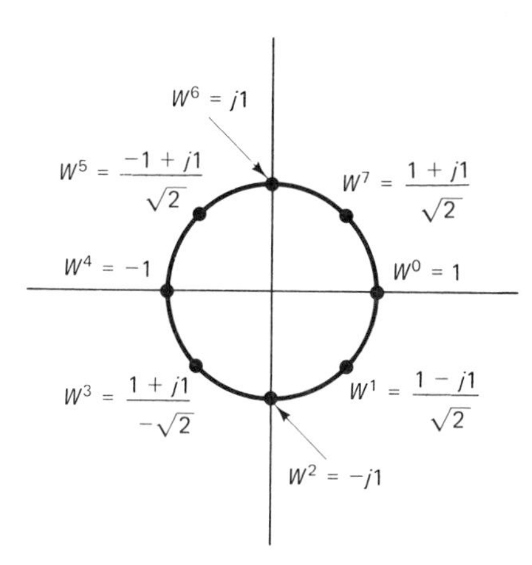

Figure 9.15

(see Fig. 9.16). In this case the input data are in proper order, but the output arrives in bit-reversed order, as was the case for the decimation-in-time FFT of Fig. 9.12.

The basic difference between decimation in time (Fig. 9.12) and decimation in frequency (Fig. 9.15) can be seen as a different method of computing the butterfly. Thus, for the decimation-in-frequency approach, the simplified butterfly pattern is given by

$$\begin{aligned} x_{m+1}(i) &= x_m(i) + x_m(j) \\ x_{m+1}(j) &= [x_m(i) - x_m(j)]W^l \end{aligned} \tag{9.34}$$

Comparing (9.34) with Fig. 9.11b serves to show the fundamental distinction. Note that the form of (9.34) is illustrated in the first stage of Fig. 9.13 since that development was carried out as decimation in frequency. Just as Fig. 9.14 indicates, the identical computation of the FFT illustrated in Fig. 9.10b, there are many ways in which the nodes may be moved around to obtain seemingly different algorithms. For instance, the output nodes may be moved to their natural order, making the corresponding flow graph appear more chaotic. In addition, there are many variations on the two basic schemes, including algorithms that do not require that the number of data points be an integer power of 2. But we will not discuss these developments. Rather, let us consider how the FFT can be used to calculate convolutional sums and correlations.

Inverse FFT

At this point we have discussed many facets of different FFT algorithms. It is useful to point out explicitly how the $[\text{FFT}]^{-1}$ can be implemented in terms of our previous discussion. Recall that the $[\text{DFT}]^{-1}$ of $\{X(n)\}$ is given by

$$x(k) = \frac{1}{N} \sum_{n=0}^{N-1} X(n) W^{-kn} \tag{9.35}$$

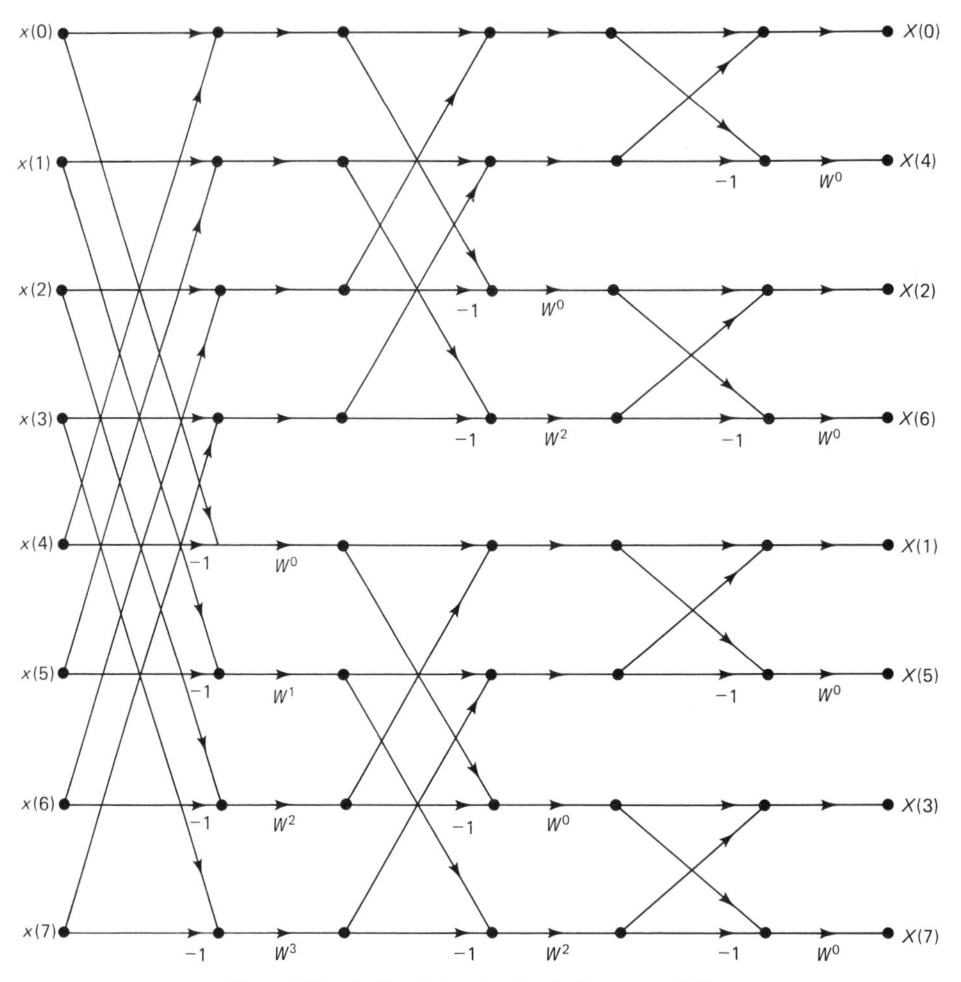

Figure 9.16 An 8-point decimation-in-frequency FFT.

for $k = 0, 1, 2, \ldots, N - 1$. Thus, to convert an FFT algorithm to an inverse FFT algorithm, we simply divide by N and use powers of W^{-1} instead of powers of W!

Alternatively, we can invert the basic decimation-in-time butterfly computation of (9.29), i.e., solve for $x_m(i)$ and $x_m(j)$, to obtain the basic butterfly pattern of an inverse FFT, that is,

$$\begin{aligned} x_m(i) &= \frac{1}{2}\,[x_{m+1}(i) + x_{m+1}(j)] \\ x_m(j) &= \frac{1}{2}\,[x_{m+1}(i) - x_{m+1}(j)]W^{-l} \end{aligned} \tag{9.36}$$

Using (9.36) for construction, we can derive the inverse FFT shown in the flow

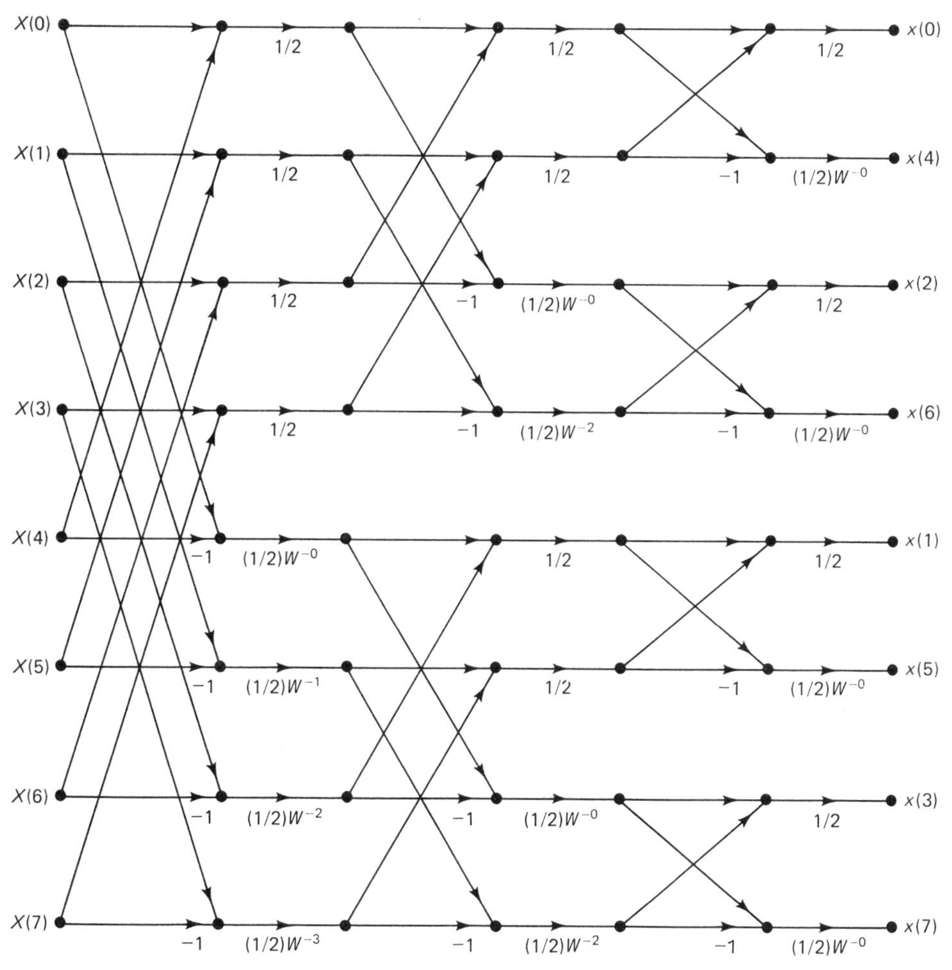

Figure 9.17 An 8-point inverse FFT.

graph of Fig. 9.17, which, when a factor of $\frac{1}{2}$ for each stage is removed, appears identical to the direct (decimation-in-frequency) FFT of Fig. 9.16 except for corresponding negative powers of W.

9.6 DFT PROPERTIES AND FAST CONVOLUTION

In this section we study how the FFT can be used to aid in the calculation of discrete-time signal processing. As we begin to think of the DFT as a "transform" that the computer can help us to apply, we become more interested in its general properties. Table 9.2 lists for reference a few of the more useful properties that can be proved for the DFT.

TABLE 9.2 SELECTED PROPERTIES OF THE DFT

Property	Time Sequence	Frequency Sequence
1. Linearity	$ax(k) + by(k)$	$aX(n) + bY(n)$
2. Time shift	$x(k - k_0)$	$X(n)W^{nk_0}$
3. Frequency shift	$x(k)W^{-kn_0}$	$X(n - n_0)$
4. Product	$x(k)y(k)$	$\frac{1}{N}\sum_{m=0}^{N-1} X(m)Y(n-m)$
5. Circular convolution	$\sum_{m=0}^{N-1} x(m)y(k-m)$	$X(n)Y^*(n)$
6. Parseval's theorem	$\sum_{k=0}^{N-1} x^2(k) = \frac{1}{N}\sum_{n=0}^{N-1} \lvert X(n)\rvert^2$	

Referring to Table 9.2, entry 5 indicates a property entitled *circular convolution*. We are very familiar with discrete convolution of two sequences (see Section 4.5). To distinguish between the two types of convolution, let us refer to ordinary convolution as *linear convolution.*

Example 9.8 (Linear Convolution)

As a quick review of discrete-time convolution, let us assume that

$$x = \{0, 1, 2, 3\}$$
$$h = \{2, 1, 1, 2\}$$

to calculate $y = h * x$. Recall that the convolution summation is given by

$$y(k) = \sum_{n=-\infty}^{\infty} h(n)x(k-n) \tag{9.37}$$

where implicit in our assumption both $h(k)$ and $x(k)$ are zero except for $k = 0, 1, 2$, and 3. Using the graphical aid of Fig. 9.18, we implement (9.37) as follows:

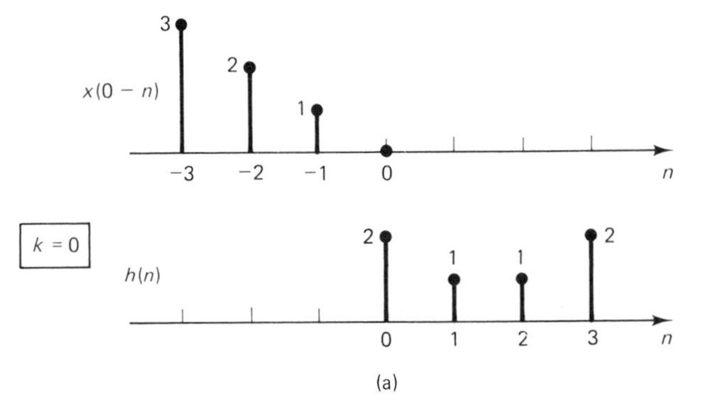

Figure 9.18

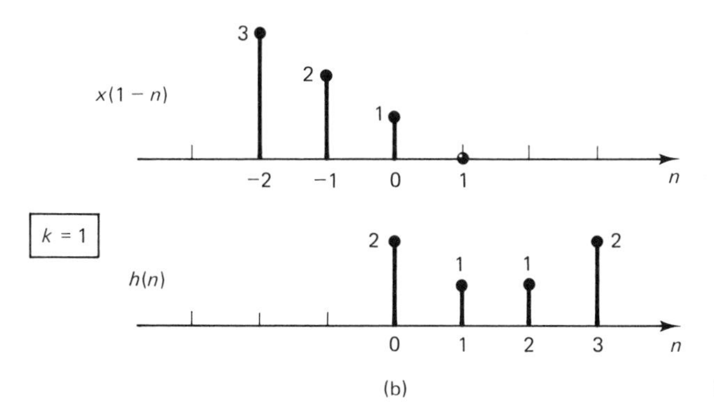

Figure 9.18 (continued)

Continuing conceptually for $k = 2, 3, \ldots, 6$, we construct the result of the linear convolution between the two finite sequences h and x:

$$y(k) = \{0, 2, 5, 9, 7, 7, 6\} \tag{9.38}$$

Circular Convolution

The z-transform discussed in Chapter 6 allowed us to bypass the convolution process by equivalently taking the product of the transforms of the two sequences in the z-domain.The corresponding property of the DFT, unfortunately, does not relate the product in frequency to the linear convolution in time, as would be desirable. The fact that the DFT's product corresponds to a circular convolution forces us to be interested in this slightly modified form of convolution.

As we have previously seen, the DFT operates as if the time and frequency sequences were both periodic. This fact is the key to circular or cyclic convolution. The *circular convolution* between h and x is given by the expression $y_c(k)$,

$$y_c(k) = \sum_{m=0}^{N-1} h(m)x(k-m) \tag{9.39}$$

where both h and x are taken to be periodic with period N. To illustrate, let us calculate the circular convolution between the two sequences given in the previous example.

Example 9.9 (Circular Convolution)

Given the two sequences of length N,

$$x = \{0, 1, 2, 3\}$$

$$h = \{2, 1, 1, 2\}$$

calculate $y_c = h \circledast x$, the circular convolution of h and x.

First we show x and h repeated periodically.

x:	$\cdots$	0	1	2	3	0	1	2	3	0	1	2	3	0	1	2	3	$\cdots$
h:	$\cdots$	2	1	1	2	2	1	1	2	2	1	1	2	2	1	1	2	$\cdots$

But implementing (9.39) requires that x be transposed or reversed in time. Thus, for various values of k, we have the following overlapping sequences to product and sum (over one period):

h:		$\cdots$				2	1	1	2	$\cdots$
x:	$(k = 0)$	$\cdots$	3	2	1	0	3	2	1	$\cdots$
x:	$(k = 1)$	$\cdots$	0	3	2	1	0	3	2	$\cdots$
x:	$(k = 2)$	$\cdots$	1	0	3	2	1	0	3	$\cdots$
x:	$(k = 3)$	$\cdots$	2	1	0	3	2	1	0	$\cdots$
x:	$(k = 4)$	$\cdots$	3	2	1	0	3	2	1	$\cdots$

Performing the calculations,

$$y_c(k) = \{7, 9, 11, 9\} \tag{9.40}$$

Note the y is of length $N = 4$ and that y is periodic. It is also clear that y_c is not the same as y in (9.38), even though x and h are the same. A conceptual way of performing circular convolution is to imagine that the two given sequences are represented (equally spaced) around the circumferences of two "wheels," one in reversed order, and to form the product and sum of adjacent sequence values as one circle is rotated relative to the other. For the previous example, this would appear as in Fig. 9.19a for $k = 0, 1, 2$.

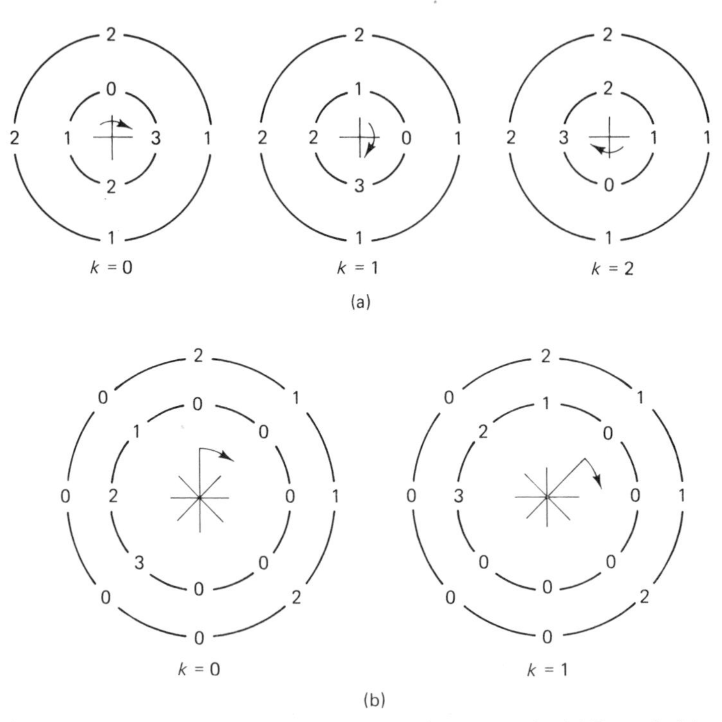

Figure 9.19 Illustration of the circular convolution process for (a) Example 9.9 and (b) Example 9.10.

Now that we have begun to understand circular convolution, we will attempt to find a way of using circular convolution to perform the more practical linear convolution. To do this, consider that two finite-length sequences of lengths M and N will (linearly) convolve into a sequence of length $M + N - 1$ elements; this larger value must be our basic sequence size. To avoid having undesired overlapping elements, we must "pad" our original sequences with zeros. To illustrate this process, consider the following example.

Example 9.10 (Linear Convolution Using Circular Convolution)

Following our previous examples, we define

$$\hat{x} = \{0 \quad 1 \quad 2 \quad 3 \quad 0 \quad 0 \quad 0 \quad 0\}$$

$$\hat{h} = \{2 \quad 1 \quad 1 \quad 2 \quad 0 \quad 0 \quad 0 \quad 0\}$$

where x and h of the previous example have had sufficient extra zeros added to prevent undesired overlapping elements. Referring to Fig. 9.19b, we obtain the circular convolution result

$$\hat{y} = \{0, 2, 5, 9, 7, 7, 6, 0\} \tag{9.41}$$

Note that the nonzero elements of (9.38) have been faithfully reproduced with circular convolution.

The above example shows us that linear convolution can be performed as circular convolution if we are careful in formulating the problem. This implies that the FFT can be used to help us calculate linear system operations. We summarize this process to calculate $y = h \circledast x$ as follows:

1. Add zeros to x and h to form $\hat{x}$ and $\hat{h}$. Enough zeros are added to each sequence to prevent "wraparound" error and to extend both sequences to a length $N = 2^n$ for some integer n.
2. Compute $\hat{X}$ and $\hat{H}$ (using an FFT algorithm).
3. Calculate the product $\hat{Y} = \hat{H} \cdot \hat{X}$ (point by point).
4. Invert $\hat{Y}$ by an FFT algorithm.

This procedure is often referred to as *fast convolution.*

To complete this development, we include a method of using fast convolution to process information that may arrive as a very long sequence.

Data Sectioning

Here we restrict our discussion to a method described as the *overlap-add method.* It is assumed that a continuing stream of data x is to be processed by a filter with a unit-pulse response of length M. Decomposing the input x into nonoverlapping sections of length L, we obtain

$$x(k) = \sum_{i=0}^{\infty} x_i(k) \tag{9.42}$$

where

$$x_i(k) = \begin{cases} x(k) & \text{for } iL \le k < (i+1)L \\ 0 & \text{otherwise} \end{cases}$$

the desired output $y = h * x$ can then be written as

$$y(k) = h * \sum_{i=0}^{\infty} x_i = \sum_{i=0}^{\infty} h * x_i = \sum_{i=0}^{\infty} y_i \qquad (9.43)$$

where

$$y_i(k) = h(k) * x_i(k)$$

Since the length of y_i is $L + M - 1$, expression (9.43) indicates that end parts of a given section must add on to preceding and succeeding sections. Figure 9.20a presents the idea conceptually, showing that the output stream y is the sum of overlapping sections, each of which is itself a convolution. If an N-point FFT algorithm is to be used to do the section convolutions, N must be at least $M + L - 1$. Typically, if the filter response M is fixed and $N = 2^n$ is a convenient

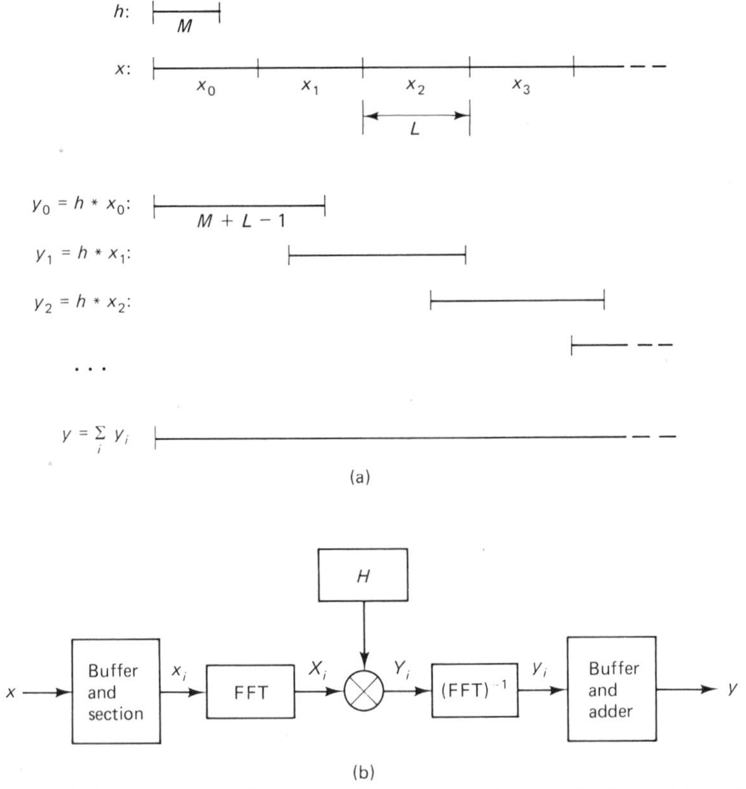

Figure 9.20 (a) Overlap-add method of sectioning and processing input data and (b) block diagram showing fast convolution used with the overlap-add method.

length for the FFT, then the input section length L is chosen to be $N - M + 1$. Consider the following example for illustration.

Example 9.11 (Overlap-Add Method of Input Sectioning)

Given the simple filter specified by

$$h = \{1, 2\} \qquad \text{for } M = 2$$

and assuming that a 4-point FFT is to be used, calculate $y = h * x$, where x is the "long" sequence

$$x = \{1, -2, 3, 0, -1, 2\}$$

We take $L = N - M + 1 = 4 - 2 + 1 = 3$, so that (extending the sections to length $N = 4$), we have

$$h = \{1, 2, 0, 0\}, \quad x_0 = \{1, -2, 3, 0\}, \quad \text{and} \quad x_1 = \{0, -1, 2, 0\}$$

The corresponding DFTs are given by

$$H = \{3, 1 - j2, -1, 1 + j2\}$$

$$X_0 = \{2, -2 + j2, 6, -2 - j2\}$$

$$X_1 = \{1, -2 + j1, 3, -2 - j1\}$$

Assuming H has been precalculated, and that the input and output are suitably buffered and synchronized, Fig. 9.20b shows the procedure in block diagram form. For our example,

$$Y_0 = \{6, 2 + j6, -6, 2 - j6\}$$

$$Y_1 = \{3, j5, -3, -j5\}$$

and

$$y_0 = \{1, 0, -1, 6\}$$

$$y_1 = \{0, -1, 0, 4\}$$

The reader should check y_0 and y_1 in two ways: as indicated by $(\text{DFT})^{-1}$ and also by convolution. Overlapping y_0 and y_1 (by one element) and adding, we obtain

$$y = \{1, 0, -1, 6, -1, 0, 4\}$$

as desired.

Although Fig. 9.20b appears quite formidable to implement, it has been found to be more efficient than straight convolution. For the remaining section of this chapter, we return to the basic ideas of data truncation.

9.7 DATA WINDOWS

Throughout our previous discussion, truncation of data always occurred in a rectangular fashion, i.e., the data was simply clipped off at each end (see the truncation window of Fig. 9.2). Such an abrupt truncation causes large "ripples" in frequency such as indicated in Fig. 9.3c. Looking back at our development, we see that the rectangular truncation to a data length of T seconds corresponds to a convolution of the frequency version of the data with a sinc Tf function. See expression (9.12). When we look in detail at the sinc function, we note that the maximum value is at $f = 0$, but that other "peaks" exist. These side peaks, often called *sidelobes*, are principally the cause of the undesired ripples in frequency mentioned before.

The key idea here is that the disturbance (in frequency) caused by truncation (in time) can be minimized by using smooth, nonrectangular windows that have no discontinuities at the beginning and end. A number of special window functions have come into use as we shall discuss subsequently. Each window design is a particular compromise between a narrow mainlobe and low sidelobes, which are inherently conflicting goals. Figure 9.21 illustrates the difference between rectangular truncation and "raised-cosine" truncation, i.e., the figure shows the two window functions and their corresponding transforms. The transform functions of Fig. 9.21

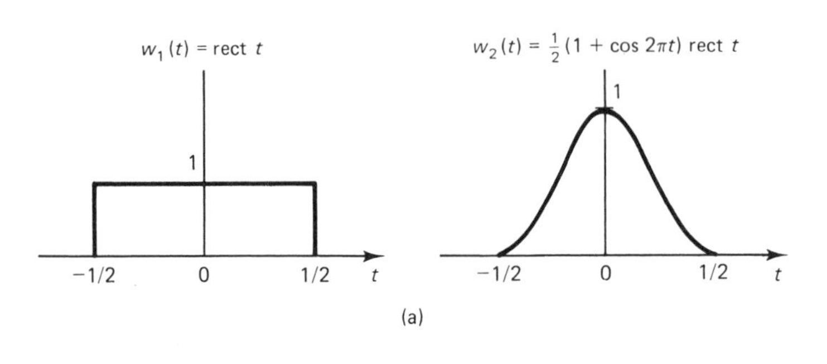

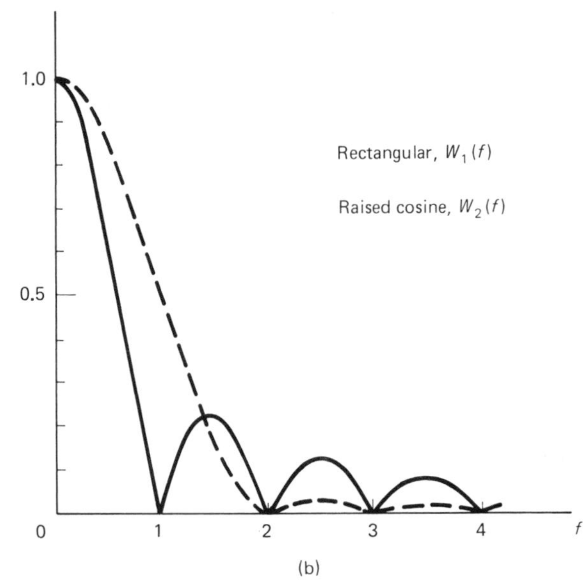

Figure 9.21 Comparison of rectangular and raised-cosine window functions: (a) windows and (b) spectra.

exhibit the effect of reducing the sidelobe level at the expense of broadening the mainlobe. Since the functions are all real and even, only the positive halves of the window spectra have been shown. It is, without doubt, an interesting phenomenon that by "chopping down" the input data in exactly the right way (in this case, multiplying by a raised cosine) that the effect of the "spectral leakage," or shifting of energy from one frequency band to another, can be reduced. However, understanding this effect is an elementary matter of considering the convolution in frequency of one function with another that exhibits large sidelobes. By using a properly designed window function on the data, then, one source of error, mentioned earlier in connection with using a finite length of data, is reduced. The penalty paid is a smearing effect (loss of frequency resolution) caused by convolution with a broad mainlobe.

In the context of windowing discrete data of length N, we list in Table 9.3 the elements of certain window functions at the discrete-time instants with $w(0)$ normalized to unity, and include a measure of the highest sidelobe, calculated in decibels (db), i.e.,

$$\text{Peak sidelobe (db)} = 20 \log_{10} \frac{W(f_p)}{W(0)} \tag{9.44}$$

where f_p is the frequency at which the highest (usually the first) sidelobe peak occurs. In addition to rectangular, triangular, and raised-cosine windows, Table 9.3 includes a popular window function called the *Hamming window* in which the two coefficients of the raised-cosine window were optimally adjusted to reduce the sidelobe level

TABLE 9.3 SELECTED WINDOW FUNCTIONS

1. Rectangular: $w_1(k) = 1 \quad \text{for } k = 0, 1, \ldots, N-1$

   Peak sidelobe = 0.2172 = −13.26 db
   Mainlobe half-amplitude width[a] = 1.0

2. Triangular: $w_2(k) = \begin{cases} \dfrac{2}{N+1}(1+k) & \text{for } 0 \le k \le \dfrac{N-1}{2} \\ \dfrac{2}{N+1}(N-k) & \text{for } \dfrac{N-1}{2} < k \le N-1 \end{cases}$

   Peak sidelobe = 0.04719 = −26.52 db
   Mainlobe half-amplitude width[a] = 1.5

3. Raised Cosine: $w_3(k) = \dfrac{1}{2} + \dfrac{1}{2}\cos\dfrac{\pi(2k-N+1)}{N+1} \quad \text{for } k = 0, 1, 2, \ldots, N-1$

   Peak sidelobe = 0.02671 = −31.47 db
   Mainlobe half-amplitude width[a] = 1.7

4. Hamming window: $w_4(k) = 0.54 + 0.46\cos\dfrac{\pi(2k-N+1)}{N+1} \quad \text{for } k = 0, 1, 2, \ldots, N-1$

   Peak sidelobe = 0.006628 = −43.57 db
   Mainlobe half-amplitude width[a] = 1.5

[a]These values are normalized to the value for rectangular truncation to permit relative comparison.

without significantly widening the mainlobe. Notice the significant reduction in sidelobe level. There are many other window functions of a more complicated form that are available but are not discussed here. For more information in this area, refer to any one of the many specialized texts on digital signal processing. We conclude with a simple example that uses the Hamming window.

Example 9.12: (Data Windowing)

In an earlier example of extracting the Fourier series of a symmetric square wave (Example 9.3), we were given the raw data

$$x = [1 \quad 1 \quad 0 \quad -1 \quad -1 \quad -1 \quad 0 \quad 1]^T$$

To show how the windowing is used, we modify the data using the Hamming window (entry 4 of Table 9.3), to obtain

$$x' = [0.188 \quad 0.460 \quad 0 \quad -0.972 \quad -0.972 \quad -0.770 \quad 0 \quad 0.188]^T \tag{9.45}$$

For instance, the first element ($k = 0$) is found as follows:

$$x'(0) = 1\left[0.54 + 0.46 \cos \frac{\pi}{9}(-7)\right] = 0.188$$

Thus (9.45) is a properly windowed set of data with which to work. It is not worthwhile to complete this example, since the windowing serves no useful purpose and, in fact, introduces some asymmetry into the data. However, as a working tool, windowing is an important element of filtering and spectral estimation. In addition, the basic concept finds application in such diverse areas as optics and antenna design.

9.8 PROBLEMS

9.1. Sketch the pulse-type signals given in Fig. P9.1 and label the origin of the t-axis so that the natural symmetry of the pulse is shown. State whether the symmetry is even or odd.

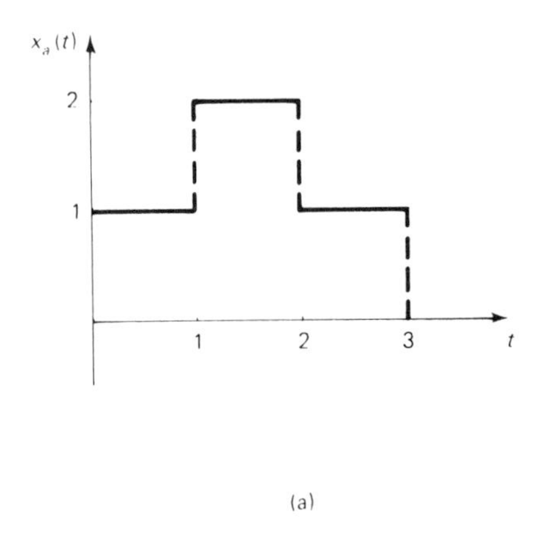

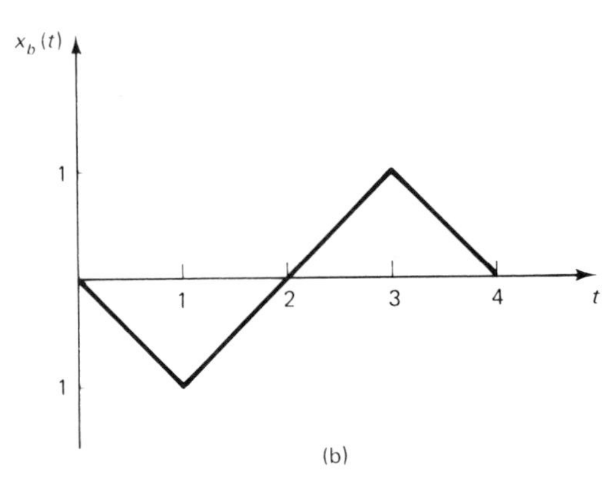

Figure P9.1

9.2. Show that if $x(t)$ has even symmetry, the Fourier transform can be written as

$$X(j\omega) = 2\int_0^{\infty} x(t)\cos\omega t\, d\omega$$

9.3. Sketch the spectrum of comb t exp $(-10|t|)$. What can be said about the spectrum of an ideally sampled time signal?

9.4. Sketch the time-domain signal that corresponds to a periodic version of $X(f)$ in entry 5 of Fig. 9.1 if the period is 20 Hz. Neglect any overlap of the tails.

9.5. Explain the result of truncation of a time-domain signal as it affects the frequency-domain description.

9.6. Describe with sketches the possible spectra resulting from a product of comb t with $\cos\omega t$ as ω varies.

9.7. Using equation (9.15), calculate the 4-point DFT of time-domain sequence $\{1, 1, -1, -1\}$.

9.8. Demonstrate the relation between the DFT and the z-transform using the 4-point sequence given in Problem 9.7.

9.9. For each pair of sequences given in Fig. P9.9, perform the ordinary discrete-time convolution $x * h$.

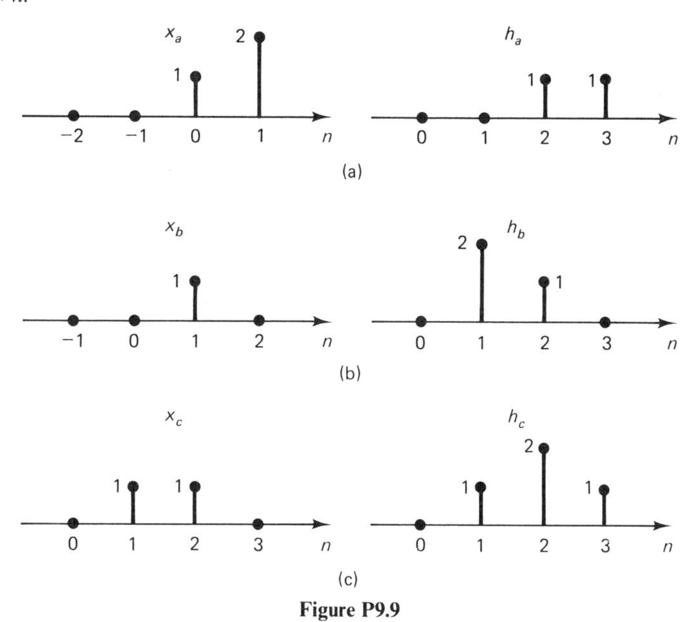

Figure P9.9

9.10. Determine the circular convolutional sum $x \circledast h$ for each pair given in Fig. P9.9.

9.11. Given that the unit-pulse response of a discrete-time system is $h(n) = (0.5)^n u(n)$, determine and sketch the unit-step response by direct convolution.

9.12. The Fourier transform of a discrete-time signal may be defined as

$$H(e^{j\omega}) = \sum_{n=-\infty}^{\infty} h(n)e^{-j\omega n}$$

with the inverse transform given by

$$h(n) = \frac{1}{2\pi} \int_{-\pi}^{\pi} H(e^{j\omega}) e^{j\omega n} \, d\omega$$

Calculate and sketch the Fourier transform of the 4-point sequence given in Problem 9.7. How does the Fourier transform of a sequence relate to the (doubled-sided) z-transform of that sequence?

9.13. Using the defined Fourier transform from Problem 9.12,

(a) Solve Problem 9.11 using the basic transform relation that $Y(e^{j\omega}) = H(e^{j\omega})X(e^{j\omega})$.

(b) Repeat the solution using z-transforms.

9.14. Plot $|X(e^{j\omega})|$ in decibels versus ω when $x(n) = 0.5[1 + \cos(\pi n/4)]$ for $|n| \leq 4$ and $x(n) = 0$ otherwise. Use

$$|X|_{db} = 20 \log_{10} |X|$$

9.15. Calculate the magnitude-squared function $|H|^2$ if $h(n) = \delta(n) - \delta(n - N)$. Sketch for $N = 4$. This filter is sometimes referred to as a *comb filter*.

9.16. Draw a block diagram of the basic butterfly computation in expression (9.19).

9.17. One method of designing a digital filter to approximate a particular frequency response is to modify the Fourier coefficients of the filter magnitude function.

(a) Calculate the first five (exponential) Fourier coefficients for the low-pass filter shown in Fig. P9.17.

(b) Multiply the coefficients obtained in part (a) by the raised-cosine window

$$W_n = \frac{1}{2}\left(1 + \cos \frac{n\pi}{\frac{N}{2} + 1}\right) \qquad \text{for } |n| \leq \frac{N}{2}$$

for $N = 4$.

(c) Shift the results of part (b) to form causal unit-pulse response sequences for the filters.

(d) Repeat parts (a), (b), and (c) for an eighth-order filter, $N = 8$.

(e) Plot the frequency responses for the two filters.

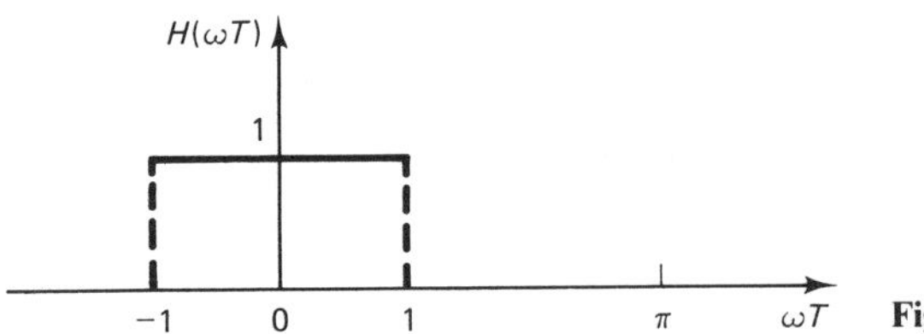

Figure P9.17

9.18. Show that the 4-point inverse DFT of $F = \{0, 2(1 - j), 0, 2(1 + j)\}$ is $f = \{1, 1, -1, -1\}$

9.19. The 8-point sampling of the filter function of Fig. P9.17 is $\{1, 1, 0, 0, 0, 0, 0, 1\}$ where $\omega T = 0$ is included, but $\omega T = 2\pi$ is not. Note that the basic function is repeated periodically with period 2π. Calculate the FFT using Fig. 9.12.

9.20. A Fortran program for implementing an FFT is given in Fig. P9.20. Determine if the algorithm is a decimation-in-time or a decimation-in-frequency algorithm.

```
100 REM      * * *   F F T  Algorithm   * * *
110 PRINT "Input the number of points,  N ."
120 PRINT " N  must be an integer power of  2 ."
130 INPUT "N = ";N
140 DIM XR(N), XI(N), XM(N), XA(N)
150 PRINT "Input the data  x(k) , real and imaginary parts:"
160 FOR I=0 TO N-1
170 PRINT "        Real part of x(";I;") = ";
180 INPUT XR(I)
190 PRINT "Imaginary part of  x(";I;") = ";
200 INPUT XI(I)
210 NEXT I
220 PI = 3.141593
230 NU = INT((LOG(N)/LOG(2)) + .5)
240 L = 1
250 N2 = N/2
260 NU1 = NU - 1
270 K = 0
280 IF L > NU THEN 610
290 I = 1
300 M = INT(K/(2^NU1) + .5)
310 GOSUB 510
320 P = IBR
330 THETA = P*2*PI/N
340 WR = COS(THETA)
350 WI = -SIN(THETA)
360 TR = WR*XR(K+N2) - WI*XI(K+N2)
370 TI = WI*XR(K+N2) + WR*XI(K+N2)
380 XR(K+N2) = XR(K) - TR
390 XI(K+N2) = XI(K) - TI
400 XR(K) = XR(K) + TR
410 XI(K) = XI(K) + TI
420 K = K + 1
430 WHILE I < N2
440 I = I + 1
450 GOTO 360
460 WEND
470 K = K + N2
480 IF K < N-1 THEN 290
490 K=0: L=L+1: N2=N2/2: NU1=NU1-1
500 GOTO 280
510 ' SUBROUTINE FOR BIT REVERSAL
520 ' INPUT M IS CONVERTED TO IBR
530 I1 = 1: IBR = 0
540 WHILE I1 <= NU
550 J2 = M/2
560 I2 = CINT(2*(J2 - INT(J2)))
570 IBR = IBR + I2*(2^(NU-I1))
580 M = INT(J2+.01)
590 I1 = I1 + 1
600 WEND: RETURN
610 M = K
620 GOSUB 510
630 WHILE IBR <= K
640 IF K = N-1 THEN 760
650 K = K + 1
660 GOTO 610
```

Figure P9.20

```
670 WEND
680 I = ISR
690 TR = XR(K)
700 TI = XI(K)
710 XR(K) = XR(I)
720 XI(K) = XI(I)
730 XR(I) = TR
740 XI(I) = TI
750 GOTO 640
760 ' DISPLAY OUTPUT
770 CLS
780 COLOR 2,0,8
790 PRINT "Index n","Re{X(n)}","Im{X(n)}","Mag{X(n)}","Ang{X(n)}"
800 PRINT "____________________________________________________________________________"
810 PRINT
820 FOR I=0 TO N-1
830 DEF FNROUND(X) = INT(X*1000 + .5)/1000
840 XR(I) = FNROUND(XR(I))
850 XI(I) = FNROUND(XI(I))
860 XM(I) = FNROUND(SQR(XR(I)^2 + XI(I)^2))
870 GOSUB 920
875 DEF FNROND(X) = INT(X*10 + .5)/10
880 XA(I) = FNROND(XA(I)*57.2958)
890 PRINT I, XR(I), XI(I), XM(I), XA(I)
900 NEXT I
910 END
920 IF (SGN(XR(I))=1 AND SGN(XI(I))=1 ) THEN 1010
930 IF (SGN(XR(I))=1 AND SGN(XI(I))=-1) THEN 1020
940 IF (SGN(XR(I))=-1 AND SGN(XI(I))=-1) THEN 1030
950 IF (SGN(XR(I))=-1 AND SGN(XI(I))=1) THEN 1040
960 IF (SGN(XR(I))=0 AND SGN(XI(I))=1) THEN 1050
970 IF (SGN(XR(I))=0 AND SGN(XI(I))=-1) THEN 1060
980 IF (SGN(XR(I))=1 AND SGN(XI(I))=0) THEN 1070
990 IF (SGN(XR(I))=-1 AND SGN(XI(I))=0) THEN 1080
1000 IF (SGN(XR(I))=0 AND SGN(XI(I))=0) THEN 1090
1010 X=ABS(XR(I)): Y=ABS(XI(I)): XA(I) =ATN(Y/X): RETURN
1020 X=ABS(XR(I)): Y=ABS(XI(I)): XA(I) =-ATN(Y/X): RETURN
1030 X=ABS(XR(I)): Y=ABS(XI(I)): XA(I) =PI + ATN(Y/X): RETURN
1040 X=ABS(XR(I)): Y=ABS(XI(I)): XA(I) =PI - ATN(Y/X): RETURN
1050 XA(I) =PI/2: RETURN
1060 XA(I) =-PI/2: RETURN
1070 XA(I) =0: RETURN
1080 XA(I) =PI: RETURN
1090 XA(I) =0: RETURN
```

Figure P9.20 (continued)

9.21. Derive a bit-reversed sequence of subscripts for a 16-point FFT, e.g., 0, 2, 1, 3 for a 4-point FFT.

9.22. Write an equivalent matrix operation for the 4-point algorithm shown in Fig. 9.14.

9.23. Show that if $\{X_n\} = \text{DFT}\ \{x_n\}$, then

$$\sum_{n=0}^{N-1} |x_n|^2 = \frac{1}{N} \sum_{n=0}^{N-1} |X_n|^2$$

9.24. What sample rate and FFT size would you recommend to perform a frequency analysis of a signal that is band-limited to 20 kHz with a resolution of 50 Hz?

9.25. Indicate how the overlap-add method could be used to convolve a 20-point unit-pulse response with a 500-point input sequence if a 64-point FFT algorithm is the largest available.

9.26. Calculate the FFT of the raised-cosine window sequence of Problem 9.17 for $N = 8$. Explain your result.

9.27. Verify the maximum sidelobe level for the rectangular and raised-cosine windows shown in Fig. 9.21.

Index

Additivity, signal operator, 104
Admittance function, 213, 215
Amortization model, 53–55
Analog network, 224–25
Analog-to-digital (A-D) conversion, 5–7
Annulus, 177
Anticausal exponential signals, 131–32, 159–61, 171–72
Approximation, signal (*see* Signal approximation)
Autocorrelation function, 147–48

Block diagrams, 215–17
Bounded-input-bounded-output (BIBO) stability, 231–33
Butterfly computation, 321–22

Cadzow, James, A., 10*n*, 14*n*, 52*n*
Causal exponential signals, 129–30, 159–61, 169–71
Causal linear operators, 110–11
Characteristic signals:
  linear continuous-time signal operator, 102, 108–10
  linear discrete-time signal operator, 100, 106–8
Circuit analysis, 9–10
Circular convolution, 329–31
Closed-form expression, 23
Closed-loop transfer function, 151, 228–30
Complex domain description (*see* Laplace transform)
Complex exponential signals (*see* Exponential signals)
Computer control, 238–40
Continuity, 67–69
Continuous-time signals, 65–95
  changing the time variable, 72–77
  complex exponential signals, 91–94
  continuity, 67–69
  conversion to discrete-time signals, 5–7
  defined, 2
  differentiability, 67, 69–72
  differentiation operation, 78–80
  examples of, 4
  illustration of, 2, 3
  linear (*see* Linear continuous-time signal operators)
  multiplication operation, 80–81
  notation of, 65
  representation of, 66–67

summation operation, 81–82
unit-impulse signal, 85–91
unit-step signal, 82–85
Continuous-time system stability, 232–33
Convergence, radius of, 177
Convergence, region of:
Laplace transform, 128, 131–33, 160–61, 168–72
single-sided Laplace transform, 203
z-transform, 177
Convolution:
circular, 329–31
linear, 328–29
Convolution operation:
causal operators, 110–11
evaluation of, 116–19
exponential response, 114–16
Fourier transform, 301
Laplace transform, 142–45
linearity of operators, 105–6
operators, 100–103
single-sided z-transform, 219–21
unit-impulse response, 106–10
z-transform, 179
Cooley–Tukey formulation of fast Fourier transform, 321–22
Cosines, 250–54
Critically damped second-order system response, 211
Cross-correlation function, 145–47
Cyclic convolution, 329–31

Damping parameter, 93
Data sectioning, 331–33
Data windows, 333–36
Decimation in frequency, 322–25
Decimation in time, 321–24
Delay operation (*see* Right-shift operation)
Derivatives, 69–72, 78–80, 91
Difference operation:
continuous-time signals, 82
discrete-time signals, 39
Differentiability, 67, 69–72
Differential equations, 14–19
Differentiation operation:
continuous-time signals, 78–80
Fourier series, 287–88
Fourier transform, 301
Laplace transform, 137–39, 142
single-sided Laplace transform, 203–5
Digital control, 238–40
Digital filtering, 9, 315–17
Dirac delta signal, 85–91
Dirichlet conditions, 269, 271–72
Discrete Fourier transform (DFT), 303–36
application of, 309–317
fast Fourier transform algorithms and (*see* Fast Fourier transform algorithms)
of a general signal, 308, 309
properties of, 327–31
Discrete-time signals, 22–59
changing the time variable, 25–29
complex exponential sequences, 45–48
conversion of continuous-time signals to, 5–7
defined, 2
digital filtering, 9
examples of, 4
exponential signal generators, 48–51
illustration of, 2–3
linear (*see* Linear discrete-time signal operators)
modeling by discrete-time operators, 52–59
multiplication operation, 36–38, 40
notation of, 3–4
operator rule, 29–31, 49–51
representation of, 23–25
sequence, concept of, 22–23
shift operation, 31–36, 40
signal size measurement, 51–52
summation operation, 38–40
unit-impulse sequence, 40–43
unit-step sequence, 43–44
Discrete-time system stability, 232
Domain of definition, 65
Dual nodes, 321–22

Electrical networks, switched, 209–13
Equivalent representations, 134
Error signal, 238
Excitation sequence, 31
Exponential bounded signals, 169–71

Exponential Fourier series, 263
periodic signal representation (*see* Periodic signal representation)
signal approximation, 270–75
Exponential response of linear operators, 114–16
Exponential signals:
anticausal, 131–32, 159–61, 171–72
causal, 129–30, 159–61, 169–71
continuous-time, 91–94
discrete-time, 45–48
generators of, 48–51
sequence, 48

Fast convolution, 331
Fast Fourier transform (FFT) algorithms, 318–36
data windows, 333–36
development approach, 318–25
discrete-time signal processing calculation, 327–33
inverse, 325–27
Feedback control systems:
block diagram, 215–16
digital, 238–40
Laplace transform system poles and, 150–51
transfer functions and, 227–31
Field-controlled d.c. motor, block diagram for, 216–17
Filtering, digital, 9, 315–17
Final signal value:
single-sided Laplace transform, 205
z-transform, 181–82
First-difference operator rule, 29–30
First-order differential equations, 14–18
Folding frequency, 308
Forced solution, 206
Fourier coefficients:
computation of, 273–75
defined, 251
discrete Fourier transform, 311–13
Fourier series representation, 246–91
discrete Fourier transform, 312–14
frequency response, 248–49
multiple sinusoidal inputs, 249–50
network theory review, 246–48
periodic signal representation (*see* Periodic signal representation)
signal approximation (*see* Signal approximation)
signal symmetry, 254–55
sine-cosine series, 250–54
Fourier transform, 299–336
discrete (*see* Discrete Fourier transform)
inverse, 314–15
Laplace transform and, 299–300
properties of, 300–302
transform pains, 302, 303
Frequency convolution, Laplace transform, 142
Frequency differentiation:
Fourier transform, 301
Laplace transform, 142
Frequency response, 248–49
Frequency shifting:
discrete Fourier transform, 328
Laplace transform, 141–43
Fundamental frequency, 253

Generalized functions, 85
Gibb's phenomenon, 253–54
Gram-Schmidt orthogonalization method, 262

Half-wave symmetry, 255
Hamming window, 335–36
Harmonic analysis, 312–14
Harmonic frequency, 253
Homogeneity, signal operator, 104
Homogeneous solution, 206, 221

Identity operator, 35
Impedance function, 213–14
Impulse response signal, 8
Impulse train, Fourier series representation of, 285–86
Initial signal value, z-transform, 181
Initial-value problems, 206–13
Input sequence, 31
Integration, numerical, 10–13
Integration operation:
continuous-time signals, 79–80

Fourier transform, 301
Laplace transform, 142
single-sided Laplace transform, 205
Interconnected systems, 215–17
Inverse fast Fourier transform, 325–27
Inverse Fourier tranform, 314–15
Inverse Laplace transform, 152–62
Inverse z-transform, 182–94

Jury array, 236–37
Jury stability test, 236–37

Kronecker delta signal, 40–43

Laplace transform, 116, 125–64
anticausal exponential signals, 131–32
causal exponential signals, 129–30
depiction of, 128
Fourier transform and, 299–300
frequency distribution property, 142
frequency shift property, 141–43
inverse, 152–62
linearity property, 134–37, 142
multiplication (frequency convolution) property, 142
notation of, 126
pair tables, 133–34
poles and zeros, 148–51
rational and stable signals, 162–64
region of absolute convergence, 128, 131–33, 160–61, 168–72
single-sided (*see* Single-sided Laplace transform)
time-convolution property, 142–45
time-correlation property, 142, 145–48
time-domain differentiation property, 137–39, 142
time integration property, 142
time scaling property, 142
time shift property, 139–42
unit-impulse signals, 132–33
value of, 126
Laplace transform pair, 126
Laplace transform pair tables, 133–34
Laurent series, 183–84
Left-shift operation, discrete-time signals, 34–36
Legendre signals, 265–66
Leibnitz differentiating theorem, 109*n*
Lighthill, M. J., 88*n*
Linear continuous-time signal operators:
causal, 111
convolution integral relationship, 102–3
convolution operation, evaluation of, 117–19
exponential response, 115–16
linearity, establishment of, 105–6
response to periodic inputs, 289–91
stability, 113, 114
time invariance, 112
unit-impulse response determination, 102, 108–10
(*See also* Laplace transform; Linear operators)
Linear discrete-time signal operators:
causal, 110–11
convolution operation, evaluation of, 116–17
convolution summation expression, 100–102
defined, 100
exponential response, 115
linearity, establishment of, 106
stability, 112–14
time invariance, 111–12
unit-impulse response determination, 106–8
(*See also* Linear operators; z-transform)
Linearity:
discrete Fourier transform, 328
Fourier transform, 301
Laplace transform, 134–37, 142
single-sided Laplace transform, 203, 205
z-transform, 177–78
Linear operators, 99–120
continuous-time signal (*see* Linear continuous-time signal operators)
discrete-time signal (*see* Linear discrete-time signal operators)
linearity concept, 103–5
time-varying, 119–20
(*See also* Laplace transform; z-transform)
Line spectrum, 278–79

Magnitude spectrum, 279–80
Mean squared error criterion, minimization of, 257–60
Measurement noise, 9
Mechanical translation system, 223–27
Models, 8
Modulation, Fourier transform, 301
Multiple sinusoidal inputs, 249–50
Multiplication operation:
 continuous-time signals, 80–81
 discrete Fourier transform, 328
 discrete-time signals, 36–38, 40
 Fourier transform, 301
 Laplace transform, 142
 z-transform, 179
Multiplying scalars, 105

Natural solution, 206
Network functions, 213–17
Networks, switched electrical, 209–13
Network theory, 209–17, 246–48
Norms, sequence, 51–52
Numerical differentiation, 14–19
Numerical integration, 10–13

One-sided exponential sequence, 48–50
Open-loop transfer function, 228, 229
Operator rule, 29–31, 49–51
Operator stability, 112–14
Orthogonal basis signals, 260–75
 changing the time interval of approximation, 266–67
 exponential Fourier series, 263, 270–75
 general Fourier series, 267–70
Orthoganity property, 307–8
Orthonormal basis signals, 262–63
Output sequence, 31
Output-to-input signals, 213
Over damped second-order system response, 211
Overlap-add method of input sectioning, 331–33

Parseval's theorem, 281–84, 328
Partial-fraction expansion method:
 inverse Laplace transform, 152–62
 z-transform, 187–94, 196–98
Particular solution, 206, 221
Periodic signal representation, 275–91
 differentiation operation, 287–88
 impulse train, 285–86
 Parseval's theorem and signal power, 281–84
 response of linear systems to periodic inputs, 289–91
 spectral content of period signals, 278–81
Phase spectrum, 279–80
Phasors, 247
Poles:
 Laplace transform, 148–51
 z-transform, 188–90, 199
Polynomial approximation of arbitrary signals, 265–66
Power, signal, 281–84
Proper rational function, 187
Proportional-integral (PI) control, 238

Rabbit population model, 55–56
Radar-tracking model, 57–59
Rademacher signals, 263–65
Radius of convergence, 177
Raised-cosine truncation, 334–35
Rational signals, 162–64
Rational transform function, 148–62
Real exponential sequences, 46–48
Real periodic signals, 280–81
Real sinusoidal sequence, 47–48
Rectangular approximation of an integral, 11–12
Rectangular truncation, 334–35
Region of convergence (*see* Convergence, region of)
Residues, method of, 185–87
Resonance, 248
Resonant frequency, 248
Response sequence, 31
Right-shift operation, discrete-time signals, 31–36, 40
Routh array, 233–34
Routh stability test, 233–36

Sampled data, z-transforms for, 195–99
Sampling period, 5–7

Sampling property, 88
Scaling the time variable (*see* Time scaling)
Second-order response, 210–13
Sequences:
  characteristic (*see* Characteristic signals)
  complex exponential (*see* Exponential signals)
  concept of, 22
  difference, 39
  excitation, 31
  representation of, 23–25
  response, 31
  size of, 51–52
  sum, 38–39
  unit-impulse (*see* Unit-impulse signals)
  unit-step (*see* Unit-step signals)
  (*See also* Discrete-time signals)
Sequential time basis, 22
Shifting the time variable (*see* Time shifting)
Sidelobes, 334–36
Sifting property, 88
Signal approximation, 256–75
  changing the time interval of, 266–67
  exponential Fourier series, 270–75
  general Fourier series, 267–70
  orthogonal basis signals, 260–66
Signals:
  concept of, 1–2
  continuous-time (*see* Continuous-time signals)
  discrete-time (*see* Discrete-time signals)
  linear operations on (*see* Linear operators)
  measurement of size, 95
  power of, 281–84
  symmetry of, 254–55
Simulation diagram, 226–27
Sines, 250–54
Single-sided Laplace transform, 202–13
  final-value theorem, 205
  initial-value problems, 206–13
  linearity property, 203, 205
  s-domain differentiation property, 205
  s-domain shift property, 205
  time differentiation property, 203–5
  time integration property, 205
  time shift property, 205
Single-sided z-transforms, 182, 217–22
Singularities, Laplace transform, 148–51
Sinusoidal inputs, multiple, 249–50
Spectral content of periodic signals, 278–81
Stability:
  operator, 112–14
  signal, 162–64
  system, 231–37
Summation operation:
  continuous-time signals, 81–82
  discrete-time signals, 38–40
Sum signals, 38–39, 81–82
Superposition property, 104
Switched electrical networks, 209–13
Symmetric square wave, 252–53
Symmetry signal, 254–55
System modeling, 52–59
System stability, 231–37
System theory, 1, 7–10

Temperature regulation system, 230–31
Time advance, z-transform, 182–83
Time-axis origin, 174
Time constants, 208–9
Time convolution (*see* Convolution operation)
Time differentiation:
  Laplace transform, 137–39, 142
  single-sided Laplace transform, 203–5
Time-domain approach, 173
Time-domain differentiation (*see* Differentiation operation)
Time integration (*see* Integration operation)
Time interval of approximation, changing, 266–67
Time invariance, 111–12
Time reversal, z-transform, 182
Time scaling, 72–74
  Fourier transform, 301
  Laplace transform, 142, 145–48
Time shifting:
  continuous-time signals, 74–75
  discrete Fourier transform, 328
  discrete-time signals, 25–27, 29, 31–36, 40
  Fourier transform, 301
  Laplace transform, 139–42
  single-sided Laplace transform, 205
  z-transform, 178–79

Time transposition:
continuous-time signals, 76–77
discrete-time signals, 27–29
Time variable changes (*see* Time scaling; Time shifting; Time transposition)
Time-varying linear operators, 119–20
Transfer functions:
continuous-time, 125–26, 144, 213–17 (*See also* Laplace transform)
discrete-time, 217–22
system applications (*see* Transfer function system applications)
Transfer function system applications, 223–40
digital control, 238–40
feedback control, 227–31
mechanical translation, 223–27
system stability, 231–37
Transfer impedance, 214
Transformation procedure, 7
Transforms (*see* Fourier transform; Laplace transform; *z*-transform)
Translation:
single-sided *z*-transform, 218–19
z-transform, 178–79
Triangular truncation, 335

Underdamped second-order system response, 211
Uniqueness, *z*-transform, 177
Unit-Dirac excitation, 109–10
Unit-doublet, 89–90
Unit-impulse signals:
continuous-time, 85–91
discrete-time, 40–43
Unit-n tuplet, 90
Unit-response signals:
Laplace transform, 132–33
linear continuous-time, 102, 108–10
linear discrete-time, 106–8
Unit singlet, 90
Unit-step signals:
continuous-time, 82–85
discrete-time, 43–44
Unity feedback system, 151

Vector-space approach, 25

Zero initial conditions, 213
Zero-input solution, 221
Zeros, Laplace transform, 148–51
z-transform, 115, 173–99
convolution property, 179
defined, 174–76
final signal value property, 181–82
initial signal value property, 181
inverse, 182–94
linearity property, 177–78
multiplication operation, 179
sampled data, 195–99
single-sided, 182, 217–22
time-advance property, 180–81
time-reversal property, 180
translation property, 178–79
uniqueness property, 177